BASIC RESEARCH AND INDUSTRIAL INNOVATION IN CHINA

Series on Innovation and Operations Management for Chinese Enterprises

Print ISSN: 2591-7188
Online ISSN: 2591-7196

Editor: Xiaobo Wu *(School of Management, Zhejiang University, China)*

Published:

Vol. 2: *Basic Research and Industrial Innovation in China*
by Xielin Liu, Yubing He and Peng Cheng

Vol. 1: *Independent Innovation in China: Theory and Cases*
by Huibo Zhong and Jiasu Lei

Forthcoming:

Vol. 3: *The Theory and Policies of Mutual Benefit and Win-Win Strategy: Research on Sustainable Development of China's Open Economy*
by Lixin Yu and Wanling Chen

Vol. 4: *The Internationalization of Technological Innovation for Chinese Enterprises*
by Jin Chen

Series on Innovation and Operations Management for Chinese Enterprises – Vol. 2

BASIC RESEARCH AND INDUSTRIAL INNOVATION IN CHINA

Liu Xielin
University of Chinese Academy of Sciences, China

He Yubing
Fuzhou University, China

Cheng Peng
Beijing Forestry University, China

World Scientific

ZHEJIANG UNIVERSITY PRESS
浙江大学出版社

Published by

World Scientific Publishing Co. Pte. Ltd.
5 Toh Tuck Link, Singapore 596224
USA office: 27 Warren Street, Suite 401-402, Hackensack, NJ 07601
UK office: 57 Shelton Street, Covent Garden, London WC2H 9HE

and

Zhejiang University Press
No. 148, Tianmushan Road
Xixi Campus of Zhejiang University
Hangzhou 310028, China

Library of Congress Cataloging-in-Publication Data
Names: Liu, Xielin, author. | He, Yubing, author. | Cheng, Peng, author.
Title: Basic research and industrial innovation in China / authored by Xielin Liu, Yubing He and Peng Cheng.
Description: [Hackensack] New Jersey : World Scientific, 2018. | Series: Series on innovation and operations management for Chinese enterprises ; vol. 2 | Includes bibliographical references and index.
Identifiers: LCCN 2017056372 | ISBN 9789813235588
Subjects: LCSH: Technological innovations--China. | Research, Industrial--China.
Classification: LCC T173.5.C5 L58 2018 | DDC 338/.0540951--dc23
LC record available at https://lccn.loc.gov/2017056372

British Library Cataloguing-in-Publication Data
A catalogue record for this book is available from the British Library.

This edition is jointly published by World Scientific Publishing Co. Pte. Ltd. and Zhejiang University Press. This edition is distributed worldwide by World Scientific Publishing Co. Pte. Ltd., except China.

For any available supplementary material, please visit
https://www.worldscientific.com/worldscibooks/10.1142/10866#t=suppl

Typeset by Zhejiang University Press

Printed in China

Preface

China has undergone rapid economic growth in the last three decades. However, from 2013 onwards, this giant economy has been slowing down and meeting with many deep challenges: The first one is a structural problem; the manufacturing industry faces an overcapacity problem and new economy is coming but slowly. Second, China is facing a heavy pollution challenge from air to water, and shortage of natural resources. Third, though the government spends huge amount of money on science and technology (S&T), and many enterprises have been listed among The *Fortune* Global 500 companies, innovation capacity at the enterprise level is still quite low. So, one key question is how a country can transform heavy S&T investment into real industrial innovation capability. In the period of long-run catch-up of countries such as Japan and the Republic of Korea (hereinafter Korea), some enterprises became forerunners in their industry globally, such as Toyota in the automobile industry in Japan and Samsung in the electronics industry in Korea. But in China, given this critical economic crisis, only Huawei has been successful in making an impact globally. So, how China can upgrade to meet the industrial innovation capacity is a great challenge for the future of China.

For a long time, Chinese enterprises have relied on technology import and imitation as their main technology strategy, and cost advantage is their base of competitiveness. Even now, their research and development (R&D) intensity is no more than 1%. So, Chinese enterprises have to learn how to build up their R&D centers and improve their innovation capability in an open and global context. At the same time, the universities and government research institutes have to change their strategy from being paper-centered to innovation-centered.

In this book, based on our cross-country case analysis and on the history of industrial innovation, we proposed the concept of industry-driven basic research, and we are convinced that both the government and enterprises should focus on this concept. We believe that there is a gap between what the government aims at and what enterprises do in China. The former loves to support major science and current key technologies but misses out on the next-generation industrial technology. The latter only focuses on production development in the short term.

Our book is structured in the following manner:

Chapter 1 is an introduction of basic research and analysis of its implication on the industrial innovation in China.

Chapter 2 examines the S&T investment in different stages of economic development; we strive to understand the pattern of the level of S&T investment in key countries following Porter's framework.

Chapter 3 suggests that there are some problems of imbalance of S&T investment in China: too little investment in basic research, too little investment in industry-driven research. These are the potential factors causing low industrial innovation capacity in China.

Chapter 4 focuses on the relationship between basic research and industrial innovation based on Pasteur's quadrant; it analyzes why the government supports basic research and enterprises do some basic research.

Chapter 5 presents different models of industry-driven basic research.

Chapter 6 is a comparative study of how to support industry-driven basic research in the leading countries and draws some implication for China.

Chapter 7 focuses on how China has invested in S&T in the last 20 years and the changing role of universities, government research institutes and enterprises during this period. We found that universities and government research institutes still dominate basic research in China.

Chapter 8 provides a deep analysis of how the national S&T programs support basic research and the role of enterprises, universities and government research institutes in these programs in China. We found that enterprises are marginal except some large state-owned enterprises (SOE).

Chapters 9 to 11 cover three industrial cases; we select biopharmaceutical industry, TFT-LED industry and carbon fiber industry to see the role of basic research in the industrial innovation against world leading countries. Case analyses show that leaders of the three industries—the United States (biological medicine and carbon fiber), Japan (LCD panel and carbon fiber) and Korea (LCD panel) share a common characteristic, which is industry-driven basic research. China's subsidies to basic research have long been constrained to conduct self-initiated, purely theoretical research or national planned research. Although some businesses have invested into application or even basic research, few Chinese businesses support basic research because it is public product and takes a long time to pay off. Large SOEs like PetroChina, Sinopec, China Mobile and State Grid have enough funds and research institutes and labs to conduct medium- and long-term research, but a majority of Chinese businesses lack the ability to do so.

Chapter 12 contains the conclusion and policy suggestion. Our policy suggestion is: China's indigenous innovation should pay more attention to strengthening industry-driven basic research, which is based on the specific needs of industrial innovation and market judgment. It serves to solve the key bottleneck of industrial development and transform scientific discovery into the core technological capacity of enterprises.

The book is a result of cooperative research. We obtained financial support from the Chinese Academy of Sciences. Our team in the School of Economics and Management of the University of Chinese Academy of Sciences (UCAS) worked jointly for many years to complete this work. Besides the authors here, Zuo Kairui (Academy of Telecommunications, Ministry of Industry and Information Technology) contributed to Chapters 2 and 3. Gao Taishan (State Information Center), Gao Wei (State Grid), Gao Yuchen (PhD candidate in UCAS), Qiu Shumin (PhD candidate in Tsinghua University), Chen Jian (PhD candidate in UCAS), Chen Ao in Shougang Group also contributed to the relevant chapters. Finally, we want to thank Wang Xi and Xu Xiaodan, who are both research assistants in our team and who also contributed greatly to the writing and editing of this book.

About the Authors

Liu Xielin is a professor at the School of Economics and Management, University of Chinese Academy of Sciences (UCAS), with a bachelor's degree, a master's degree and a PhD from Peking University, the Chinese Academy of Sciences and Tsinghua University respectively. His research areas mainly cover innovation policy, technology and innovation management and entrepreneurship. He has played a prominent role in advising the Chinese government on the national technology development strategies. He has served as Vice President of the Chinese Association for Science of Science and S&T Policy and editor of the *Journal of Science and Technology Management* (since 2003). His research has been published in leading journals including *Research Policy*, *Technovation*, and *International Journal of Management Studies*. According to Elsevier, he is among the most cited Chinese researchers.

He Yubing is a professor at the School of Economics & Management, Fuzhou University, with a PhD from Zhejiang University and he did post-doctoral research at the Chinese Academy of Sciences. He is mainly engaged in technological innovation management, strategic management, technology and innovation policy research, and serves as a director member of the Chinese Association for Science of Science and S&T Policy.

Cheng Peng is an associate professor at the School of Economics and Management, Beijing Forestry University, the director of Property Management, with a PhD from Jilin University and he did post-doctoral research at the Chinese Academy of Sciences. He is mainly engaged in innovation management, S&T policy and industrial policy.

Contents

Chapter 1

Introduction

1.1 Research background

In the past 10 years, China has made significant achievements in independent innovations well-known to the world. In respect of S&T investment, from 1998 to 2013, China's R&D expenditure had increased rapidly from 55.112 billion RMB to 1,184.66 billion RMB, the proportion of R&D investment in GDP (R&D intensity) had risen from 0.65% to 2.01%, and the full-time equivalent of R&D personnel had increased from 755.2 thousand to 3.53

million ① (Figure 1-1).

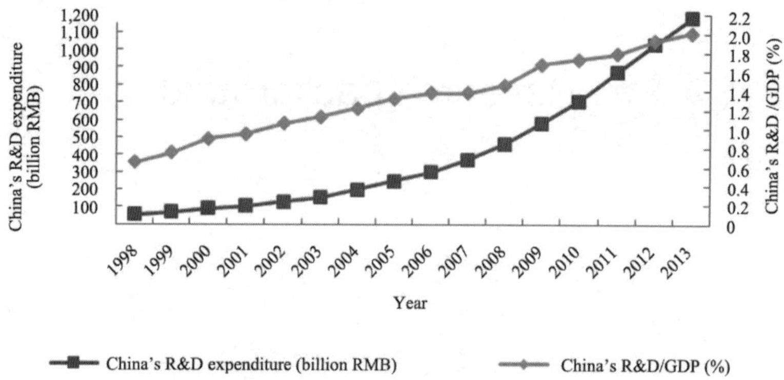

Figure 1-1 Share of China's R&D expenditure in GDP by year

Data source: *China Statistical Yearbook on Science and Technology 2014*, the website of the Ministry of Science and Technology of China.

In terms of S&T output, China has already been an important player in the world, and its numbers of R&D staff, scientific papers and patent applications have all witnessed rapid growth (OECD, 2008). Its research capacity in many disciplines has already reached an

① 3.53 million is data collected in 2013. In recent years, the total volume of China's R&D human resources has witnessed rapid development. According to relevant statistics of the Ministry of Science and Technology of China, the number of China's full-time equivalent of R&D staff has surpassed those of Japan, Russia, and the United States in 2000, 2002, 2008. China has become a country of the largest human resources of R&D staff in the world. In 2010, the total of China's R&D staff (3.542 million) accounted for 23.6% of the world's total and almost doubled that of the United States, which ranked second (1.93 million, accounting for 18% of the world's total). However, China's proportion of R&D staff in one thousand labor workers is only 1/10 that of Japan's. In respect of R&D expenditure, the US National Science Board released a report *Science and Engineering Indicators 2014*, which claimed that China has become the world's second largest R&D spender just second to the United States. In 2011, its R&D expenditure accounted for 15% of world's total, while the R&D expenditure of the United States accounted for 1/3. According to reported data, up to 2013, China's scientific R&D funding share of GDP has doubled since 1998, reaching 2.01%, while the real annual growth rate of China's R&D expenditure remained at about 20%.

advanced level in the world. ① In some key fields of technology (such as nanotechnology, some fields of life sciences), China has already achieved leading-edge advantages (Zhou & Leydesdorff, 2006). In terms of high-tech output, this trend is quite obvious (Figure 1-2).

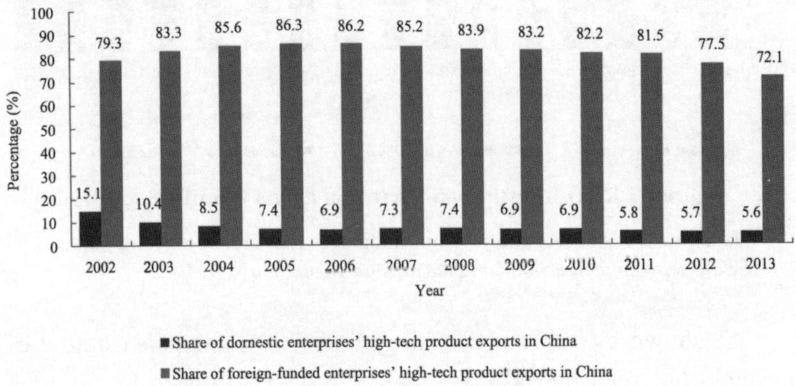

Figure 1-2 Shares of domestic and foreign-funded enterprises' high-tech product exports in China by year

Data source: *China Statistical Yearbook on Science and Technology 2014*, the website of China Customs, *China Statistics Yearbook on High Technology Industry*, the website of China Science and Technology Statistics (http://www.sts.org.cn/sjkl/gjscy/data2014/data14.pdf).

However, there is another important trend. In the past 10 years, the proportion of China's investment in basic research in the gross domestic expenditure on R&D (GERD) has not improved; on the contrary, the proportion declined from 5.96% in 2004 to 4.70% in 2013 (Figure 1-3).

① According to the data released by Basic Research Management Center of the Ministry of Science and Technology of China, in 2013, the total number of China's scientific papers published in international journals amounted to 232,070, accounting for 13.5% of the world's total and jumping to the second position in the world. The reference count of its papers ranked fourth in the world. The number of Chinese scientists' papers has also seen a significant increase on authoritative journals such as *Nature*, *Science* and *Proceedings of the National Academy of Sciences*.

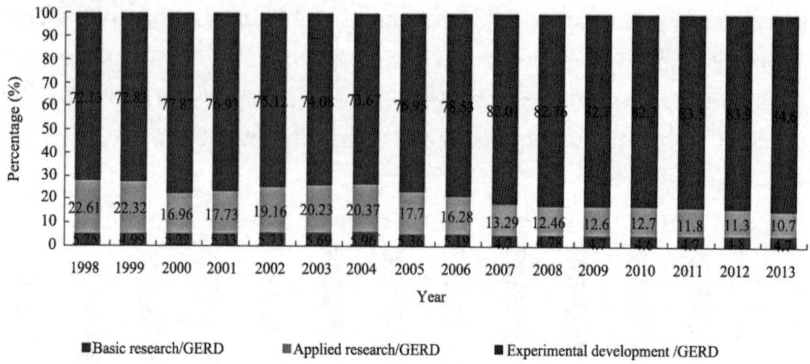

Figure 1-3 R&D Investment Structure of China from 1998 to 2013

Data source: The website of China Science and Technology Statistics, *China Statistical Yearbook on Science and Technology 2014*.

As shown by Table 1-1, in the past 9 years, the paper output of business has seen the fastest growth (78.6%), followed by research institutes (57.9%) and medical institutions (50.0%), while the lowest is higher education (40.9%).

Table 1-1 Domestic S&T papers by type of institution (1,000 papers)

Type	2005	2006	2007	2008	2009	2010	2011	2012	2013
Higher education	235	243	306	318	342	343	336	337	331
Research institutes	38	42	47	50	56	57	58	56	60
Business	14	13	15	16	18	20	21	33	25
Medical institutions	52	91	76	71	87	89	92	79	78
Others	16	14	19	17	18	21	23	19	23
Total	355	403	463	472	521	530	530	524	517

Data source: *China Statistical Yearbook on Science and Technology 2014*, the website of the Ministry of Science and Technology of China.

The number of SCI papers has witnessed the fastest growth in the past 9 years from 2005 to 2013, which has increased by 2.4 times (Table

1-2). China has now become the second largest country of SCI papers (Table 1-3).

Table 1-2 Chinese S&T papers indexed by SCI, EI and CPCI-S (1,000 papers)

Type	2005	2006	2007	2008	2009	2010	2011	2012	2013
SCI	68	71	89	96	109	122	144	193	232
EI	54	65	76	85	93	112	124	124	164
CPCI-S	31	36	43	59	52	38	52	78	69

Data source: *China Statistical Yearbook on Science and Technology 2014*, the website of the Ministry of Science and Technology of China.

Table 1-3 S&T papers indexed by SCI, EI and CPCI-S in selected countries (2013)

Country	SCI		EI		CPCI-S	
	Number (1,000 papers)	Rank	Number (1,000 papers)	Rank	Number (1,000 papers)	Rank
China	232	2	164	1	69	2
USA	474	1	103	2	11	1
UK	133	3	29	5	20	4
Germany	119	4	34	3	20	5
Japan	96	5	32	4	22	3
France	83	6	26	7	13	6
Italy	74	7	2	9	13	7
Canada	72	8	19	11	12	8
India	60	10	28	6	11	9
Korea	57	12	25	8	9	10
Brazil	44	13	9	16	6	14
Russia	32	15	14	13	5	18
Total	1,476		503		310	

Data source: *China Statistical Yearbook on Science and Technology 2014*, the website of the Ministry of Science and Technology of China.

However, in China, the ratio of investment in basic research actually has not risen in the past 9 years. A large amount of S&T funds has been invested in application research and product development. Thus, it leads to an interesting result. It is an indisputable fact that although China's scientific ability has improved, the improvement of this ability might be attributed to the research on a lot of application and product development. It is such research that has been transformed into papers instead of new technologies and products that can be then commercialized.

The status of patents can also prove that industrial innovation capacity has not been effectively improved. In terms of China-based USPTO (the United States Patent and Trademark Office) patents, a large number of these patents are achieved by multinational corporations in China (Figure 1-4). From the indicator of China-based USPTO patents, China's industrial innovation capability has not improved as fast as its scientific ability. In other words, China's scientific capacity has improved rapidly, but at the same time, its industrial innovation capability has not improved accordingly. Therefore, it becomes difficult for industries in China to develop leadership in the world.

From 1981 to 2010, China had a total of 12,419 USPTO patents, among which 78% were granted to firms, 12% to individuals and 9% to universities and institutions (Figure 1-4).

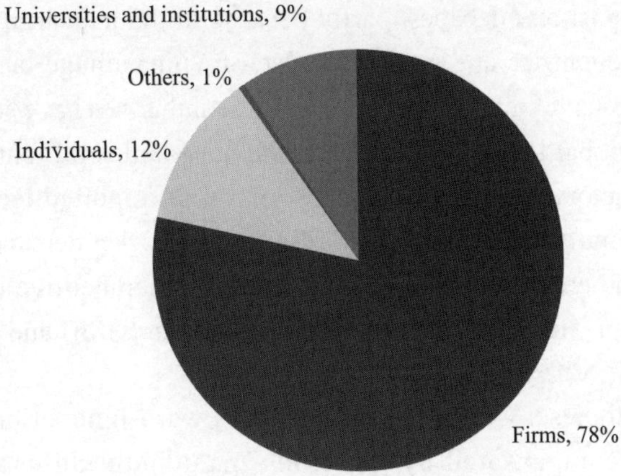

Figure 1-4 Share of China-based USPTO patents by assignee type

However, among the firms, 35% are Taiwan-invested companies registered in the Chinese mainland and 29% are foreign-funded enterprises of the Unites States in China, and the number of enterprises based in the Chinese mainland in a real sense occupies only 23% (Figure 1-5).

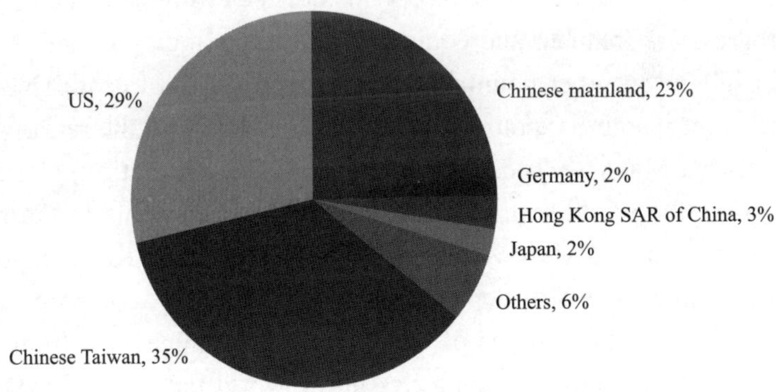

Figure 1-5 Share of firm-owned China-based USPTO patents by assignee origin

Data source: Branstetter, L., Li, G. & Veloso, F. The globalization of R&D: China, India and the rise of international co-invention. Working paper of Carnegie Mellon University, 2013.

In the past few decades, partly because of the industrial transfer of developed countries and partly because of China's huge labor market, which contains infinite momentum, Chinese industries have successfully become a global innovation follower and have laid solid foundations as a "world factory", relying on its absorption of imported technologies, low-cost manufacturing, and rapid response to market demand. In many fields, it has created a number of world-class competitive enterprises such as Haier, Huawei, Lenovo, Founder, Galanz, CIMC and so on.

Nevertheless, China's weak and backward innovation capacity cannot be compensated by its strong manufacturing capability of industries. Industry development in China mostly relies on the advantages of low labor cost and natural resources, which are unreliable, instead of core technological innovation (Liu & Zhao, 2001; OECD, 2008). Because of this, a lot of core high-tech knowledge in China needs to be introduced from other countries. China lacks proprietary intellectual property rights and most industries are much less competitive than those of other countries in the world (Chen, 2005). In many fundamental industries (such as IC, supporting software, automobile engine, LCD panel) where technological changes and progress are frequent and common, core technologies are still heavily dependent on other countries. The economy is still cost-driven and the profit margin is quite low in China. According to the released data of China Machine Tool & Tool Builders' Association, China has become the world's largest producer of machine tools in the past 3 years in a row. However, it is not a powerful producer. Key components of machine tools still depend on imports. In 2011, compared to the previous year, the imports of metal working machine tools from Japan and Germany, powerful countries in the world industry, increased by 50.3% and 40.0% respectively, and the imports of intermediate machine tools from Chinese Taiwan and Korea grew by 23.7% and 24.3% respectively (Li, 2012).

These phenomena constitute a paradox of China's innovation

process. Chinese enterprises have not embarked on the road of innovation even though China has increased R&D investment and intensified introduction of technologies for a long time. The technological innovation capability of industries has not grown in pace with S&T investment. High R&D investment of the country contrasts sharply with the low innovation capability of enterprises.

Questions 1: Is China's investment in S&T resources still far from enough? Or is it already adequate but a matter of resources allocation?

In the current context of market economy, S&T resources have become strategic resources that determine economic and social development. The method applied to optimize the allocation of resources is critical for the improvement of a country's economic efficiency. Since China proposed a national strategy of independent innovation in 2006, its S&T investment has been rising and has maintained an average annual growth rate of 22%, much higher than the GDP growth rate. Therefore, China's R&D investment level has increased from a relatively lower level to the second largest in the world. The R&D share of GDP has also increased from barely meeting the standards to 2.01% in 2013. China is already a big country of S&T resources. However, in terms of innovation ability, China still needs to import core industrial technologies. Although it has the second largest number of patents in the world, its disruptive innovation is low and at an average level of developing countries. First, economic investments have stimulated large purchases of foreign equipment. Therefore, we can say that the efficiency of China's S&T resources allocation and innovation system is at a relatively low level. This is the reason why in the 1980s and 1990s, after the peak of the S&T system reform, China proposed another round of S&T structural reforms to solve the mismatching problems of science and economy. On July 17, 2013, Xi Jinping visited the Chinese Academy of Sciences, and once again pointed out that "we must resolutely eliminate the institutional barriers that influence the improvement of technological innovation capacity, build a sound channel for technological and economic transformation, optimize S&T policy support, and improve the S&T evaluation system. We should prioritize S&T fields that can promote the transformation of economic development modes and create new economic growth points, focus on bottlenecks that restrict China's sustainable development of economy and society and strengthen the input in burgeoning, cutting-edge and crossing fields" (Li, 2013).

Question 2: Considering the innovation system, has a country's investment of S&T resources really driven the S&T investment of enterprises? Can enterprises continuously acquire new knowledge from universities and institutions?

Since innovation is large-scale commercialization of new ideas, enterprises, as the subject of economy, are inevitably the subject of innovation. But how can we measure whether an enterprise is a subject of innovation or not? Does it depend on the enterprise's R&D share of a country's total, or the enterprise's performance of innovation output as a subject? Is there any innovative impetus for enterprises? Is the government impairing enterprises' innovative impetus or guiding their innovative direction? Is the enterprise in a subordinate position or dominant position in terms of the government's innovation strategy? What is the appropriate role for universities and institutions in an innovation system?

The national innovation system is a very useful framework for analyzing the efficiency of a country's S&T resources allocation (Freeman, 1987; Liu & White, 2001). The core concept of this framework is as follows: The efficiency of a country's innovation does not simply depend on its investment level of resources, although all countries are seeking to improve their S&T investment level. For example, the European Union's goal is to reach 3% by 2010, the United States to reach 3% in the future, and China to reach 2.5% by 2020. However, politicians often forget that the efficiency of the innovation system is the determinant factor of a country's innovation ability.

The efficiency of innovation system also refers to the efficiency of collaborative innovation between enterprises, universities and research institutes, who, based on their respective strategic objectives and strategic intent, establish a complementary, risk-and-profit sharing partnership with each other in order to seize new market opportunities, enhance independent innovation capability, and realize their common interests with their respective advantages of resources.

Leydesdorff & Etzkowitz (1996) explained the strategic issues of government-industry-university-research institute cooperation based on their famous Triple Helix Theory, which is considered to be a new paradigm of innovation research: Universities (scientific research institutes and universities are both seen as the knowledge production sector), industry, and the government strengthen resources sharing and information communication between each other and improve the efficiency of S&T resources by structural arrangements and institutional design of organizations.

The cooperative relationship between the government, industry, universities and research institutes stated by the Triple Helix Theory can be expressed using a government-industry-university-research institute Triple Helix Model to each participant party with strong "interactive and reflexive" effects. "Interactive" refers to the hybrid organization with a tripartite network created in the process of their tripartite interaction. "Reflexive" means that each participant also plays the role of other participants while playing its own role. The Triple Helix Theory emphasizes the cooperative relationship between industry, government and academic circle, whose common interests are directed towards the social value they create. Its key point is that the boundaries between the public and the private, science and technology, universities and industry are accessible. Each party also takes the tasks that are used to be taken only by other parties. From the government point of view, the theme of its work has been increasingly oriented towards building these relationships at different levels of S&T policies.

Its core value is that it combines the government, enterprises and universities that have their own respective value systems in terms of S&T resources, turns the three forces of knowledge field, administrative field and production field into one, and builds a solid foundation for economic and social development. In order to create such a resultant force, it is important to break traditional boundaries, such as disciplinary boundaries, industrial boundaries, geographical boundaries and conceptual boundaries and establish new research management systems and market operation mechanisms based on these boundaries. This requires us to develop the cooperative organization mode and various mechanisms that can truly enable the integration of the government, industry and universities/research institutes.

The allocation ratio of R&D funds for government, industry and universities/research institutes changes every year. The investment of enterprises has increased for consecutive years (Figure 1-6), and has increased by 16% in the past 13 years, which reflects that China's S&T resources are gradually and increasingly allocated to enterprises, the subject of innovation; the share of research institutes has declined year by year, and declined by about 13% in the past 13 years; the share of universities has slightly decreased, while the share of other public institutions has first decreased, but then increased after a while and remained at a steady level as a whole. According to the Triple Helix Model proposed by Etzkowitz and Leydesdorff, it is the "overlapping" part of universities, enterprises and research institutes or their mutually penetrating part that comprises of the core element of the innovation system, and it is the relations among the three parties that are the key factors pushing the production and dissemination of knowledge. In the process of transforming knowledge into productivity, participants interact and influence each other; their structural proportion constantly adjusts along with the changes of accumulated knowledge and external environment, and it promotes R&D capability in a spiral way during this dynamic process.

Figure 1-6 shows that China's S&T resources allocation model becomes gradually similar to that of developed countries: Enterprises have gradually become the subject of S&T innovation, and universities and research institutes are playing an increasingly important role despite their decreased share of funds. The relationship between enterprises, universities and research institutes has become a complementary, trade-off relationship during this dynamically changing process. We believe that at a certain stage, with the support of the government, the funds allocation of enterprises, universities and research institutes will be set at a specific ratio. This ratio is not static. With the changes of economic cycles, market forces will facilitate the interaction among them, so the ratio will undergo dynamic adjustment after a period of time and set at a new one in the end, which is just the dynamic process of the Triple Helix Model.

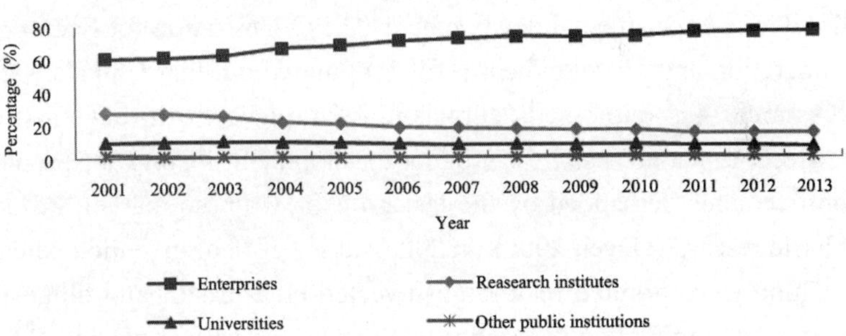

Figure 1-6 The three main parties' R&D expenditure in recent 13 years

At present, the proportion of R&D funds for S&T resources generally maintains at the following level: enterprises : universities : research institutes : other public institutions = 76.6 : 15.0 : 7.2 : 1.1 (2013). Based on this ratio, many people believe that China's S&T investment structure has already proven that enterprises are the main actor of innovation. If enterprises are still not the main actor of innovation, it should be caused by other reasons, instead of government investment. As a result, the cause for the paradox seems not to be the lack of R&D investment from enterprises, but the structure of investment.

The common feature of many world-leading innovation countries and regions (US, EU, Japan, Korea, etc.) is that enterprises' R&D investment is in a dominant position. Actually, in recent years, Chinese enterprises have gradually transferred their innovation input from importing technologies to technology research and development, and have become the main force of China's R&D investment. In 2013, the total of industrial enterprises' R&D funds was 907.6 billion RMB, accounting for 76.6% of China's total R&D investment of 1,184.7 billion RMB (Figure 1-7). However, an outstanding feature of China's research and development is that "development" is more than "research", research and development mainly focus on experimental development, only about 15% of domestic R&D funds have been invested in basic research and applied research, and over 80% of the funds have been allocated to experimental development. The Organization for Economic Cooperation and Development (OECD) pointed out that China's R&D investment has witnessed "surprisingly" rapid growth, but most of the R&D funds are used for product development, mainly for product improvements demanded by the fast-growing Chinese market, instead of basic research (Dyer, 2006). In 2006, nearly 60% of invention patents in China were granted to foreign-invested enterprises, and domestic enterprises' innovation capability had not grown as fast as their R&D investment (OECD, 2011).

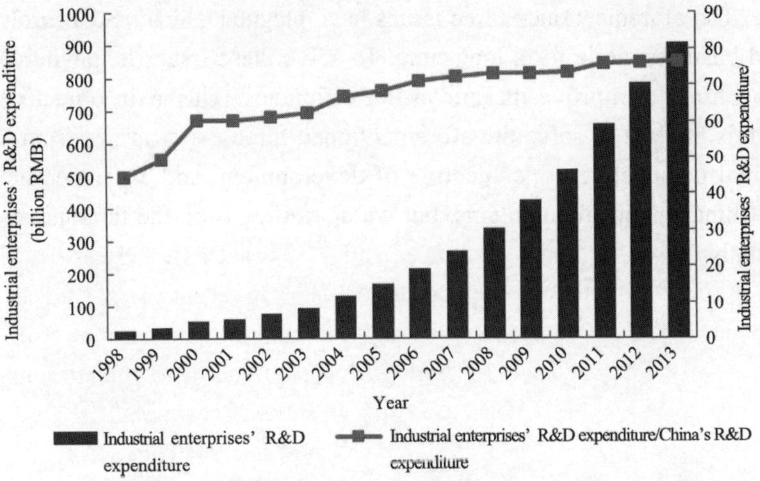

Figure 1-7 Share of enterprises' R&D funds in China by year

Data source: *China Statistical Yearbook on Science and Technology 2014*, the website of the Ministry of Science and Technology of China.

Question 3: As China is a developing country, should we focus on basic research or applied research while allocating innovation resources, if we want to promote enterprise innovation? In other words, what is the appropriate proportion of basic research, applied research and experimental development at different stages of development? What are the appropriate roles for enterprises, universities and research institutes at different stages of development? Is it the result of or the reason for a particular stage of development that enterprises become the main actor of innovation? How should China, as a developing country, allocate these S&T resources at the three different stages?

Our basic conclusion is this: The real demand-oriented scientific research has been ignored in China. Therefore, we propose a research on the serious shortage of basic industrial research. And this shortage has resulted in the inefficiency of a large number of S&T funds nowadays, which is an important reason for the serious shortage of industrial innovation.

The aforementioned three issues have plagued China's S&T circle and business circle for a long time. To solve these issues is inevitable for China to improve its innovation efficiency. The main objective of this book is to solve the aforementioned three major issues from a global perspective, a perspective of development, and a perspective of China's national conditions, but we do not focus on the three issues equally.

1.2 Main ideas

We propose that the important reason for the weak international competitiveness of Chinese industries is China's long-term low investment in research and its imbalanced structure, specifically the lack of basic research in terms of core industrial technological innovation, which is also a major challenge faced by the technology investment system in our country. Scientific (basic) research is an investment of the future. Influenced by catching-up strategy, the Chinese government and corporations tend to invest more in experimental development and applied research for a long period of time. As a result, our original knowledge innovation lags seriously behind the industrial technological innovation and there is a great need to enhance international competitiveness, due to this kind of overemphasis on secondary innovation strategy. Until now, when the domestic demand is on an explosive increase, many Chinese corporations, driven by interim profits, are still concentrating on a large amount of incremental innovation on the industrial technological tracks, which have already been explored by foreign companies. Once a foreign company achieves an innovative breakthrough in a certain industrial field and enters a new technological track, Chinese corporations will fall in dire straits of "introducing—falling behind—introducing again". The industry of household appliances is a typical case in point. Although Chinese corporations got the opportunity to learn the advanced household appliance technology in the world when the international industry was transferring to developing countries, they generally focused on market segmentation and price competition, aimlessly becoming original equipment manufacturers (OEMs) of other brands, thus overlooking the basic research in the semiconductor chip, magnetron, compressor and so on. In consequence, a new round of significant breakthrough in household appliance technology caused serious excess capacity[①], and the corporations entered the stage of introducing and learning again.

The research of the OECD also indicates that low investment in research, lack of basic research, and deficiency of government support are the major bottlenecks in innovation for Chinese enterprises (OECD, 2008). On the whole, the innovation of Chinese corporations is still following the market, which lacks industry-driven basic research[②] to cultivate a new technological industry or R&D planning of corporations to lead the next generation of technology. One typical example is the overall decline of China's LCD industry during the upgrading and updating of technology.

It is of great significance to correctly understand the role of basic research in increasing the core industrial technological capacity (Liu & He, 2011). During the past 30 years, policy-makers in China tended to separate basic research from industrial innovation. They took a rather

① China industry annual report series: Home appliances, http; //www.cei.gov.cn/, 2007.

② In the book, "industry-driven basic research" and "industrial basic research" have the same meaning.

narrow view of basic research, holding that the results of basic research in universities and research institutes could be smoothly transferred to corporations. In other words, they believed that the country should allocate all basic research resources to public research institutes and universities, while enterprises only needed to transform the research results into products, which limited basic research to the national level. Therefore, China attaches great importance to the transformation of S&T achievements, with a political guide of "basic research (universities and research institutes) → applied research (enterprises, universities, and research institutes) → experimental development (enterprises)". As the transformation of S&T achievements is related to a lot of problems such as intellectual property, benefits distribution and market demand, the basic research achievements of universities and research institutes have not contributed too much to industrial S&T innovation. As a result, universities, research institutes and enterprises focus on their own part in R&D activities, thus leading to low efficiency of resources allocation. Chinese enterprises are unaware of the importance of basic research, so they have not concentrated enough in it, while the basic research achievements made by universities and research institutes are also difficult to be applied to industry.

Many scholars believe that as a public good, basic scientific knowledge can be transmitted on a global scale, so developing countries do not need to conduct their own basic research and can make their secondary innovation by absorbing research achievements of developed countries (Pavitt, 2001). Nevertheless, basic research is not for free; scientific knowledge is not a kind of information that can be easily transmitted and if there is no corresponding background knowledge or skill resources, enterprises are unable to understand and use the basic research achievements of other countries, especially when tacit knowledge is contained (Callon, 1994). The attempt of enterprises to get a free ride on scientific knowledge of developed countries will not be conducive to core technological breakthroughs of China's industry. There is another view that the key to China's industrial innovation is confidence, and innovation can be achieved as long as Chinese enterprises have confidence (Gao, 2006; Lu, 2006; Ge et al., 2010). However, in fact, the core technological innovation of industries has its own objective laws; it is an accumulation process of knowledge, and its success cannot just rely on confidence and enthusiasm. Differing from business model innovation, breakthroughs and innovation of core industrial technologies are mostly based on scientific principles, so technological breakthroughs can be achieved only when enterprises have mastered natural phenomena and core principles of industrial technology. Since Chinese enterprises have ignored basic research for a long term, Chinese enterprises' innovation ability remains low when the industrial technology (such as biomedicine) is increasingly related to basic research. The key to improving the technological innovation capability of China is to strengthen basic research. Enterprises have been the weakest part of China's innovation system for a long time (OECD, 2008). In addition, China's innovation system has low efficiency of innovation output, and there is a huge gap between the research projects of public research institutes and enterprises' R&D demand (Liu & White, 2001), which has forced many enterprises to meet their own development needs by importing foreign technologies.

The relationship between basic research and industrial technological innovation is not a new topic. Many documents have pointed out that basic research is the prerequisite and catalyst of applied research and the fundamental driving force of technological innovation (Bush, 1945). Studies show that a lot of core technologies are long-term accumulation of basic research. Technologies imported from other countries have been increasingly more advanced and complicated, while most advanced technologies are based on basic principles. If China cannot solve the problem of long-term neglect on basic research investment or change investment and employment mechanisms of basic research, the problem of low industrial innovation capacity will be difficult to be solved completely. Although in the past 10 years, China has increased its investment in basic research by implementing special projects such as National Program on Key Basic Research Project of China (Program 973), National Natural Science Foundation of China, Knowledge Innovation Project, "Project 211" and "Project 985", State Key Scientific Research Program and so on, but compared with the leading innovation countries, China is still in a disadvantageous position in basic research investment (Figure 1-8).

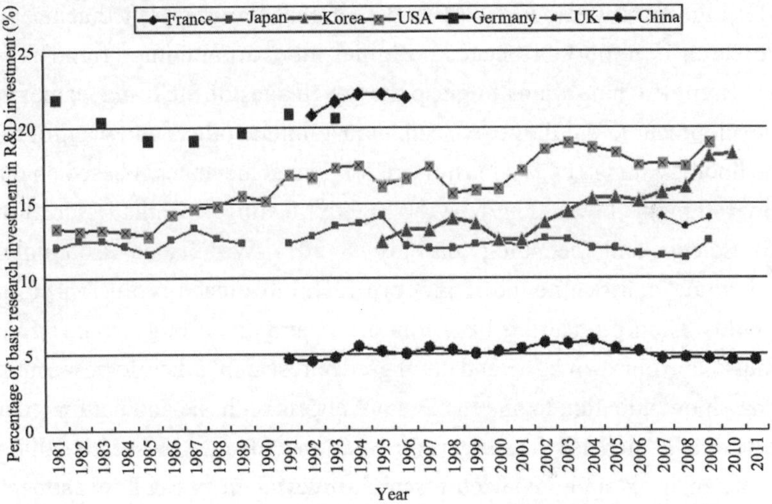

Figure 1-8 Comparison of basic research investment between China and leading innovation countries in the world

Data source: The website of the Ministry of Science and Technology of China, OECD's Main Science and Technology Indicators (2011).
Note: Some countries' data in a certain period are unavailable.

China's industries cannot achieve more breakthrough innovations without investment in basic research. Even though China's R&D funds structure is quite similar to that of the United States, EU countries, Japan and Korea (enterprises' R&D investment is the main part), in terms of research types, quite a number of R&D funds have been invested in experimental development, followed by applied research, so investment in basic research is low. This is prone to result in low-level equilibrium dilemma.

In recent years, the Chinese government has come to realize the critical importance of basic research. Hu Jintao repeatedly pointed out that "we need to pay enough attention to basic research", because "a lot of major revolutionary scientific achievements have originated from basic research" (Hu, 2004). China should "strive to make an original breakthrough in basic research and leap-type breakthrough in high-tech sector" (Hu, 2011). In his speech to the S&T circle of Beijing on November 3, 2009, Wen Jiabao pointed out that "basic research is important because original innovation comes from basic research. If China wants to grasp the advantages of S&T and economic development in the future, we cannot just imitate others and wait for the technology transfers. We must make original achievements based on our own strength" (Wen, 2009). At the Eighth Meeting of China Association for Science and Technology on May 28, 2011, Wen Jiabao also pointed out that "achievements of modern technological revolution come mainly from pioneering basic research" and "as a large country, we must have our own basic and cutting-edge research"; therefore, we must "pay more attention to and make more efforts on basic and cutting-edge research", "increase the support for basic and advanced research and the share of basic and advanced research investment in S&T investment, and provide long-term steady support for such research via our funds in order to maintain its continuity and stability"(Wen, 2011).

As a developing country, China should particularly attach importance to the impact of basic research on independent innovation strategies. A country's funding for basic research mainly comes from the government's R&D investment. Although the share of China's investment in experimental development has increased significantly over the past 10 years, the share of applied research, especially basic research, has been declining continuously. One challenge for China to effectively implement independent innovation policies is how to allocate government's R&D resources and how to increase the investment of the government and enterprises in basic research activities. In light of this, this book mainly focuses on how to improve the basic research driven by industry, promote the contribution of basic research to industrial technological innovation, and explore the investment mechanism and mode of basic research in industrial technological innovation; in other words, this book researches on "why" we should and "how" we could carry out basic research for industrial technological innovation. This issue involves three interrelated aspects:

(I) How is the basic research investment in China's independent innovation? How is the basic research for enterprises and what are the problems? Why should we attach importance to basic research for industrial technological innovation? Why is it that China's overall R&D investment and enterprises' R&D investment are growing, but its industrial core technological ability is still weak, especially for those high-tech industrial innovations that depend on scientific discovery? Why is "technological imitation—technological catch-up—another technological catch-up" a recurring process for Chinese enterprises?

(II) China has made significant achievements in terms of basic scientific research, but a lot of these achievements are results of national projects and have not been taken into the consideration of industrial needs. How do the investment in and allocation of resources from national S&T programs balance with the demands of industrial technological innovation? Under the background of open innovation, how can we balance the relationships between enterprises, universities and research institutes on basic research? How can we balance the relationship between internal basic research and external knowledge acquirement of enterprises? How can we promote the collaboration between the government and market on the allocation of basic research resources?

(III) In order to accelerate the independent innovation breakthroughs, what should China's S&T policies, especially basic research policies, focus on? For example, what adjustments should be made for existing institutional mechanisms of China's basic research funding and allocation? How can we design more effective incentive policies to guide enterprises to increase investment in basic research or to promote research collaboration between industry, universities/research institutes, and enterprise on basic research for industrial technological innovation?

Basic research is at the front end of the whole innovation chain and the focus of modern industry competition. In the 12th Five-Year Plan (FYP) for National Economic and Social Development of the People's Republic of China, China clearly required that the share of China's R&D funds in GDP should reach 2.2% by 2015, and the number of patents for every 10 thousand people should reach 3.3. The major problem for S&T management is how to provide continuous and steady support mechanisms and institutions for basic research, so as to increase the contribution of basic research to industrial technological innovation. In recent years, some scholars have carried out some studies on the basic research of Chinese enterprises (Wen & Li, 2007; Liu & He, 2011; Cheng et al., 2011), but a complete analysis framework has not been established yet. Currently, the focus of international industrial competition has increasingly transferred to the front end. Under this background, in order to improve its innovation ability of core technologies, it is inevitable for China to increase its investment in basic research, promote the rational allocation of resources among enterprises, universities and research institutes, and enhance the contribution of basic research to industrial development. For the Chinese government, it is not easy to solve the allocation problem of R&D funds. Increasing the investment of the government and enterprises in basic research activities will not only be a critical long-term task but also be a major challenge for the government to effectively implement independent innovation policies. Achievements from research on the investment pattern of basic research resources for industrial technological innovation and S&T policies of basic research for industrial development will be of great academic value and profound practical significance.

On the one hand, enterprises are reluctant to invest in basic research because the output of basic research has strong nature of externality. On the other hand, lacking basic research capacity, enterprises are unable to launch new products with outstanding competitiveness. In theory, there are two ways to increase the basic research capacity of industries:

(I) Enterprises initiatively realize the long-term benefits of basic research, and actively invest resources in the form of the basic research laboratory, which may bring about exclusive gains from innovation breakthroughs years later. For example, in the United States, there is the first internal laboratory established by DuPont, the Bell Labs of AT&T, and the laboratories of Xerox, which have been guiding the direction of industrial development for decades and even a century and have obtained monopolistic innovations. However, this mode requires enterprises to have both strong financial strength and innovative spirit to be able to see the technological development trends of industries, and to have corporate scientists whose abilities are at the same level as that of scientists in universities and research institutes. Japan once imitated this mode while catching up, based on the fact that they had many outstanding research personnel in some disciplines and fields, but we cannot fully imitate it under the present situations in China.

(II) If enterprises cannot initiatively recognize the importance of basic research and also lack outstanding scientists who can see the future direction of industrial technology, how can scientists in universities and research institutes assume this responsibility? Traditionally, basic research of universities and research institutes are mainly interest-oriented projects (such as most NSFC projects) and state-planned projects (such as Program 973), whose main objective is to solve the deep-seated fundamental problems in theories, improve China's S&T level and solve the major problems of China's economy. However, what these two models invest in is too far away from the needs of enterprises, so it is hard for them to make targeted basic research. If it is difficult for enterprises to get basic research topics and spend more funds on basic research related to industrial development bottlenecks, should the government reform the resources allocation? If so, how can that be done? Should the government pay more attention to solving these basic research (i.e., industrial basic research) problems?

The starting point of this book is to analyze the relationship between basic research and innovation capability of industrial core technologies, so its core idea is that if we want to make breakthroughs and innovation of industrial core technologies, we need to invest in basic research continuously. This book also suggests that the overall weakness of basic research capability has already become a bottleneck for China's industrial innovation. Therefore, it is a great challenge for China to build its industrial core technological innovations on more effective scientific knowledge on its way to an innovation-originated country. At the end, this book makes suggestions on relevant policies.

1.3 Data sources

In terms of data sources, we have mainly referred to and cited various kinds of journals and international literature and books (including some raw data for statistical purpose). We have tracked and retrieved many domestic and international academic databases (such as ISI Knowledge database developed by Thomson Reuters Corporation). International databases mainly include websites of important working papers for innovation management, such as EBSCO, JSTOR, PQDD, SDOS, Wiley Interscience, Emerald, SSRN and SPRU. Domestic databases are mainly CNKI and CQVIP. At the same time, we have also attached great importance to the tracking of several important authoritative journals on this aspect, including *Research Policy*, *Strategic Management Journal*, *Academy of Management Journal*, *Administrative Science Quarterly*, *Technovation*, *Science and Public Policy* and so on.

In terms of statistical analysis of the data, we have adopted domestically and internationally authoritative scientific and patent statistics databases, such as OECD's Main Science & Technology Indicators, Research and Development Statistics, Research and Development Database, Science, Technology and Industry Scoreboard, Science, Technology and R&D Statistics, and Analytical Business Enterprise Research and Development at different years and periods; NSB's *Science and Engineering Indicators* and *National Patterns of R&D Resources* at different periods; as well as public data released by governmental S&T statistics departments of Germany, the United Kingdom, Japan, Korea and so on. Domestic data mainly include data from the series of *China Statistical Yearbook on Science and Technology*, *International Statistical Yearbook*, *National Science and Technology Funds Statistical Bulletin* and *China Science and Technology Indicators* in different years and periods, and data of national research on R&D resources inventory, China Science and Technology Statistics website of the Ministry of Science and Technology of China, and relevant statistical data of National Bureau of Statistics of China.

In terms of cases, we have extensively collected information from domestic and international literature, industrial research reports, publicly available information and websites of industrial research organizations on LCD panels, biomedicine and carbon fiber. We have also conducted field interviews and research on some enterprises.

References

Branstetter, L., Li, G. & Veloso, F. The globalization of R&D: China, India and the rise of international co-invention. Working paper of Carnegie Mellon University, 2013.

Bush, V. *Science: The Endless Frontier*. Washington DC: US Government Printing Office, 1945.

Callon, M. Is science a public good?. *Science, Technology, & Human Values*, 1994 (19): 395-424.

Chen, J. Basic research: The foundation of independent innovation. *Anhui Science and Technology*, 2005 (6): 4-6. [陈佳洱. 基础研究：自主创新的基石. 安徽科技，2005（6）: 4-6.]

Cheng, P., et al. Basic research on and the catch-up of China's high-speed rail industry. *Management Review*, 2011 (12): 46-55. [程鹏，等. 基础研究与中国产业技术追赶：以高铁产业为案例. 管理评论，2011（12）: 46-55.]

Dyer, G. China has surpassed Japan and become the second largest country of R&D investment. *China Reform Daily*, 2006-12-06.

Freeman, C. *Technology Policy and Economic Performance*. London: Printer, 1987.

Gao, X. Independent innovation: A wrong way for enterprises?. *Science & Technology for Development*, 2006 (10): 24-31. [高旭东. 科技创新：企业迷途？. 科技促进发展，2006（10）: 24-31.]

Ge, H., Gao, X. & Tong Y. Empirical research about the firm-level endogenous self-confidence of the indigenous R&D. *Science of Science and Management of S. & T.*, 2010 (4): 14-20. [葛沪飞，高旭东，仝允桓. 我国企业自主研发内生信心因素实证研究. 科学学与科学技术管理，2010（4）: 14-20.]

Hu, J. Talent is the key to the S&T innovation. *People's Daily* (Overseas Edition), 2004-12-15. [胡锦涛. 人才是科技创新的关键. 人民日报（海外版），2004-12-15.]

Hu, J. Protecting intellectual property and making innovation achievements. State Intellectual Property Office of the P.R.C., 2011-12-

19. [胡锦涛．保护知识产权，使一切创新成果得到肯定．中华人民共和国国家知识产权局，2011-12-19．]

Leydesdorff, L. & Etzkowitz, H. Emergence of a triple helix of university-industry-government relations. *Science and Public Policy*, 1996, 23: 279-286.

Li, B. http://www.cas.ac.cn/xw/zyxw/yw/201307/t20130717_3900830.shtml, 2013-07-17.

Li, D. Machine tool industry: Facing the challenge of producing good products. *Science and Technology Daily*, 2012-03-26. [李大庆．机床工具工业：面临"做"好产品的挑战．科技日报,2012-03-26．]

Liu, X. & He, Y. Basic research is the source of industrial core technological innovation in China. *China Soft Science*, 2011 (4): 104-117. [柳卸林，何郁冰．基础研究是中国产业核心技术创新的源泉．中国软科学，2011（4）：104-117．]

Liu, X. & White, S. Comparing innovation systems: A framework and application to China's transition context. *Research Policy*, 2001, 30: 1091-1114.

Liu, X. & Zhao, J. 10 aspects of degradation of the environment of basic research in China. *Forum on Science and Technology in China*, 2001 (6): 16-19, 41. [柳卸林，赵捷．我国基础研究环境恶化的 10 个方面．中国科技论坛，2001（6）：16-19，41．]

Lu, F. Independent innovation needs courage. *Decision and Information*, 2006 (6): 15-19. [路风．自主创新需要勇气．决策与信息，2006（6）：15-19．]

OECD. Reviews of innovation policy: China, 2008.

OECD. Science, technology and R&D statistics, 2011.

Pavitt, K. Public policies to support basic research: What can the rest of the world learn from US theory and practice? (And what they should learn). *Industrial and Corporate Change*, 2001, 10 (3): 761-779.

Wen, K. & Li, L. Analysis of basic research in domestic enterprises in order to enhance indigenous innovation capability. *Science of Science and Management of S. & T.*, 2007 (2): 1-9, 20. [温珂，李乐旋．从提升自主创新能力视角分析国内企业基础研究现状．科学学与科学技术管理，2007（2）：1-9，20．]

Wen, J. S&T: Leading the sustainable development of China. China Radio International, 2009-11-03. [温家宝．让科技引领中国可持续发展．中国广播网，2009-11-03．]

Wen, J. S&T personnel should undertake the mission of promoting social development. Xinhuanet, 2011-05-28. [温家宝．科技工作者要担起推动社会发展使命．新华网，2011-05-28．]

Zhou, P. & Leydesdorff, L. The emergence of China as a leading nation in Science. *Research Policy*, 2006, 35: 83-104.

Chapter 2

Study on the Relationship of S&T Investment and Economic Development Stages

There is no doubt that S&T can promote economic development. The conclusion that "science and technology is the primary productive force", put forward by Deng Xiaoping, is an important part of the socialist theory with Chinese characteristics. As rapid S&T development has been quickly converted into real productivity in the last one hundred years, S&T has increasingly become the most active factor and main driving force of modern productive forces. S&T advancement has increasingly played a decisive role in a country's economic and social development.

But what is the relationship between S&T investment, which is an indispensable factor of S&T advancement, and economic development? It has been a research focus of scholars for a long time. On the one hand, some believe that industrial innovation capacity is low, because China's S&T investment is rather low compared with those of developed countries. On the other hand, some believe that China's S&T investment has reached a rather high level; therefore, if industrial innovation capacity is low, it should be caused by low efficiency of resources allocation, instead of low investment level. Another important question is: As China is a developing country, if we want to build an innovative country on this basis, should China re-emphasize the need for S&T investment, or, to what extent should we invest to ensure that China will become an innovative country? What is the relationship between S&T investment level and economic development stages? In other words, if we invest more than what the economic development stage needs, will it become inefficient, as the whole economy lacks the corresponding absorption capacity?

In this chapter, we first review scholars' classical economic theories on S&T investment and economic development stages. Then, we apply the regression model and the logistic curve model for the study of the relationship between S&T investment and economic development stages, and compare and analyze the results of the United States and China so as to answer the questions aforementioned.

2.1 Theories of S&T investment and economic development stages

2.1.1 Porter's theory of economic development stages

Michael Porter (2002) used the diamond theory as a research tool to examine economic performance via competitive advantage, and analyzed the process of economic development through the phenomenon of competition. He proposed that there are different driving forces behind the four stages of national economic development; hence, he described that a country will have to go through the following stages: factor-driven stage, investment-driven stage, innovation-driven stage and wealth-driven stage. Figure 2-1 shows the driving forces in the four stages. The first three stages involve successive upgrading of a nation's competitive advantages and will normally be associated with progressively rising economic prosperity. The fourth stage is one of drift and ultimate decline.

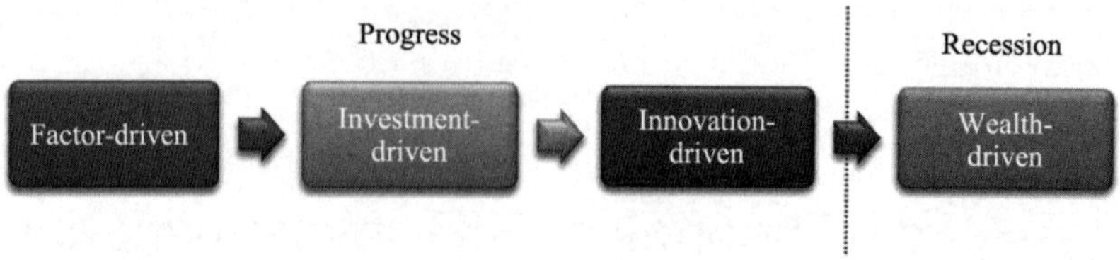

Figure 2-1 Four development stages of national economic competitiveness

1 Factor-driven stage

In the national level and at this initial stage, virtually all internationally successful industries in the nation draw their advantage almost solely from basic production factors, whether they are natural resources such as coal, oil and so on, or an abundant and inexpensive semi-skilled labor pool. In the "diamond" system of this stage, only factor conditions are an advantage.

The factor-driven stage has the following characteristics:

(1) A nation's indigenous enterprises in such an economy compete solely on the basis of price in industries that require either little product or process technology or technology that is inexpensive and widely available. Technology is sourced largely from other nations and not created. This occurs in some industries through imitation or more often through the acquisition of foreign capital goods. Industrial competitiveness is primarily competitiveness of low cost or geographic location of production factors.

(2) Products are mainly manufactured to meet low-level consumer demand. This is a stage characterized by food, as its aim is to provide enough food and clothing.

(3) Related industries and upstream industries lack international competitiveness, meanwhile forward, backward and sideways effects

of industries cannot be created. Also, industries lack communication between each other, and their interaction is limited.

(4) Very few enterprises at the national level at this stage have direct contact with end users. Foreign enterprises provide most of the access to foreign markets. Overseas trading opportunities are also completely in the hands of foreign agents. As a result, enterprises don't have international competitiveness. They mostly lack innovation ability and don't have the ability to participate in international competition.

2 Investment-driven stage

At the investment-driven stage, national competitive advantage is based on the willingness and ability of a nation and its enterprises to invest aggressively. Compared with those of the factor-driven stage, its production factors, corporate strategies, corporate structure, competitive environment and other factors have seen a marked improvement.

(1) A large part of competitive advantage is derived from the improvement of production factors. For example, enterprises increasingly employ skilled workers and technical personnel, establish some international marketing channels and even run educational institutions and research institutes.

(2) At this stage, competitive advantages come more from supply push than demand pull. Nations that are most likely to be successful in the investment-driven stage are those where their home market demand is relatively large due to local circumstances.

(3) Compared to the factor-driven stage, nations at this stage have been able to compete in a wider range of industries and industrial sectors, of which some have high entry barriers. At this stage, related and supporting industries are largely undeveloped in the nation. The production of related industries is almost all done by technologies,

equipment or components of other countries.

(4) Competitive advantages also come from a series of improvements from corporate strategy, corporate structure and competitive environment.

3 Innovation-driven stage

Many industries have already had a complete diamond system. All key elements of the diamond system not only play their roles well but also exhibit the strongest interaction effects. This stage has the following characteristics:

(1) Competitive advantages due to factor costs become increasingly rare. This stage is called innovation-driven because enterprises not only appropriate and improve technology and methods from other nations but create them. A great many industries expedite their pace of improvement and innovation thanks to the increasing new players in the industry. Important industrial clusters begin to emerge in world-class supporting industries. In respect of the overall environment, a more advanced system of infrastructure, research institutes and universities is forming.

(2) It has developed from a stage characterized by food and with an aim of providing enough food and clothing to the well-off stage attaching equal importance to other material consumption.

(3) The competitiveness of downstream industries promotes the development of upstream and supply industries; downstream industries might do the same activities that are usually done by upstream industries. Competitive advantages spread among industries. At this time, improvement of national productivity is not dependent on the export performance of a handful of industries; instead, it depends on the market performance of various industries.

(4) Enterprises will compete in the international market via

differentiation. Although their competition is indispensable from cost, they focus more on the performance of advanced and high-level technologies. The differentiation of technology and product becomes the focus of competition. Stimulated by the development of the manufacturing industry, domestic service industries will also be internationalized. The most important role of the government is to create advanced production factors and improve the quality of demand, for example, setting strict product standards, improving health insurance and environmental protection and so on.

4 Wealth-driven stage

The driving force of the wealth-driven stage is the wealth that has been accumulated. This stage is a turning point, after which the economy might develop from progress to recession. The main characteristics of this stage are as follows:

(1) Successful wealth accumulation in the past has changed the domestic capital market structure. Investors begin to retain wealth instead of accumulating wealth.

(2) People become more interested in work in other fields than work in industry.

(3) Economic viability begins to decline, enterprises begin to lose their international competitive advantages, and mergers and acquisitions become frequent. A large number of enterprises begin to reduce their corporate size.

2.1.2 Classification of economic development stages in global competitiveness report

Since 1979, the World Economic Forum, headquartered in Geneva, Switzerland, has published a Global Competitiveness Report annually, which divides economic development stages into three: factor-driven stage, efficiency-driven stage and innovation-driven stage.

The factor-driven stage is the first stage of economic development, at which countries rely on their production factors (mainly unskilled labor and natural resources) to compete. Enterprises sell basic products and commodities and rely on price to compete; therefore, their return is low because of their low productivity. Competitive advantages at the factor-driven stage mainly rely on well-functioning systems, well-developed infrastructure, stable macroeconomic environment, as well as healthy labor force with basic education.

With the development and progress of the country and the increase of productivity, return will also increase. As a result, the country's competitiveness will be improved continuously. The economic development of the country will get into the efficiency-driven stage from the factor-driven stage. At the efficiency-driven stage, the country must begin to develop more efficient production processes and improve product quality. At this time, enhancement of competitiveness mainly relies on higher education and training, highly efficient commodity market, well-functioning labor market, developed financial market, ability to earn profit from existing technologies and large domestic or foreign markets.

Ultimately, the country's economic development will enter the innovation-driven stage; at this stage, only businesses that can produce new and unique products will survive. The main source of national competitiveness will be constant development of new products and utilizing complicated manufacturing processes to create new and unique products. Determining factors of national competitiveness at different stages of economic development are shown in Figure 2-2.

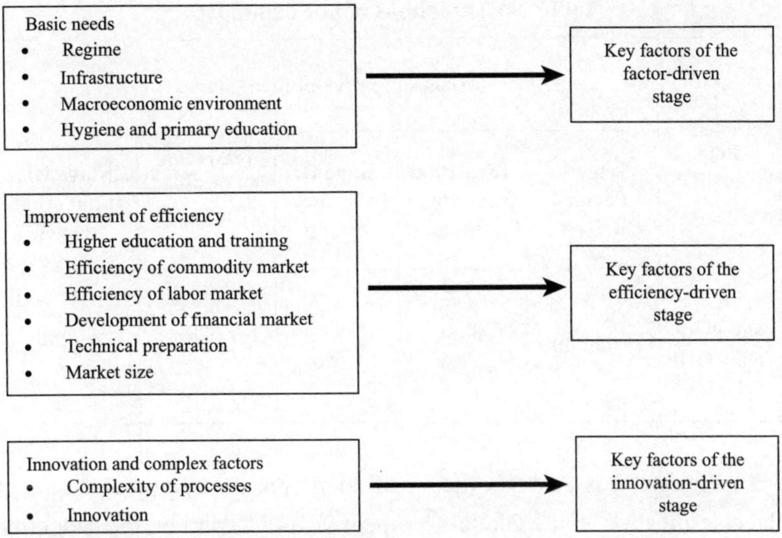

Figure 2-2 Determining factors of national competitiveness at different stages of economic development

The Global Competitiveness Report mainly uses two indicators to classify the major economic development stages of countries in the world. One is per capita GDP at the market exchange rate. This indicator is considered a proxy variable for wage, because there is no internationally comparable data on wages (threshold of per capita GDP at each stage is shown in Table 2-1). Another indicator is the total export/import ratio of mineral products. If this ratio exceeds 70%, it is believed that the country's economy is at the factor-driven stage.

Table 2-1 Threshold of per capita GDP

Item	Economic development stages				
	Stage I: Factor-driven	Transition from Stage I to Stage II	Stage II: Efficiency-driven	Transition from Stage II to Stage III	Stage III: Innovation-driven
Threshold of per capita GDP (USD)	< 2,000	2,000–2,999	3,000–8,999	9,000–17,000	> 17,000

Whether it is Porter's theory of economic development stages or the classification of the Global Competitiveness Report on the economic development stages, both have adequately focused on the promoting role of technologies on economic growth and believed that different stages of economic development have different needs for technological advancement. Also, the investment intensity of S&T resources needs to match with the needs of economic development stages in order to give full play to these resources. For example, Porter believed that at the factor-driven stage, the technologies used by domestic enterprises are widely spread and common technologies that are easy to get. The needs for technologies are rather low, and the enterprises themselves lack innovation ability; so too much S&T investment will not result in beneficial technological progress. Instead, it will occupy the resources that should be used for expanded reproduction and adversely influence the development of economy. However, at the innovation-driven stage, enterprises will compete in the international market via differentiation. Although their competition is still indispensable from cost, they focus more on the performance of advanced and high-level technologies. The differentiation of technology and product becomes the focus of competition. At the innovation-driven stage, technological innovation has become the main source of competitive advantages, and corresponding adequate technological resources have become a necessary requirement for maintaining competitive advantages.

2.1.3 Neoclassical theory of economic growth on technological progress

In the 1950s and 1960s, the neoclassical theory of growth with representatives such as Solow (1956) and Swan (1956) began to rethink the possibility of long-term economic growth. This theory is based on the model of Cobb-Douglas production function. Based on the assumption of diminishing returns, the theory believes that without introduction of some exogenous factors, the neoclassical growth model that is based on the neoclassical production function will incvitably go into an equilibrium state of zero growth. Solow (1957) has proposed the residual method under the assumption of neutral technical progress, which is known as Solow Residual, and calculated the contribution rate of technological progress to economic growth. However, this model only shows the dynamic change process of per capita capital's accumulation, and cannot explain the technological progress mechanism that can lead to long-term economic growth; therefore, this model cannot explain the root cause of long-term economic growth.

2.1.4 New economic growth theory on technological progress

The new economic growth theory with representatives such as Lucas (1988) and Romer (1986) regards technological progress as endogenous variable and believes that the law of diminishing marginal benefits of capital can be avoided based on the background of technological progress. Becker & Murphy (1992) have compared the S&T investment share of GDP at OECD countries and believe there is a close relationship between S&T investment and economic growth. Guellec & de la Potterie (2001) have compared the long-term impacts of different types of R&D investment on the growth of productivity. All of these studies have confirmed the facilitating role of S&T investment for economic development. However, some scholars also believe the economic development will in turn influence the advancement of technology, and the more economically developed a country is, the more S&T investment it will make and as a result its S&T progress will be promoted.

2.2 S&T investment and economic development in the United States

We believe that there should be a matching relationship between S&T investment and economic development stages: At a certain stage of economic development, there should be a "most appropriate" intensity of S&T investment; if the intensity of S&T investment is too low, the long-term economic growth potential will be affected, while if the intensity of S&T investment is too high, the share of investment for expanded reproduction will be occupied, thus adversely influencing economic growth. In the following sections, we study the relationship between S&T resources investment and economic development stages in China and the United States with R&D investment as an important characterization of S&T investment, and per capita GDP as a measurement method of economic development stages.

The R&D investment and per capita GDP of the

United States from 1981 to 2008 are shown in Figure 2-3. From this figure, we can see that the R&D investment and per capita GDP of the United States from 1981 to 2008 have maintained a rather consistent growth momentum. Meanwhile, is there a long-term and stable relationship between the two? We have conducted further studies on the relationship between the two via a time series model.

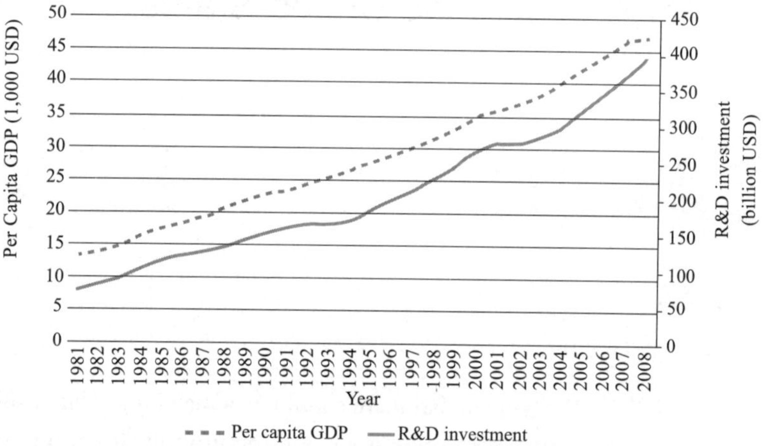

Figure 2-3 R&D investment and per capita GDP in the United States

2.2.1 Co-integration test of R&D investment and per capita GDP in the United States

If the means or auto covariance functions of time series changes over time, the series is a non-stationary time series. Time series analysis of non-stationary time series will result in a "spurious regression" problem. In order to get meaningful comparison, we can stabilize it and do a regression for the difference in sequence. However, this will miss some useful information contained in the original time series, which is quite necessary. For this, we can adopt the co-integration theory and conduct co-integration analysis. The prerequisite of the co-integration test is that the time series is non-stationary and unstable and has a unit root. Therefore, we must first conduct a unit root test for relevant data. We study and then establish a regression model of non-stationary time series.

The time series diagram of the United States' per capita GDP is shown in Figure 2-4, and the results of the unit root test are shown in Table 2-2 and Table 2-3, of which Table 2-2 is the results of ADF test and Table 2-3 is the results of PP test. Test results all show that there is no unit root for the time series data of per capita GDP in the United States from 1981 to 2008 (Figure 2-4).

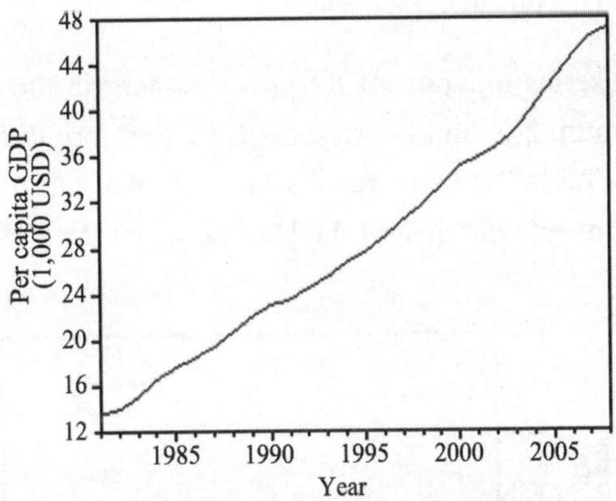

Figure 2-4 Time series chart of per capita GDP in the United States

Table 2-2 Unit root test of per capita GDP in the United States (ADF test)

Item		*t*-Statistic	Prob.*
Augmented Dickey-Fuller test statistic		2.621486	1.0000
Test critical values	1% level	−3.737853	
	5% level	−2.991878	
	10% level	−2.635542	

Null Hypothesis: GDPPC has a unit root.
Lag Length: 3 (Automatic based on SIC, MAXLAG = 6).
*MacKinnon (1996) one-sided *p*-values.

Table 2-3 Unit root test of per capita GDP in the United States (PP test)

Item		Adj. *t*-Stat	Prob.*
Phillips-Perron test statistic		−3.138929	0.0355
Test critical values	1% level	−3.699871	
	5% level	−2.976263	
	10% level	−2.627420	
Residual variance (no correction)		0.003720	
HAC corrected variance (Bartlett kernel)		0.005784	

Null Hypothesis: GDPPC has a unit root.
Bandwidth: 2.
*MacKinnon (1996) one-sided *p*-values.

The time series diagram of R&D investment in the United States is shown in Figure 2-5, and the results of the unit root test are shown in Table 2-4 and Table 2-5. Test results have shown that there is no unit root for R&D investment data of the United States from 1981 to 2008.

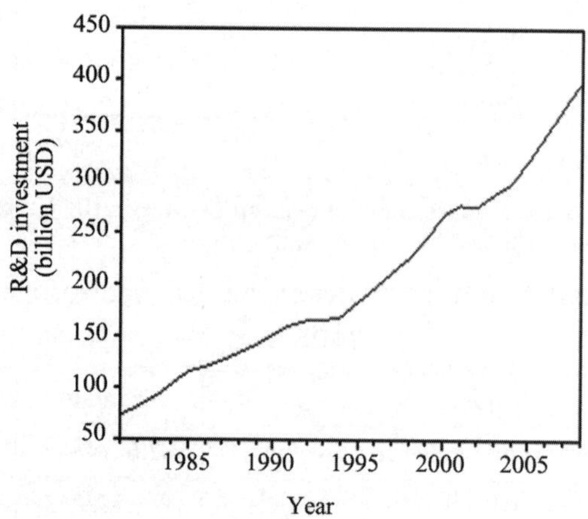

Figure 2-5 Time series chart of R&D investment in the United States

Table 2-4 Unit root test of R&D investment in the United States (ADF test)

Item		*t*-Statistic	Prob.*
Augmented Dickey-Fuller test statistic		1.769936	0.9995
Test critical values	1% level	−3.711457	
	5% level	−2.981038	
	10% level	−2.629906	

Null Hypothesis: RD has a unit root.
Lag Length: 3 (Automatic based on SIC, MAXLAG = 6).
*MacKinnon (1996) one-sided *p*-values.

Table 2-5 Unit root test of R&D investment in the United States (PP test)

Item		Adj. *t*-Stat	Prob.*
Phillips-Perron test statistic		2.852695	1.0000
Test critical values	1% level	−3.699871	
	5% level	−2.976263	
	10% level	−2.627420	

Null Hypothesis: RD has a unit root.
Bandwidth: 2.
*MacKinnon (1996) one-sided *p*-values.

2.2.2 OSL regression analysis of R&D investment and per capita GDP in the United States

Both the results of ADF test and PP test have shown that the United States' R&D investment and per capita GDP data from 1981 to 2008 have remained stationary. According to the theories of time series analysis, we can use OLS regression to analyze stationary data directly. The results of OLS regression analysis are shown in Table 2-6.

Table 2-6 OSL regression analysis results of R&D investment and per capita GDP in the United States

Item	Coefficient	Std. Error	t-Statistic	Prob.
C	–512.1321	51.43474	–9.956930	0.0000
GDPPC	896.4930	17.01483	52.68893	0.0000
R^2	0.990721	Mean dependent var		2044.136
Adjusted R^2	0.990364	Durbin-Watson Test		0.468693
F-statistic	2776.123	Prob. (F-statistic)		0.000000

Dependent Variable: RD.
Method: Least Squares.
Sample: From 1981 to 2008.
Included observations: 28.

From the OSL regression analysis results, we have found that there is a significant positive correlation between R&D investment and per capita GDP in the United States. This shows that there is a long-term stable relationship between R&D investment and economic development stages and the same relationship between S&T investment and economic development stages in the United States.

We have also studied the R&D investment and per capita GDP data of Japan, Germany and Korea and found similar conclusions with that of the United States, whose regression results are not shown in this book due to the constraint of limited pages. Therefore, we believe the assumption that there is a matching relationship between S&T investment and economic development stages is right.

2.2.3 Logistic curve fitting of R&D investment and per capita GDP in the United States

By analyzing the data of the United States, Japan and other countries, we have been able to conclude that the efficiency can be maximized when there is a matching relationship between S&T investment and economic development stages. Now, let's analyze the matching relationship between S&T investment and economic development stages in the

United States and China via the Logistic curve.

The Logistic function, also known as the growth curve function, was first found in the biological reproduction study by the American biologist and demographer Raymond Pearl (1879–1940) and Lowell Reed (1888–1966), and was then widely applied to describing the processes of biological growth and industrial growth. Economic growth and S&T investment, in essence, also a gradual process of growth, so there will be rather high similarity and feasibility if we use the Logistic growth curve function to analyze the characteristics of its stages. Developed countries such as the United States and Japan have quite a slow rate of economic growth in recent years and have entered their mature stage, while China's economy has enjoyed a rapid growth ever since its reform and opening-up and is now in its growth stage.

The Logistic function curve is kind of an elongated S-shaped curve as shown in Figure 2-6. The part before point *A* $(0 - t_1)$ is called an introductory phase, which is characterized by slow growth. The part between point *A* and point *C* $(t_1 - t_3)$ is called the growth phase, featured by rapid growth; there is an inflection point *B* between points *A* and *C*. From point *A* to point *B*, it has an increasing growth rate; while from point B to point C, it has a decreasing growth rate. Both growth rates are rather high, so point *B* is the prime period of growth stage. After point *C* $(t_3 - \infty)$, the growth slows down.

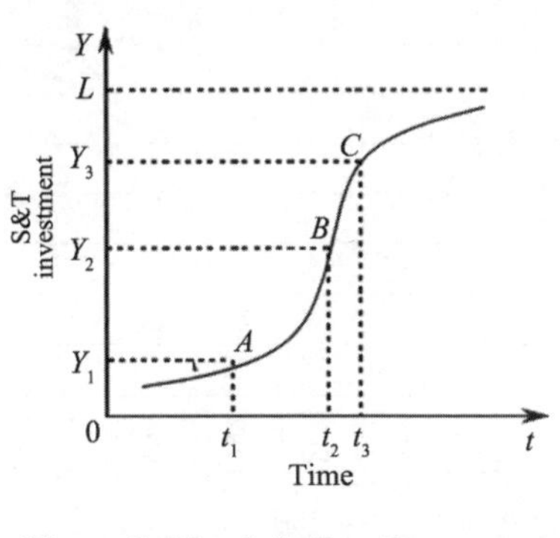

Figure 2-6 Logistic function curve

Generally, the Logistic function equation is $Y=\frac{L}{1+ae^{-bt}}$, of which t is time and Y is a dependent variable which represents S&T investment or per capita GDP, L is maximum limit, and a and b are controllable parameters. Their values are determined by the characteristics of growth itself and the spreading environment.

If we calculate the first derivative of the Logistic function and make it 0, we get $t=\frac{\ln a}{b}$ and point B. We continue to calculate the second derivative of the Logistic function and make it 0, and then we get $t_1=\frac{\ln a-1.317}{b}$ and $t_3=\frac{\ln a+1.317}{b}$, that is, the values of time t on horizontal axis for point A and point C.

If we calculate the time values of point A, point B and point C on the economic development stage curve and S&T investment curve, we can compare and get the characteristics of the two curves' stages, and know whether they are matching.

We have obtained the United States' per capita GDP data from 1970 to 2010 and R&D investment data from 1981 to 2008 through OECD. StatExtracts Database is shown in Figure 2-7.

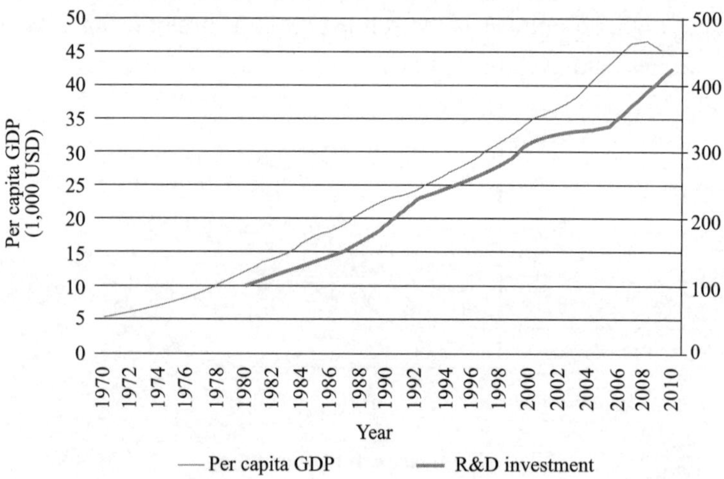

Figure 2-7 Per capita GDP and R&D investment in the United States

First, we need to determine the ideal value of the United States' per capita GDP and R&D investment: Normally, it is the L in the Logistic function equation of $\frac{L}{1+ae^{-bt}}$. The data shows that the annual compound growth rate of the United States' per capita GDP from 2005 to 2010 is 1.89%. If we assume that the United States will maintain this growth rate, the per capita GDP of the United States will reach 56,206.95 USD 10 years later; so we set this value as the ideal value of the United States' 10 per capita GDP. Similarly, we set 477.804 billion USD as the ideal value of the United States' R&D investment.

Logistic curve fitting results of the United States' per capita GDP and R&D investment are shown in Table 2-7, Figure 2-8, and Figure 2-9. The fitting results show that the United States' per capita GDP has entered its growth stage since 1981, reached its peak in 1995 with the highest growth rate and slowed down after entering into a mature stage in 2009. The United States' R&D investment started its growth stage from 1985, with growth rate at its highest point in 1997 and slowed down since 2010.

Table 2-7 The United States' per capita GDP growth curve index

Item	Per capita GDP	R&D investment
a	11.373	5.567
b	0.097	0.104
R^2	0.993	0.967
t_1	11.48	3.82
t_2	24.98	15.46
t_3	38.64	29.1
Start year of growth stage	1981	1985
Year of prime stage	1995	1997
Start year of mature stage	2009	2010

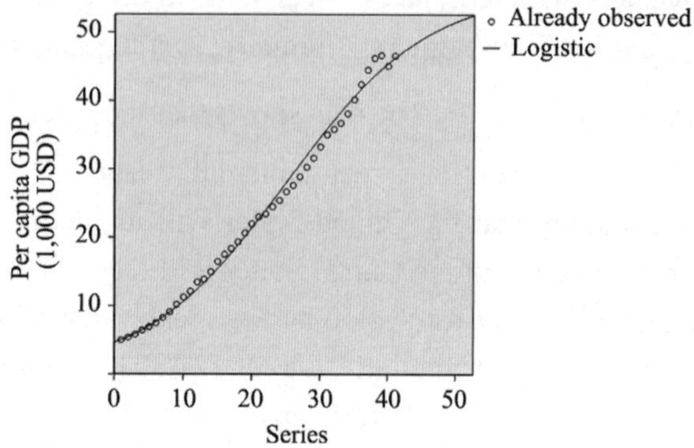

Figure 2-8 Logistic curve fitting results of the United States' per capita GDP

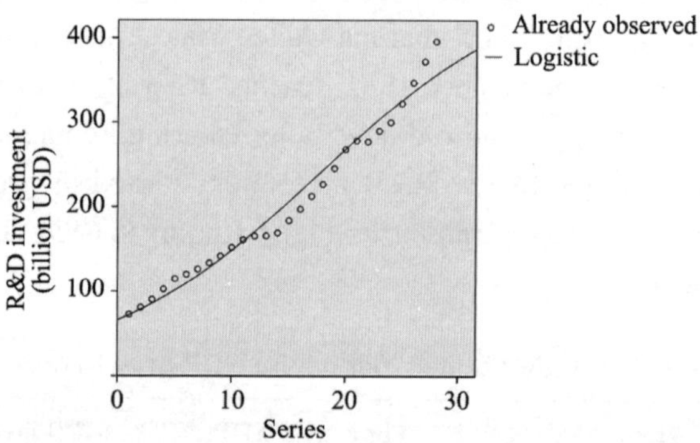

Figure 2-9 Logistic curve fitting results of the United States' R&D investment

From the perspective of economics, industrial structure will also change along with economic growth. The shares of the United States' three industries in GDP from 1970 to 2010 are shown in Table 2-8. With the comparison of Table 2-7 and Table 2-8, we can see that the growth of the United States' per capita GDP entered its growth stage in 1981; from 1970 to 1980, the United States' per capita GDP has had a relatively slow growth rate. The share of the United States' tertiary industry almost remained at the same level, while the share of the primary industry decreased rapidly from 7.3% to 2.5%, while the share of the secondary industry increased to 33.5%. After entering the growth stage, its primary industry's share changed a little, only gradually declining from 2% to about 1%; its secondary industry's share declined year by year from 33.5% in 1980 to 18.6% in 2009; its tertiary industry's share increased year by year from 64% in 1980 to 80.4% in 2009. These data indicate that before the United States entered the growth stage, economic growth mainly depended on the growth of the secondary industry. But after entering the growth stage, the main driving force of economic growth became the tertiary industry. However, as the development of the tertiary industry slowed down since 2008, the United States' economic growth entered its mature stage and its growth began to slow down.

Table 2-8 Shares of the United States' three industries in GDP from 1970 to 2010

Year	Primary industry (%)	Secondary industry (%)	Tertiary industry (%)
1970	7.3	28.7	64.0
1980	2.5	33.5	64.0
1985	2.1	31.1	66.8
1990	2.1	28.0	69.9
1995	1.6	26.8	71.7
1997	1.7	26.2	72.1
2000	1.2	24.2	74.6
2004	1.3	22.0	76.7
2008	1.1	19.8	79.0
2009	0.9	18.6	80.4
2010	1.1	18.9	80.0

Analyzing the relationship between the United States' R&D investment and structure of three industries, we find that along with the rapid development of its secondary industry, the United States' R&D investment had increased and entered into a growth stage in 1985. After this, the United States' economy developed rapidly, its share of the tertiary industry upgraded quickly, and its tertiary industry became the second driving force of its rapid growth of R&D investment. However, the S&T need generated by the development of the tertiary industry was relatively low compared to that of the secondary industry; so its promotion for the increase in R&D investment was weak. The growth of R&D investment began to slow down quickly when the development pace of the tertiary industry started to slow down. The growth stage of the United States' per capita GDP is from 1981 to 2009; the growth stage of its R&D investment is from 1985 to 2010; it is in 2008 that its tertiary industry started to slow down.

2.3 Logistic curve fitting of R&D investment and per capita GDP in China

If we simulate growth curves of China's R&D investment and per capita GDP development, we can determine the characteristics and current status of their life cycle curves and compare their priorities horizontally. The growth curve can be easily estimated using the SPSS software, but before estimating, we need to determine the ideal values of China's R&D investment and per capita GDP. In 2008, the United States' GDP was 14.33 trillion USD, about three times that of China's. Therefore, we set the ideal values of China's R&D investment and per capita GDP as three times of that in 2008. We used EVIEWS6.0 to simulate the growth curves of China's per capita GDP and R&D investment, got the values of parameters, and then calculated the main time feature points. Overall, the best fit with R^2 is very high, with all above 0.98, as shown in Table 2-9.

Table 2-9 Growth curve indexes of China's S&T investment and economic development stages

Item	Per capita GDP	R&D investment
a	47.498	46.063
b	0.161	0.147
R^2	0.983	0.995
t_1	15.47	17.12
t_2	23.98	26.09
t_3	32.49	33.05
Start year of growth stage	2005	2007
Year of prime stage	2014	2016
Start year of mature stage	2022	2023

Based on our analysis, China's economic development entered the growth stage in 2005 and will reach its inflection point in 2014, before which the growth rate of economy will increase and after which will decrease. Then, in 2022, it will enter the mature stage, after which economic growth will slow down. China's S&T investment entered the growth stage in 2007, two years later than the economic growth. It will reach its inflection point in 2016 by our forecasting, two years later than the economic growth, and then enter its mature stage in 2023.

China's structure of three industries from 2001 to 2010 is as shown in Table 2-10. Comparing Table 2-10 with Table 2-8, we can find that the stage China is now in is similar to that of the United States during 1970 to 1985: Both have their share of the primary industry declining and share of the secondary industry and the tertiary industry increasing, and per capita GDP and R&D investment entering their growth stage successively with accelerated growth.

Table 2-10 China's structure of three industries from 2001 to 2010

Year	Primary industry (%)	Secondary industry (%)	Tertiary industry (%)
2001	14.4	45.1	40.5
2002	13.7	44.8	41.5
2003	12.8	46.0	41.2
2004	13.4	46.2	40.4
2005	12.1	47.4	40.5
2006	11.1	48.0	40.9
2007	10.8	47.3	41.9
2008	10.7	47.5	41.8
2009	10.3	46.3	43.4
2010	10.1	46.8	43.1

Nowadays, China is in a stage where its secondary industry and tertiary industry are developing rapidly, and a stage with the highest need for S&T. Hence, the state needs to increase its S&T investment. Compared with that of the United States, the intensity of China's R&D investment was obviously far from enough. The share of R&D investment in GDP in the United States has remained above 2% since 1981, while that of China did not exceed 1% until 2002 (Figure 2-10).

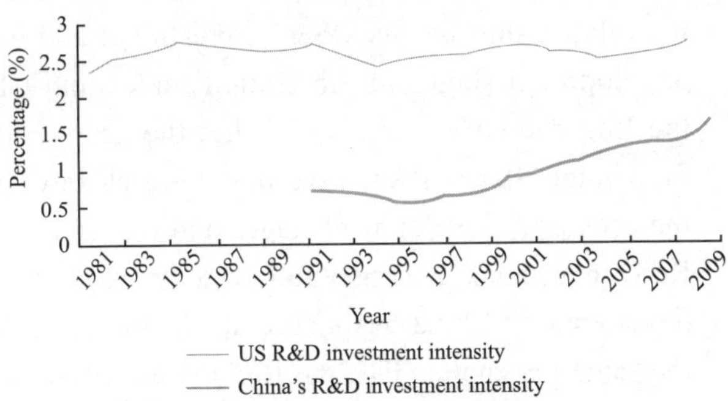

Figure 2-10 R&D investment intensity of the United States and China

2.4 Summary

This chapter has first defined the connotation and denotation of S&T resources and then it has reviewed existing research results of the relationship between S&T resources investment and economic development stages. We found that there is a long-term stable correlative relationship between R&D investment and per capita GDP in countries such as the United States, Japan and Germany via co-integration analysis. This chapter has also studied the relationship between S&T investment and economic development stages in the United States and China via the Logistic curve, and found that the growth stages of the United States' R&D investment are closely related to the economic development stages and the growth trend of R&D investment obviously followed the trend of economic development. China has a similar phenomenon, but with the same per capita GDP, the R&D intensity of the United States is obviously higher than that of China, reflecting that China should further strengthen its R&D investment.

References

Becker, G. S. & Murphy, K. M. The division of labor, coordination costs, and knowledge. *The Quarterly Journal of Economics*, 1992, 107 (4): 1137-1160.

Guellec, D. & De la Potterie, B. P. R&D and productivity growth: Panel data analysis of 16 OECD countries. *OECD Economic Studies*, 2001/II (33): 103-126.

Lucas, R. E. On the mechanics of economic development. *Journal of Monetary Economics*, 1988, 22 (1): 3-42.

Porter, M. *The Competitive Advantage of Nations*. New York: The Free Press, 2002.

Romer, P. M. Increasing returns and long-run growth. *The Journal of Political Economic Performance*, 1986: 1002-1037.

Solow, R. M. A contribution to the theory of economic growth. *The Quarterly Journal of Economics*, 1956, 70 (1): 65-94.

Solow, R. M. Technical change and the aggregate production function. *The Review of Economics and Statistics*, 1957, 39 (3): 312-320.

Swan, T. W. Economic growth and capital accumulation. *Economic Record*, 1956, 32 (2): 334-361.

Chapter 3

S&T Resources Allocation and Corresponding Efficiency of China

In Chapter 2, we have concluded that although China's S&T investment has been increasing for several decades, it still needs to be further increased when compared with those of some developed countries in the world, in order to meet the needs of China to build an innovative country.

Another important question is: Why is there less significant innovation for China even though it has invested in S&T for over 10 consecutive years with a high growth rate? This relates to the efficiency of S&T investment. So, it is better to analyze the output efficiency of China's S&T input-output efficiency in domestic universities, research institutes and enterprises, in an attempt to draw some useful conclusions.

3.1 China from a global perspective: International comparison of major S&T indexes

3.1.1 Investment in S&T resources increasing significantly

In 2010, R&D expenditure of the world (40 countries, similarly hereinafter) reached 1.2 trillion USD, 1.5 times that in 2000. From a global perspective, from 2000 to 2010, R&D expenditure of emerging countries such as China, Korea, Brazil, India and Russia have maintained a good momentum of growth, with an average annual growth rate of 18.2%, 9.4%, 8.7%, 8.6% and 5.9% respectively. However, the R&D expenditure of developed countries such as the United States, Japan, Germany and the United Kingdom have a relatively low growth rate, less than the world's average (4.1%). Switzerland even had a negative growth rate. In the past 10 years, China's R&D expenditure has enjoyed a rapid growth rate in the world, reaching 104.32 billion USD in 2010, when for the first time its R&D expenditure exceeded 100 billion USD and surpassed Germany and ranked third in the world. Its share of global total also increased from 1.7% in 2000 to 8.6% in 2010, with a narrower gap with those of the United States and Japan. Even though its share of global total R&D expenditure is decreasing year by year, the United States, as a powerful S&T country, still enjoys a leading position in respect of total R&D expenditure. In 2009, its total R&D expenditure reached 401.58 billion USD, 2.4 times that of Japan, which was ranked second in the world.

In 2010, R&D personnel in the world reached 10.614 million, 1.4 times that of 2000. From 2000 to 2010, R&D personnel of most countries, except countries such as Romania and Russia, were on the rise, especially obvious for emerging economies. For example, average annual growth rates of R&D staff in countries such as Turkey, China, India and the Czech Republic reached 11.7%, 10.7%, 8.8% and 8.0% respectively, much higher than the global average (3.4%). In 2010, China's total R&D staff reached 2.554 million, ranking top in the world and increasing from 12.1% of the world's total in 2000 to 24.1% in 2010.

3.1.2 Knowledge output reaching a leading position in the world

From 2000 to 2010, the number of SCI papers of the world has an annual growth rate of as high as 4.6%, with a total of 1.367 million papers in 2010. Overview of the world shows that SCI papers of countries such as China, Brazil, Korea and India have a higher growth rate than those of developed countries. In 2010, China had 125,000 papers collected by SCI database, ranking second only to the United States in the world. China's share of total SCI papers rose from 2.8% in 2000 to 9.1% in 2010. In the past decade, the number of China's SCI papers has enjoyed an average annual growth rate of 17.7%, with the highest growth rate in the world. In 2010, the United States' SCI papers collected by SCI database was 330,000, accounting for 24.2% of the world's total, dropping by 5 percentage points compared with that of 2000.

During 2006 to 2010, SCI papers have been cited 36,262,000 times in the world, two times that during 1996 to 2000. While the number of China's SCI papers is in rapid growth, the quality of these papers is also improving. Citation times of China's SCI papers have increased from 128,000 times during 1996 to 2000 to 1,669,000 times during 2006 to 2010, with an increase of about 12 times, ranking sixth in the world. Top five countries of citation times are the United States, the United Kingdom, Germany, Japan and France, with 11.418 million times, 2.982 million times, 2.7 million times, 1.849 million times and 1.787 million times respectively.

In 2010, the quantities of global patent applications and grants reached 1.743 million and 0.504 million respectively, with an increase of 44.1% and 76.7% respectively compared with those of 2000. China's growth of patents has played an important role in the global growth, with contribution as high as 63.6% and 33.6% respectively. From 2000 to 2010, the quantities of global patent applications and grants in China have had an average annual growth rate of 22.4% and 29.2% respectively, much higher than those of the world's average (3.7% and 5.9% respectively). In 2010, the application quantity of China's patents reached 391,000, accounting for 22.4% of the world's total, surpassing Japan and ranking second in the world for the first time. Its grants quantity reached 80,000, accounting for 15.8% of the world's total, ranking only behind Japan and the United States. In addition, quantities of patent applications and grants in India also maintained a sound growth momentum, with an average annual growth rate of 14.1% and 15.7% respectively, in the past 10 years.

3.1.3 High-tech industry growing rapidly

In 2010, the total added value of high-tech industry in the world reached 1.3 trillion USD, 1.3 times that of 2000, with an average annual growth rate of 2.7% in the past decade. In 2010, the added value of high-tech industry of the United States reached 385.94 billion USD, accounting for 29.3% of the world's total, ranking top in the world. During 2000 to 2010, China's added value of high-tech industry had an average annual growth rate of 19.6%, reaching 263.02 billion USD in 2010, accounting for 20.0% of the world's total. After surpassing Japan in 2007, China had its added value of high-tech industry ranking second in the world for four consecutive years, indicating increasingly narrowing gaps between China and the United States.

3.1.4 Exports of high-tech industry maintaining the top position in the world

In view of the world, high-tech industry exports of countries, except Belgium, India, Israel, Luxembourg and Spain, have all declined in 2009 to different extents compared with those of the prior year. The value of total high-tech industry export was 1.5 trillion USD in 2009, with a decrease of 15.4% compared with that of the prior year. In the same year, China's high-tech industry exports reached 348.29 billion USD, with a decrease of 8.7% compared with that of the prior year, ranking first in the world for five consecutive years. China's share of high-tech industry exports of the global total continued to increase rapidly, from 4.2% in 2000 to 23.7% in 2009. In 2009, Germany's high-tech industry exports surpassed that of the United States and ranked second in the world for the first time, accounting for 9.7% of the global total. Other three in the top five countries of high-tech industry exports were the United States, Korea and Japan, with a share of 9.6%, 7.0% and 6.7% in the world's total respectively.

3.2 International comparison of S&T resources allocation efficiency

S&T resources allocation efficiency reflects a country's ability to utilize and integrate S&T resources and represents the overall functionality and efficiency of this country's S&T system. So it determines the strength of its S&T capability to some extent. As a developing country, China has a more urgent need for economical, rational and highly-efficient allocation of limited S&T resources so as to get as much output as possible from limited financial and human resources.

How is the efficiency of China's S&T resources allocation? In what aspects is the efficiency low? This chapter has tried to answer these questions through an international comparative study of S&T resources allocation efficiency. At the same time, we also calculate the S&T resources allocation efficiency based on principal component analysis and data envelopment analysis (DEA).

3.2.1 Data sources

Data of this study mainly consist of two aspects: academic indicators and economic indicators. Academic indicators are primarily obtained from the Web of Knowledge retrieval system of ISI, while economic indicators are mainly obtained from "OECD main science and technology indicators" of Issue 1 of 2011 of OECD. Through a variety of channels, we have obtained OECD countries' data such as the quantity of SCI papers, nominal GDP, quantity of three-in-one patents, enterprises' R&D input, full-time equivalent of S&T staff in enterprises, universities' R&D input, full-time equivalent of S&T staff in universities, public research institutes' R&D input, full-time equivalent of S&T staff in public research institutes during 2005 to 2009, as well as countries and regions such as China and Russia.

3.2.2 Calculation of S&T resources allocation efficiency based on principal component analysis

We believe that the efficiency of S&T resources allocation can be calculated via the calculation method of production efficiency. We have integrated S&T investment indicators such as enterprises' R&D input, full-time equivalent of S&T staff in enterprises, universities' R&D input, full-time equivalent of S&T staff in universities, public research institutes' R&D input, and full-time equivalent of S&T staff in public research institutes into one unified S&T investment indicator via principal component analysis. Also, we combined the two indicators of SCI papers quantity, and three-in-one patents quantity into one unified S&T output indicator, that is, the efficiency of S&T resources allocation.

First of all, we have normalized all data:

$$x_i' = \frac{x_i - \bar{x}}{s}$$

where x_i' is the data after being normalized, $\bar{x}$ is the mean value of the variable x, while s is the standard deviation of variable x.

Then, we conducted KMO and Bartlett tests for input data of each country in 2005. The results are as shown in Table 3-1. Its test value is 0.732, and it is obvious that the data is suitable for principal component analysis.

Table 3-1 KMO and Bartlett tests

Item		Value
Kaiser-Meyer-Olkin measure of sampling adequacy		0.732
Bartlett's test of sphericity	Approx. chi-square	448.806
	Df	15
	Sig.	0.000

After principal component analysis, we have one principal component, which can explain 87.492% of the total variance, so the information extraction is very good (Table 3-2).

Table 3-2 Total variance explained

Component	Initial eigenvalues			Extraction sums of squared loadings		
	Total	Proportion in variance (%)	Accumulated proportion (%)	Total	Proportion in variance (%)	Accumulated proportion (%)
1	5.249	87.492	87.492	5.249	87.492	87.492
2	0.540	9.001	96.493			
3	0.156	2.608	99.100			
4	0.032	0.531	99.632			
5	0.016	0.273	99.904			
6	0.006	0.096	100.000			

Extraction method: principal component analysis.

The component score coefficient matrix of indexes is shown in Table 3-3. The variance contribution factors of universities' R&D input, full-time equivalent of S&T staff in universities, public research institutes' R&D input, full-time equivalent of S&T staff in public research institutes, enterprises' R&D input and full-time equivalent of S&T staff in enterprises are 0.176, 0.186, 0.189, 0.159, 0.181 and 0.177 respectively. Based on the weight of variance contribution factors, we have integrated the six input indicators into one unified indicator.

Table 3-3 Component score coefficient matrix of input index

Item	Component 1
Universities' R&D input	0.176
Full-time equivalent of S&T staff in universities	0.186
Public research institutes' R&D input	0.189
Full-time equivalent of S&T staff in public research institutes	0.159
Enterprises' R&D input	0.181
Full-time equivalent of S&T staff in enterprises	0.177

Extraction method: principal component analysis.
Rotation method: Varimax with Kaiser Normalization.

Similarly, the component score coefficient matrix of each country's output index is shown in Table 3-4. With the unified S&T output indicator, we have further calculated the indicators of S&T resources allocation efficiency for each country and region in 2005, as shown in Table 3-5.

Table 3-4 Component score coefficient matrix of output index

Item	Component 1
Papers	0.528
Patents	0.528

Extraction method: principal component analysis.
Rotation method: Varimax with Kaiser Normalization.

Table 3-5 S&T resources allocation efficiency in different countries and regions in 2005

Rank	Country/region	Input	Output	Efficiency
1	Korea	0.057579	0.123425	2.143578
2	Chinese Taiwan	–0.17067	–0.30383	1.780181
3	Spain	–0.10387	–0.16199	1.55956
4	Japan	1.709903	2.637896	1.542717
5	Germany	0.704089	1.046715	1.486624
6	Poland	–0.31859	–0.34519	1.083498
7	France	0.338815	0.341039	1.006563
8	Czech Republic	–0.419	–0.41702	0.995282
9	Finland	–0.37973	–0.37512	0.987841
10	Singapore	–0.41882	–0.40609	0.969597
11	New Zealand	–0.45207	–0.4333	0.958478
12	Hungary	–0.44525	–0.42119	0.945952
13	Norway	–0.42003	–0.39714	0.945491
14	Argentina	–0.39193	–0.36993	0.943858
15	Romania	–0.44646	–0.41969	0.94005
16	Slovak Republic	–0.47494	–0.44644	0.939984
17	Slovenia	–0.48449	–0.45312	0.935254
18	Iceland	–0.49675	–0.46351	0.933089
19	Portugal	–0.44406	–0.41367	0.931554
20	Estonia	–0.49573	–0.46122	0.930379
21	Luxembourg	–0.49686	–0.45608	0.917941
22	Greece	–0.4413	–0.40226	0.911527
23	Denmark	–0.41411	–0.37598	0.90793
24	Ireland	–0.46424	–0.41802	0.900438
25	South Africa	–0.42323	–0.36892	0.871678
26	Austria	–0.39032	–0.33563	0.859895

Continued

Rank	Country/region	Input	Output	Efficiency
27	Belgium	–0.38334	–0.32419	0.845701
28	Sweden	–0.31709	–0.26625	0.839679
29	Turkey	–0.35573	–0.28753	0.808264
30	United Kingdom	0.314193	0.246895	0.785805
31	Netherlands	–0.26955	–0.18047	0.669531
32	Mexico	–0.34502	–0.17105	0.495762
33	China	2.250637	0.798625	0.354844
34	Russian Federation	0.640651	–0.07136	–0.11138
35	Italy	–0.09425	0.026621	–0.28245
36	Canada	0.00168	–0.10288	–61.2383

From Table 3-5, we can see that, in respect of indicators of S&T resources allocation efficiency for each country in 2005, the top 10 countries or regions were Korea, Chinese Taiwan, Spain, Japan, Germany, Poland, France, Czech Republic, Finland and Singapore. But China only ranked 33rd, showing a rather low allocation efficiency.

With the same method, we have also calculated the indicators of S&T resources allocation efficiency for each country and region in 2009, as shown in Table 3-6. The top 10 countries or regions were Japan, Spain, Chinese Taiwan, Germany, Portugal, Poland, Czech Republic, Singapore, Finland and Norway. The overall ranking had not changed, and China still lagged behind[①]. From the two tables, we can see that compared to the world, China's S&T output efficiency had increased, with ranking climbing from the 33rd position in 2005 to the 25th position in 2009.

---

① Due to data availability constraints, this chapter can only select data relatively complete in 2005 and 2009 for comparison, for data of 2006–2008 are missing seriously. In 2005, data were available from 36 countries and regions; in 2009, data were obtained from 27 countries and regions.

Table 3-6 S&T resources allocation efficiency in different countries and regions in 2009

Rank	Country/region	Input	Output	Efficiency
1	Japan	1.122279	2.142787	1.909317
2	Spain	–0.11791	–0.20987	1.779959
3	Chinese Taiwan	–0.19389	–0.33145	1.709529
4	Germany	0.565618	0.845276	1.494431
5	Portugal	–0.42116	–0.43693	1.037457
6	Poland	–0.35313	–0.36276	1.027267
7	Czech Republic	–0.4348	–0.43665	1.004247
8	Singapore	–0.43109	–0.42598	0.98815
9	Finland	–0.4096	–0.4034	0.984852
10	Norway	–0.43323	–0.42168	0.973339
11	Hungary	–0.46286	–0.44381	0.958832
12	Slovenia	–0.49372	–0.4719	0.95582
13	Slovak Republic	–0.4859	–0.46269	0.952236
14	Denmark	–0.42968	–0.40852	0.950754
15	Estonia	–0.50514	–0.47774	0.945761
16	Luxembourg	–0.50561	–0.47403	0.937532
17	Ireland	–0.47334	–0.4434	0.936743
18	Romania	–0.4653	–0.43252	0.929554
19	Turkey	–0.3486	–0.31392	0.900531
20	Austria	–0.41255	–0.36952	0.895691
21	Belgium	–0.4097	–0.36451	0.889694
22	United Kingdom	0.125453	0.106517	0.849059
23	Sweden	–0.37404	–0.29508	0.788893
24	Netherlands	–0.33283	–0.23793	0.714888
25	China	2.273659	1.081357	0.475602
26	Italy	–0.15412	–0.06144	0.398624

Continued

Rank	Country/region	Input	Output	Efficiency
27	Russian Federation	0.525417	–0.03858	–0.07343

3.2.3 Valuation of S&T resources allocation efficiency based on DEA

In Section 3.2.2, we used the weighting method of principal component analysis to measure the indicators of S&T resources allocation efficiency for OECD countries in 2005 and 2009, as well as countries and regions such as China and Russia. However, this method can only get the ranking of efficiency, and is unable to study the relationship between input and output. In this section, we have used the DEA method to further analyze the S&T resources allocation efficiency of each country or region in 2005 and 2009, and compared its results with those of principal component analysis in order to find the root cause for China's low S&T resources allocation efficiency.

1 DEA method

Data envelopment analysis, also called DEA, is a popular method in the cross-boundary research of operations, management science and mathematical economics, which was founded by Charnes, Cooper and others in 1978. DEA is used to evaluate the relative effectiveness (or DEA effectiveness) of "departments" or "units" (or decision-making unit, abbreviated as DMU) that have multiple inputs and outputs. If we evaluate whether DEA is effective according to the observed data of each DMU, in essence, we are evaluating whether the DMU is located at the "frontier" of the production possibility set. Production frontier is a way of production function in economics to get more output. The use of DEA method and model can determine the structure of the production frontier, so we can regard DEA as a non-parametric method of statistical estimation. When we use DEA to evaluate the efficiency of DMU, we can get a lot of management information that contain profound economic implications and background information in economics.

In the scientific research and practice of economics and management, organization is a system that converts input into output. An enterprise is a system that converts human resources, material resources and financial resources into profits and employment; colleges and universities are a system that converts teachers, research expenditure and test facilities into graduate students, scientific research achievements and social services; while the innovation system is a system that converts R&D expenditure, R&D staff and R&D equipment into patents, new products and new services. All of the above systems are the research subject of DEA model: decision-making unit (DMU).

As shown in Table 3-7, we have studied the relative efficiency of n DMU, of which $X_j=(x_{1j}, x_{2j}, ..., x_{mj})$ represents m kinds of input of DMU_j, while $Y_j=(y_{1j}, y_{2j}, ..., y_{sj})$ is s kinds of output, with $x_j > 0$ and $y_j > 0$. DEA model is used to study the relative efficiency of this set of DMU.

The main idea of DEA model is to find two sets of weight coefficients $(\mu_1, \mu_2, ..., \mu_j)$, and $(v_1, v_2, ..., v_m)$, and then divide the weighted output by input, thus getting the relative efficiency value of DMU_{j0} .

Table 3-7 The relative efficiency of n DMU

Item	DMU_1	DMU_2	...	...	$DMUn$
m kinds of input	X_1	X_2	...	...	X_n
s kinds of output	Y_1	Y_2	...	...	Y_n

$$h_{j0}=\frac{\sum_{r=1}^{s}\mu_r y_{rj0}}{\sum_{x=1}^{m}v_k x_{kj0}}$$

As a result, the problem of relative effectiveness becomes how to determine the best set of weighting coefficients so as to maximize h_{j0}. The maximum efficiency value will be DMU_{j0}, and other DMU will have no higher relative efficiency evaluation value than this. We define that all $h_j(j=1,2,...,n)$ will be no more than 1. If the relative efficiency value of DMU_{j0} equals 1, it means that DMU_{j0} is effective if compared with other DMU; if it is less than 1, it means that its efficiency needs to be improved. Based on the above analysis, we can establish measurement model of relative efficiency value for the DMU ranking at j_0:

$$\max H = h_{j0}$$

$$\text{s.t.}\begin{cases} h_j \leqslant 1, j=1, 2, ..., n \\ \mu_r \geqslant 0 \\ v_k \geqslant 0 \end{cases}$$

The formula can be converted to :

$$\max h_{j0} = \sum_{r=1}^{5} \mu_r y_{rj0}$$

$$\text{s.t.}\begin{cases} \sum_{x=1}^{m} v_k x_{kj0} = 1 \\ \sum_{r=1}^{s} \mu_r y_{rj} - \sum_{k=1}^{m} v_k x_{kj} \leqslant 0, j=1, 2, ..., n \\ \mu_r \geqslant 0, r=1, 2, ..., s \\ v_k \geqslant 0, k=1, 2, ..., m \end{cases}$$

This is a simple linear programming problem with independent variable ($\mu_1, \mu_2, ..., \mu_3, v_1, v_2, ..., v_m$). We can get the relative efficiency evaluation value of this set of DMU by solving corresponding linear programming of each DMU.

Traditionally, DEA regards the whole system as a "black box", and only uses the effectiveness of evaluation decision-making units of initial investment and final output. As it does not take into consideration the production process, traditional DEA is unable to get the efficiency of the intermediate stage of the production process and the impacts of each stage on the overall efficiency of the production process.

We set the first stage's input as *X*, output as *W*, intermediate input as *Z*, and final output as *Y*, so this model can be shown as follows (Figure 3-1 and Table 3-8):

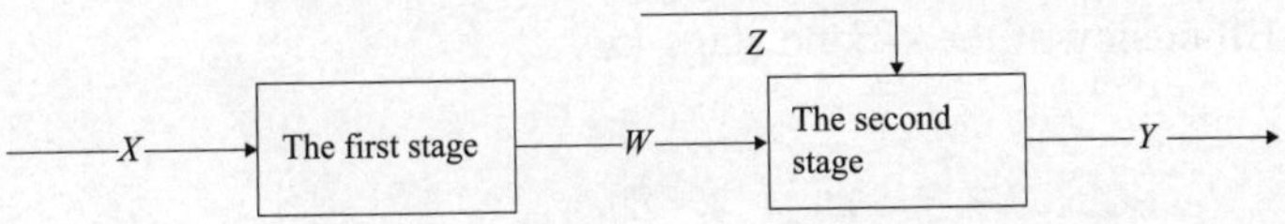

Figure 3-1 Model of input-output

Table 3-8 Model of Input-Output

Item	DMU_1	DMU_2	…	DMU_n
Input of the first stage	X_1	X_2	…	X_n
Output of the first stage	W_1	W_2	…	W_n
Intermediate input	Z_1	Z_2	…	Z_n
Final output	Y_1	Y_2	…	Y_n

In this table, $X_j = (x_{1j}, x_{2j}, \dots, x_{mj})$, $W_j = (w_{1j}, w_{2j}, \dots, w_{pj})$, $Z_j = (z_{1j}, z_{2j}, \dots, z_{qj})$, and $Y_j = (y_{1j}, y_{2j}, \dots, y_{sj})$, and corresponding weighting coefficients are $(v_1, v_2, \dots, v_m)$, $(\varpi_1, \varpi_2, \dots, \varpi_p)$, $(\lambda_1, \lambda_2, \dots, \lambda_p)$, $(\mu_1, \mu_2, \dots, \mu_s)$, respectively, so the calculation model of relative efficiency values is:

$$\max h_{j0} = \sum_{r=1}^{s} \mu_r y_{rj0}$$

$$\text{s.t.} \begin{cases} \sum\limits_{x=1}^{m} v_k x_{kj0} + \sum\limits_{i=1}^{q} \lambda_i z_{ij0} = 1 \\ \sum\limits_{r=1}^{s} \mu_r y_{rj} - \sum\limits_{k=1}^{m} v_k x_{kj} - \sum\limits_{i=1}^{q} \lambda_i z_{ij} \leqslant 0,\ j=1, 2, \dots, n \\ \sum\limits_{r=1}^{s} \mu_r y_{rj} - \sum\limits_{l=1}^{p} \lambda_l w_{lj} - \sum\limits_{i=1}^{q} \lambda_i z_{ij} \leqslant 0,\ j=1, 2, \dots, n \\ \sum\limits_{l=1}^{p} \lambda_l w_{lj} - \sum\limits_{k=1}^{m} v_k x_{kj} \leqslant 0,\ j=1, 2, \dots, n \\ \mu, v, \varpi, \lambda > \varepsilon \end{cases}$$

Efficiency of the first stage is:

$$h^1_{j0} = \frac{\sum\limits_{l=1}^{p} \lambda_l w_{lj0}}{\sum\limits_{k=1}^{m} v_k x_{kj0}}$$

Efficiency of the second stage is:

$$h^2_{j0}=\frac{\sum_{r=1}^{s}\mu_l y_{lj0}}{\sum_{l=1}^{p}\lambda_l w_{lj0}+\sum_{i=1}^{q}\lambda_i z_{ij0}}$$

We have built the DEA model for the measurement of S&T resources allocation efficiency, with six input variables such as enterprises' R&D input, full-time equivalent of S&T staff in enterprises, universities' R&D input, full-time equivalent of S&T staff in universities, public research institutes' R&D input, full-time equivalent of S&T staff in public research institutes, intermediate variables such as the quantity of three-in-one patents and the quantity of SCI papers, and final output as nominal GDP. Specific variables and indicators are as shown in Table 3-9.

Table 3-9 Indicator system of S&T resources allocation efficiency measurement

Item	Variable	Indicator description
Input variable	Enterprises' R&D input	Measurement of enterprises' input of S&T financial resources
	Full-time equivalent of S&T staff in enterprises	Measurement of enterprises' input of S&T human resources
	Universities' R&D input	Measurement of universities' input of S&T financial resources
	Full-time equivalent of S&T staff in universities	Measurement of universities' input of S&T human resources
	Public research institutes' R&D input	Measurement of public research institutes' input of S&T financial resources
Input variable	Full-time equivalent of S&T staff in public research institutes	Measurement of public research institutes' input of S&T human resources

Continued

Item	Variable	Indicator description
Intermediate variable	Quantity of three-in-one patents	Quantity of patent output by S&T input
	Quantity of SCI papers	Quantity of paper output by S&T input
Output	Nominal GDP	Economic benefit brought by S&T resources input

2 Results and analysis of DEA model

With the aforementioned DEA model, we have measured the S&T resources allocation efficiency of OECD countries in 2005 and 2009, as well as countries and regions such as China and Russia, with results as shown in Table 3-10 and Table 3-11. In 2005, the top 10 countries were Germany, Japan, Luxembourg, the Netherlands, Sweden, the United States, Austria, Korea, Denmark and Belgium, which is almost the same as indicated by principal component analysis. China's S&T resources allocation efficiency ranked 35th in 2005 with a DEA efficiency value of 0.0518. In 2009, the top 10 countries were Germany, Japan, the Netherlands, Sweden, Denmark, the United States, Belgium, Austria, the United Kingdom and Finland without much change in ranking. China's efficiency value was 0.0744 and ranked 28th, so China had improved in this aspect compared with that of 2005. This reflects that not only has China's S&T investment increased continuously in recent years, but its S&T resources allocation efficiency and innovation efficiency have also improved.

Table 3-10 Measurement results of DEA model in 2005

DUM	Country/ region	Technical efficiency	Ranking	Scale efficiency	Ranking
$DUM_{\_09}$	Germany	1	1	1	1
$DUM_{\_15}$	Japan	1	2	1	2
$DUM_{\_17}$	Luxembourg	1	3	1	3

Continued

DUM	Country/ region	Technical efficiency	Ranking	Scale efficiency	Ranking
$DUM_{\_19}$	Netherlands	1	4	1	4
$DUM_{\_27}$	Sweden	1	5	1	5
$DUM_{\_30}$	United States	1	6	1	6
$DUM_{\_01}$	Austria	0.905656	7	1	7
$DUM_{\_16}$	Korea	0.833532	8	0.87067	12
$DUM_{\_05}$	Denmark	0.791964	9	0.839722	14
$DUM_{\_02}$	Belgium	0.766737	10	0.822665	15
$DUM_{\_20}$	New Zealand	0.708619	11	0.869121	13
$DUM_{\_08}$	France	0.679134	12	0.680187	18
$DUM_{\_31}$	EU (27 countries)	0.668784	13	1	8
$DUM_{\_14}$	Italy	0.626085	14	0.633258	19
$DUM_{\_07}$	Finland	0.589014	15	0.613862	20
$DUM_{\_29}$	United Kingdom	0.581749	16	0.591439	21
$DUM_{\_13}$	Ireland	0.440569	17	1	9
$DUM_{\_21}$	Norway	0.420985	18	0.471069	24
$DUM_{\_11}$	Hungary	0.358419	19	0.491352	22
$DUM_{\_03}$	Canada	0.34341	20	0.353393	26
$DUM_{\_25}$	Slovenia	0.29155	21	0.719801	16
$DUM_{\_36}$	Singapore	0.274997	22	0.424113	25
$DUM_{\_26}$	Spain	0.189087	23	0.200258	31
$DUM_{\_10}$	Greece	0.178214	24	0.344653	27
$DUM_{\_12}$	Iceland	0.174515	25	1	10
$DUM_{\_37}$	South Africa	0.152839	26	0.306939	28
$DUM_{\_38}$	Chinese Taiwan	0.116411	27	0.133136	34
$DUM_{\_23}$	Portugal	0.105698	28	0.279441	30
$DUM_{\_32}$	Argentina	0.09554	29	0.285999	29

Continued

DUM	Country/ region	Technical efficiency	Ranking	Scale efficiency	Ranking
$DUM_{\_06}$	Estonia	0.095499	30	1	11
$DUM_{\_34}$	Romania	0.08266	31	0.471865	23
$DUM_{\_22}$	Poland	0.081629	32	0.169779	32
$DUM_{\_35}$	Russian Federation	0.061695	33	0.06201	37
$DUM_{\_24}$	Slovak Republic	0.059177	34	0.692091	17
$DUM_{\_33}$	China	0.051848	35	0.052245	38
$DUM_{\_04}$	Czech Republic	0.051721	36	0.134085	33
$DUM_{\_28}$	Turkey	0.045677	37	0.122862	35
$DUM_{\_18}$	Mexico	0.033662	38	0.07954	36

Table 3-11 Measurement results of DEA model in 2009

DUM	Country/region	Technical efficiency	Rank- ing	Scale efficiency	Ranking
$DUM_{\_07}$	Germany	1	1	1	3
$DUM_{\_11}$	Japan	1	2	1	5
$DUM_{\_13}$	Netherlands	1	3	1	7
$DUM_{\_20}$	Sweden	1	4	1	8
$DUM_{\_04}$	Denmark	0.953542	5	1	1
$DUM_{\_23}$	United States	0.952679	6	1	9
$DUM_{\_02}$	Belgium	0.660783	7	0.699589	12
$DUM_{\_01}$	Austria	0.625308	8	0.685543	13
$DUM_{\_24}$	EU (27 countries)	0.618834	9	1	10
$DUM_{\_22}$	United Kingdom	0.579869	10	0.615164	15
$DUM_{\_06}$	Finland	0.495402	11	0.538446	16
$DUM_{\_10}$	Italy	0.444915	12	0.449904	17

Continued

DUM	Country/region	Technical efficiency	Rank-ing	Scale efficiency	Ranking
$DUM_{\_12}$	Luxembourg	0.422784	13	1	6
$DUM_{\_14}$	Norway	0.33531	14	0.405982	18
$DUM_{\_09}$	Ireland	0.32996	15	1	4
$DUM_{\_05}$	Estonia	0.300952	16	1	2
$DUM_{\_28}$	Singapore	0.265567	17	0.363761	21
$DUM_{\_08}$	Hungary	0.241055	18	0.383586	20
$DUM_{\_18}$	Slovenia	0.171317	19	0.629552	14
$DUM_{\_19}$	Spain	0.146624	20	0.159835	24
$DUM_{\_29}$	Chinese Taiwan	0.1144	21	0.135816	26
$DUM_{\_15}$	Poland	0.103314	22	0.188488	22
$DUM_{\_17}$	Slovak Republic	0.082696	23	0.956717	11
$DUM_{\_16}$	Portugal	0.075272	24	0.158035	25
$DUM_{\_25}$	China	0.074444	25	0.077616	28
$DUM_{\_03}$	Czech Republic	0.063992	26	0.168732	23
$DUM_{\_21}$	Turkey	0.043125	27	0.09322	27
$DUM_{\_27}$	Russian Federation	0.036758	28	0.054134	29
$DUM_{\_26}$	Romania	0.02274	29	0.392613	19

There are two reasons why China ranked low. First, in order to unify the statistical caliber, we have selected three-in-one patents data from OECD statistics. This only includes data of the European Patent Office (EPO), the Japanese Patent Office (JPO) and the US Patent and Trademark Office (USPTO), so data pertaining to domestically applied patents in China have not been taken into account. Although the use of three-in-one patents data can better reflect the value of patents, in consideration of the large number of patent applications in China, this might lead to undervalued S&T output of China. Second, the data of papers we have chosen is also from the Web of Knowledge retrieval system of ISI (the quantities of SCI and SSCI papers) that can be compared internationally. So the papers published on domestic journals in China have not been included, thus leading to relatively undervalued output of China's papers and as a result relatively low efficiency of China's S&T resources allocation.

In order to analyze the S&T input-output efficiency of enterprises, universities and public research institutes, Table 3-12 and Table 3-13 have listed China's results of DEA slack range analysis in 2005 and 2009. Slack range is the amount that input can be reduced, when China's S&T resources allocation efficiency becomes the most efficient, that is, DEA efficiency value equals 1, while still keeping the output unchanged.

Table 3-12 shows that when the full-time equivalent of S&T staff in universities has a slack range of –8105.36, if China's S&T resources allocation efficiency value reaches 1, we only need to input 213,802.64 (221,908 – 8,105.36) of full-time equivalent of S&T staff in universities and the redundant input will account for 3.65% of the total. If the redundancy of universities' R&D input is 0, it means that the R&D input of universities has been fully utilized. It is similar for enterprises. If enterprises' R&D input has no redundancy, the redundant full-time equivalent of S&T staff in enterprises will be 3.48%, which means that even though enterprises' R&D input has an increasingly higher share in the total R&D input and enterprises' dominant role in innovation has been enhanced, enterprises' R&D input is still not enough and should be further increased. The indicators of public research institutes' R&D input and full-time equivalent of S&T staff both have redundancy, so its efficiency needs to be further improved.

Table 3-12 DEA slack range analysis in 2005

Variable	Slack range	Input	Share	Proportion
$Input_1$	Universities' R&D input	0	7,027.4	0.00%

Continued

$Input_2$	Full-time equivalent of S&T staff in universities	–8,105.36	221,908	–3.65%
$Input_3$	Public research institutes' R&D input	–514.72	15,483.1	–3.32%
$Input_4$	Full-time equivalent of S&T staff in public research institutes	–9,261.26	200,377	–4.62%
$Input_5$	Enterprises' R&D input	0	48,544.5	0.00%
$Input_6$	Full-time equivalent of S&T staff in enterprises	–24,244.8	696,413	–3.48%

Comparison of Table 3-12 and Table 3-13 shows that in 2009, although the efficiency of China's S&T resources allocation was improved, redundancy also increased. Redundancy has increased for the five indicators such as full-time equivalent of S&T staff in universities, public research institutes' R&D input, full-time equivalent of S&T staff in public research institutes, enterprises' R&D input and full-time equivalent of S&T staff in enterprises, suggesting that China's S&T resources allocation efficiency is still not high and the efficiency of public research institutes in particular needs to be improved. Meanwhile, we also need to further optimize S&T resources allocation and strengthen the evaluation and monitoring of R&D activities, in order to enhance the efficiency of resources use.

Table 3-13 DEA slack range analysis in 2009

Variable	Slack range	Input	Share	Proportion
$Input_1$	Universities' R&D input	0	12,438.2	0.00%
$Input_2$	Full-time equivalent of S&T staff in universities	–10,534.6	225,057	–4.68%
$Input_3$	Public research institutes' R&D input	–1,510.31	28,834.5	–5.24%
$Input_4$	Full-time equivalent of S&T staff in public research institutes	–14,701.2	219,483	–6.70%
$Input_5$	Enterprises' R&D input	–3,171.66	112,874.7	–2.81%

Continued

Variable	Slack range	Input	Share	Proportion
$Input_6$	Full-time equivalent of S&T staff in enterprises	–28,131.5	707,771	–3.97%

3.3 Study on the factors influencing S&T resources allocation efficiency

In order to better understand the issue of China's S&T resources allocation, we need to further analyze the factors influencing S&T resources allocation efficiency, so as to understand S&T resources allocation in its real sense and find the deep cause for such an allocation structure. Then, we can adopt appropriate policies to promote the improvement of resources allocation efficiency. Based on this, we will conduct the Tobit regression analysis for the factors influencing S&T resources allocation efficiency.

We select indicators such as imports and exports share of GDP, educational expenditure share of GDP, government's support for universities and public research institutes, quantity of examined patent applications per person, and share of full-time equivalent of R&D staff in total population. We will conduct regression analysis of factors influencing S&T resources allocation efficiency from the perspectives of openness, education level, government's financial support, dynamics of R&D activities and human capital. The model is as follows:

$$y=\beta_b+\beta_1\text{Open}+\beta_2\text{Edulv}+\beta_3\text{Rdg}+\beta_4\text{Ppp}+\beta_5\text{Hpl}+\varepsilon_i$$

y represents the efficiency value calculated via the DEA model; Open represents openness; Edulv represents education level; Rdg represents government's support for S&T input; Ppp represents intellectual property rights; Hpl represents human capital. The regression results are shown in Table 3-14.

Table 3-14 Results of Tobit regression analysis

Variable	Regression coefficient	Standard deviation	t-Statistic
β_0	2.4830	0.2409	10.3100***
Open	0.5341	0.2029	2.6300**
Edulv	27.1717	5.8370	4.6600**
Rdg	0.5180	0.0643	3.1500**
Ppp	0.5702	0.0683	8.3500**
Hpl	0.2248	0.0366	6.1400*

*, **, *** represents being significant at 0.05, 0.01, 0.001 respectively.

(1) Open regression coefficient is significantly positive, indicating that there is a significant positive correlation between a country's openness and its S&T resources allocation efficiency. Openness will promote a country's economic growth, improve the level of demand for S&T and thus increase the demand for efficiency improvement of S&T resources allocation.

(2) Edulv regression coefficient is significantly positive. It reflects that the increase of education investment will promote the improvement of S&T resources allocation efficiency. Increase of education investment, to some extent, indicates the improvement of education

level and the increase of human capital inventory, so it is beneficial to enhancing the overall quality of R&D staff and improving the efficiency of S&T resources allocation. Moreover, the coefficient of education investment is the largest of all regression coefficients, indicating that education investment has the greatest impacts on the improvement of S&T resources allocation efficiency, which is consistent with the result of the DEA model: There is redundancy for full-time equivalent of R&D staff in universities, public research institutes and enterprises.

(3) Rdg regression coefficient is positive, indicating that the government's financial support for universities and public research institutes will be beneficial to the improvement of S&T resources allocation efficiency. Basic research and applied research are important ways for a country to enhance its S&T competitiveness and capacity. Therefore, the government's financial support for universities and research institutes, the main force that engages in basic research and applied research, will promote the output of basic and core S&T achievements. This helps in knowledge accumulation for leapfrog development and S&T innovation, thus improving the efficiency of a country's S&T resources allocation. This is consistent with the result of the DEA model: There is no redundancy for universities' R&D input indicator in 2005 and 2009.

(4) The regression coefficient of the quantity of examined patent applications per person is also significantly positive. The quantity of examined patent applications per person is a proxy variable of a country's dynamics of R&D activities, so the higher the quantity is, the more R&D activities are conducted by the government, universities and enterprises. This coefficient is positive. It means that the government, universities and enterprises are attaching increasing importance to patent application and R&D activities; therefore, the efficiency of S&T resources allocation will be on the increase. This might be because of the contradiction between limited S&T resources and unlimited demand for patents and R&D achievements that has prompted universities, the

government and enterprises to continuously improve the efficiency of S&T resources allocation.

(5) The share of full-time equivalent of R&D staff in total population reflects a country's level of human capital to some extent. Its regression coefficient is significantly positive. It means that the increase of human capital level will help enhance the efficiency of S&T resources allocation.

Based on the results of Tobit regression analysis and DEA model, we find that the most effective ways to improve the efficiency of S&T resources allocation are to increase the investment in education, and train and introduce more highly-qualified R&D staff, so as to raise the level of a country's human capital; increase government's R&D investment, and establish an efficiency evaluation and monitoring mechanism for S&T resources allocation.

3.4 Summary

In this chapter, we have compared China's S&T resources allocation efficiency with those of OECD countries. We find that China's efficiency is low, and all indicators except the indicator of universities' R&D input have redundancy, but the redundancy is not high, with no redundancies exceeding 7%. Therefore, overall, China has low efficiency of S&T resources allocation. Based on the results of the DEA model, we have studied the factors influencing S&T resources allocation efficiency, and found that the most effective way to increase S&T resources allocation efficiency is to improve the level of human capital (increase the investment in education, and train and introduce more R&D staff).

Chapter 4

Study on the Relationship Between Basic Research and Industrial Innovation

In Chapter 3, we said that the intensity of China's R&D investment has been increasing continuously. In this chapter, we analyze the relationship between basic research and industrial innovation. We like to see how strong inputs to research and development can spur industrial innovation.

According to the three standards classified by Calvert (2007) to distinguish basic research, applied research and experimental development, we believe that current basic research (science) in China is mainly driven by three forces: (1) Curiosity-driven basic research. It is done by scientists out of their own research interests and for the sake of gaining new knowledge and this is mainly funded by the National Natural Science Foundation of China. (2) Government-funded basic research. This is basic research supported by the State from the perspective of strategic development in the form of S&T plans. Such plans focus on the national requirements and are designed to meet national demand. These are top-down programs, such as Program 973 and Program 863, mega projects and so on. (3) Industry demand-driven basic research or mission-oriented basic research based on market demand and addressing critical bottlenecks in industrial development. National Natural Science Foundation also supports some applied basic research. But most of the research is intended to address universal scientific problems, not based on the specific needs for industrial innovation.

This chapter defines the scope of basic research, analyzes the relationship between basic research and industrial innovation, and analyzes the ways the government can utilize to support basic research and the reasons why enterprises want to invest in basic research, so as to identify the problems existing in China's investment in basic research.

4.1 The concept of basic research

Generally, R&D activities include three aspects: basic research, applied research and experimental development. Logically, basic research is not only the foundation and source of research and development but also the foundation and source of the entire S&T activities.

It is generally believed that basic research (or fundamental research) is a scientific research field that has no national boundaries, which is driven by curiosity and not influenced by supporting funds or prospects, has no definite, predictable utility for products or processes, and is intended to explore new scientific principles and to understand the potential basis of a phenomenon or test (Dearborn et al, 1953; Calvert, 2007). The essence of basic research is to increase people's understanding of phenomenon in certain S&T fields, while applied research is intended to meet the needs of individuals, groups or society directly and so the most valuable feature is to reduce the empirical elements in actual skills (Stokes, 1997). US National Science Foundation (1959) believes that basic research is a kind of original exploration of advances in S&T knowledge and is research that seeks new knowledge without any direct commercial objectives, while applied research is research that results in gaining new knowledge for certain applications. Different from basic research that has a low appropriable degree and aims at long-term returns (Schumpeterian effect), applied research aims to apply industrial generic S&T knowledge to specific market, has a high degree of exclusivity and can generate short-term returns (Darwin Effect) (Calderini & Garrone, 2001).

Bush (1945) defines basic research in *Science: The Endless Frontier* that basic research is a scientific activity that does not consider application goals and results in generalized knowledge and understanding of nature and its laws, while applied research is an activity that solves a practical problem. Basic research and applied research, as "two ends of a lever", should be independently undertaken by different agencies. Bush puts forward the famous "linear mode of S&T development" (Figure 4-1), that is, a sequence mode that begins from basic research and ends with innovations. It believes that, from the perspective of technological advance, basic research precedes applied research and through further development can be converted into technological inventions in the end; therefore, basic research is intrinsic motivation of technological progress.

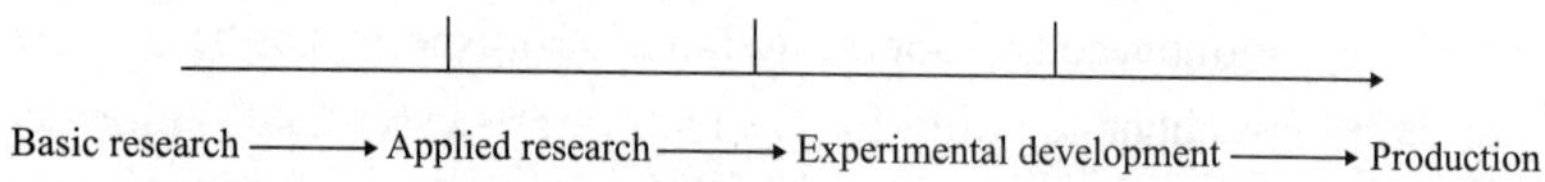

Figure 4-1 Linear mode of S&T development proposed by Bush (1945)

Bush's views have had significant impacts on the policies of basic research and directly contributed to the establishment of US National Science Foundation (NSF). This mode has been promoted by NSF as a policy guideline. In the periods before and after the Cold War, it has greatly influenced the S&T policies stipulated by the United States government and has been gradually adopted by other countries around the world (Zhou, 2008).

However, the Bush mode has also been questioned frequently. Some scholars believe that it simplifies the relationship between basic research and technological inventions in theory, suggesting that with the purpose of "knowledge seeking", basic research determines the development of applied research and technological development in the first place, but it has neglected the feedback effects of applied research on basic research. Some scholars believe that basic research and applied research cannot be absolutely distinguished in practice: Some research tagged by basic research in scientific history has obvious practical purposes, while some research regarded as applied research has elements of basic research (Cheng & Sun, 2000).

An American scholar Nelson has proposed quite influencing thoughts on industry-driven basic research from his early economic analysis of basic research (Nelson, 1959), "Yale Survey on R&D Organization" (Levin et al., 1988), research on the role taken by American universities in economic development (Rosenberg & Nelson, 1994), technology licensing development of American research universities conducted by him and others such as Mowery in recent years. Nelson proposed that it is necessary to analyze the reasonableness of basic research from the perspective of economics. Why should we invest in basic research? How much should we invest in basic research? Why should we invest this amount? Under what circumstances will the investment laws change? Can basic research capability be demonstrated and explained via quantitative analyses? There are other questions too. In these studies, Nelson's thoughts on the economic and social value of basic research help us understand these problems.

The OECD definition of basic research is often quoted by scholars and policy-makers. *Frascati Manual* compiled by the OECD in 1963 and its revised editions are intended to standardize the measurement of

S&T activities. It defines basic research as experimental or theoretical work aiming to gain fundamental, basic or new knowledge about the phenomenon that can be observed (revealing the nature and motion law of objective things; accessing new development and new doctrines). Applied research is defined as creative research carried out to acquire new knowledge, but it is mainly aimed at a specific practical purpose or objective (OECD, 2002). Obviously, this definition, similar to that of Bush (1945), still emphasizes the scientific nature of basic research, which aims to explore the nature of objective things, reveal the natural laws, or propose and validate various ideas, theories or laws, so its outcomes are usually general principles, theories or laws published in the form of papers in scientific journals or exchanged during academic conferences. Normally, it is conducted by scientists and relevant organizations (Zhou, 2008). Of course, the OECD also realizes that its definition of basic research is not perfect and admits that there is some sequence and separation (of basic research and applied research) in the real world. Some scholars have tried to introduce more precise definitions for basic research. For example, Stokes (1997) has proposed that basic research should be divided into "pure basic research" and "application-oriented basic research". What's more, some new expressions such as "translational research" and "basic technology" are trying to bridge the gap between basic research and applied research. In fact, the concept of basic research is inherently quite ambiguous (Table 4-1).

Table 4-1 Multi-angle analysis on the definitions of basic research

Item	Main content of basic research	Opposing opinions and contradictions
Epistemological definition	Unpredictability or novelty: discover a new concept or extend the boundary of existing knowledge	Researchers with "universal" concept
	Universality: research achievements can be extended to a wide range of fields	Molecular biologists of specific gene function research

Continued

Item	Main content of basic research	Opposing opinions and contradictions
Epistemological definition	Construction of theoretical framework: explain the principle of universality	Experimental physicists
	Reductionism: understand and explain phenomenon based on the most basic entities, such as researching behaviors of life at molecular level	
Objectives and purposes	Basic research is curiosity-driven research	The same research with different purposes may be classified in different ways
Distance from application	Research has no purpose for application	Futility, but useful in practice; it is not basic research, if it has no application purpose
Disclosing forms of research achievements	Disclosing forms of research achievements, including publications and reports	It is confined by location and disclosing scope; basic research of private companies that hasn't been disclosed publicly is excluded

Data source: Calvert, 2007.

In order to define the concept of basic research, Calvert & Martin (2001) have interviewed 49 scientists and policy-makers in the United Kingdom and the United States. The results show that the interviewees' understanding about basic research is significantly overlapping, but no one can accurately express the characteristics of basic research's connotations. Calvert and Martin also believe that it is because of such ambiguity and complexity of basic research's connotations that scientists (such as Pasteur) can liberate themselves from the settled system and mode and make research based on actualities. Salter and Martin also believe that excessively precise definition of basic research will limit the scope of research and restrict the diversity and heterogeneity directly related to research and innovation. So the more appropriate method is to regard different kinds of research and development activities as continuous, cross-boundary activities, without essential differences.

In fact, it is hard to distinguish basic research from applied research in the real world. It also has great limitation if we determine the nature of research based on its disclosing forms. Basic research and applied research often have quite vague boundaries. For example, the achievements of a transistor research project include both progress of basic physics knowledge and inventions and improvements of devices in practice (Hao, 2006). Influenced by factors such as disclosing scope and geographical locations, many research achievements are actually both basic and applied according to the disclosing scope. Salter & Martin (2001) think that basic research includes "curiosity-driven" basic research for the purpose of acquiring new knowledge and strategic research that has some practical applications but has no clear purposes for specific processes and products. This definition about basic research is good for explaining why some companies are engaged in basic research rather than acquiring knowledge externally.

How can we understand basic research more comprehensively? First, we believe that it is important to see this from the perspective of research purpose and objective. For example, if the purpose of research is to get applicable outcomes, no matter what its level as basic research is (e.g., unpredictability and universality), it can still be seen as applied research. What's more, the research objective is also very important. One research might be basic research for researchers, but applied research for its funders. Therefore, in the conceptual framework of research objectives, the problem we have is whose "research objective" it is. Second, if it is classified by application logically, most researchers would claim that their research has no application purpose and belongs to "basic research". However, when the researchers need public funds to finance their study, they would then claim their research has application potential. Thus, the concept of "basic research" has heterogeneity when it is used by researchers. Furthermore, it is not convincible to say it is basic research just because it has no direct, obvious application purposes.

The OECD classification of basic research has a profound impact on the policy-making process of many countries. In *Frascati Manual* (1970 revised edition) of the OECD, it divides basic research into pure research and mission-oriented research, of which the latter obviously caters to the interest of funders. However, mission-oriented research was replaced by strategic research in the 1980 revised edition. In the sixth edition of *Frascati Manual* (2002 revised edition), the OECD divides basic research into: (1) Pure basic research. Its purpose is to promote the advancement of knowledge. It does not take into consideration long-term economic or social interests nor get committed to solving practical problems or transforming research results to applicable outcomes. (2) Mission-oriented basic research. It is aimed at broadening knowledge range in order to address known or anticipated or potential research issues. Mission-oriented basic research is a concept enriched by the OECD, because the OECD found that there is a kind of basic research, which itself has no "special" purposes but is committed to fields that people are interested in and regards wide range of application as its purpose. For example, some public research projects of nanotechnologies are selected by the governments in some countries. As these researches are intended to lay a solid foundation for the next generation of technology, some private enterprises might also undertake some basic research (such as research of fuel cell technology). Of course, in consideration of potential commercial applications, private companies usually conduct basic research that can be favorable to themselves (OECD, 2002).

In the research report Governance of Public Research released by the OECD Science and Technology Policy Committee in 2003, basic research is divided into curiosity-oriented basic research (pure academic research), which mainly comprises research conducted by scientists to acquire new knowledge out of their own interests (i.e. the ancient Grecian style of "for the sake of science and technology"), and application-oriented basic research, namely, basic research prompted by commercial applications (although its exact product or process purpose is not known yet). The advantage of this classification is that it has included strategic application-oriented scientific activities for the sake of "knowledge" in basic research. This reflects the evolution of basic research's connotation. However, as it lacks statistics to support this classification, it is not put into use (OECD, 2003).

In addition, some scholars have proposed innovative classification. For example, Conant (1951) is not in agreement with the traditional definition that divides "applied research" and "basic research". He has defined basic research as all research that seeks to expand the range of knowledge in all scientific fields and put forward two types of "uncommitted research" and "programmatic research". Waterman (1965) thinks that Conant's "programmatic research" is actually "mission-oriented research", which aims to help solve some practical problems. However, it is different from applied research because its researchers are not constrained by specific practical goals but can conduct research according to their own design schemes. It is also different from "unconstrained" basic research because research funders have the right to use and dispose these research results. ①

Irvine et al. (1987) put forward two types of basic research, curiosity-oriented research and strategic research, of which the latter can either be done by researchers of basic research or researchers of applied research. The key is whether its findings have potential applications and can last for a long term. Calvert (2007) proposed three criteria to distinguish basic research from other types of research: the purpose of the research, the autonomy of researchers and the epistemological features of the research itself. With these three criteria, we can roughly define basic research in three ways: One is research without specific purpose or curiosity-driven research; the second is research with its

① "Application-oriented basic research" is different from other research, as it emphasizes the dual role of research: knowledge-oriented and application-oriented. However, "planned research", "task-oriented research" and "strategic research" emphasize the knowledge-oriented role of research under the background of strategy.

objectives and directions decided by researchers independently; the third is research that aims to explore new principles, new laws and new methods, or research whose results are unpredictable.

Stokes, former Science and Technology Policy Advisor of the President of the United States, has also greatly influenced the definition of basic research. In *Pasteur's Quadrant: Basic Science and Technological Innovation* published in 1997, Stokes has brought forward opinion that is different from the Bush's (1945). Supported by a large amount of historical data and realities, Stokes believes that separating the definition of basic research and applied research is not consistent with the development history of science. A large number of important studies have shown that the selection process of research is often influenced by two goals at the same time, such as the microbiology research pioneered by Pasteur in the 19th century, the research of thermal physics in the United Kingdom, and the research of organic chemistry in Germany. He has combined the two dimensions of basic research (application-oriented and knowledge-driven), proposed a two-dimensional quadrant mode of scientific research and classified basic research into three types that are currently popular: pure basic research, pure applied research and use-inspired basic research (see Figure 4-2).

As a result, Stokes draws a new conclusion: S&T policies should focus on the Pasteur's quadrant, namely, use-inspired basic research rather than the traditional Bohr's quadrant or the Edison's quadrant. The use-inspired research (Pasteur's quadrant [①]) proposed by Stokes is different from the "targeted basic research" proposed by the OECD and the "strategic basic research" proposed by the Chinese government,

① Pasteur is a great French microbiologist; his research methods between Bohr (pure basic research) and Edison (pure applied research), not only has a strong interest on science itself, but also concerns the practical application of scientific knowledge. He was committed to making contributions to the improvement of human health through scientific research. Therefore, a lot of basic research of Pasteur is entrusted with businesses or government agencies.

because knowledge seeking is not its sole purpose and its core purpose is to seek application-oriented knowledge. Therefore, this kind of basic research cannot be conducted out of scientists' interests, or government's plans. It is the industries and those who are keen to commercialize knowledge that are most likely to promote such kind of basic research.

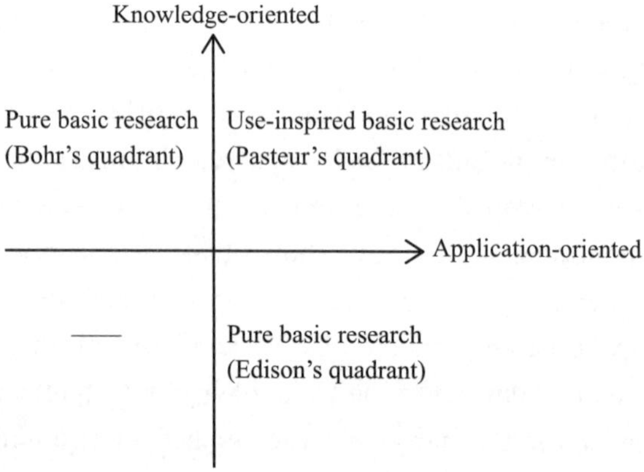

Figure 4-2 Two-Dimensional quadrant mode of scientific research

Data source: Stokes, 1997.

Representative definitions of basic research in foreign countries can be summarized as shown in Table 4-2.

Table 4-2 Types of basic research classified by scholars in foreign countries

Types of basic research	Author
Basic research (BR)	Bush (1945)/NSF
Uncommitted research, programmatic research	Conant (1951)
Curiosity-oriented BR, mission-oriented BR	Waterman (1964)
Curiosity-oriented research, strategic research	Irvine & Martin (1984)
Pure BR, use-inspired BR	Stokes (1997)

Data source: Zhou, 2008.

Generally, China follows the OECD's classification of basic research, which is well shown in related government policies and documents for basic research. For example, in the policy document Suggestions on Strengthening Basic Research jointly issued by the Ministry of Science and Technology, the Ministry of Education, Chinese Academy of Sciences, Chinese Academy of Engineering and the National Natural Science Foundation of China in 2001, it divides basic research into three types: (1) exploratory research (i.e. pure basic research or unconstrained exploratory basic research), with the main goal of understanding natural phenomena and revealing objective laws; (2) oriented research (i.e. strategic basic research) that aims to solve important scientific problems of national economy, social development and science; (3) basic research (i.e. supportive basic research) that examines, collects and evaluates fundamental scientific data, materials and information systematically, and accesses and analyzes them comprehensively in order to find basic laws. In Science and Technology Assessment Measures issued in 2003, the Ministry of Science and Technology clearly divides basic research into two categories, unconstrained exploratory program and strategic program, of which the former is not restrained by funders and is aimed at understanding phenomenon and accumulating knowledge, while the latter's goal must be consistent with that of funders' (national development strategy) (Zhou, 2008). However, in the Outline of National Plan for Long- and Medium-Term Scientific and Technological Development (2006–2020), it emphasizes that we must both adhere to national goals and encourage unconstrained exploration while developing basic research.

4.2 Impact of basic research on industrial innovation and economic development

The relationship between basic research and economic development has long been emphasized by innovation economics, industrial economics and S&T policy. More and more scholars have agreed that basic research has significant contribution to economic development, either directly or indirectly (Salter & Martin, 2001).

Economists who focus on economic growth theory first pay attention to the economic effects of basic research. Solow (1957) introduced S&T progress as "Solow residuals" into production function for the first time. In the Solow model, besides labor and capital, technological progress is another exogenous variable that has great impact on economic growth. Economists such as Arrow, Uzawa, Griliches, Romer, Grossman and Helpman all agree that the key driver of economic growth comes from technological progress and human capital accumulation, and have changed Solow's "exogenous variable" to an endogenous one. Such early macroeconomic research (such as knowledge production function proposed by Griliches) has mostly focused on the rate of return on common stockholders' equity (ROE) of public research investment, sketched out the functional relationship between basic research and economic growth, and drawn an empirical conclusion that is generally positively correlated.

To what extent can basic research promote industrial innovation and economic growth? There have been a large number of research achievements about these aspects since the 1980s. Adams (1990) believes that academic research can significantly improve the productivity of manufacturing enterprises, but there is also a lag phase of at least 30 years. De Marchi & Rocchi (2000) point out that basic research is an important factor in the rise of Italian manufacturing industry. Pavitt (2001) believes, in practice, the ways basic research contributes to industrial development are complicated: They may be direct or indirect, implicit or explicit. In addition, the contribution varies in different industries and at different periods. Salter & Martin (2001) have summarized six kinds of contributions of public R&D to economic growth: increase stock of useful knowledge, train skilled students, create new scientific rules and methods, form a network and promote social interaction, improve S&T problem-solving ability and create new businesses. They also believe that basic research, either directly or indirectly, has significant contribution to economic development. These studies have all shown that basic research is not only the source of dominant information but also the source of new technological opportunities; so it is very important for promoting the performance of industrial innovation. If enterprises invest too little in R&D activities, especially in basic research, its innovation output and performance will be weak.

In order to explore the impact basic research has on industrial innovation and economic development and the ways it impacts them, we will analyze three different approaches: First is a linear model approach that focuses on the impact of basic research on innovation as its source and other indirect impacts; second is a radical innovation approach, which emphasizes basic research's fundamental role in breakthroughs of core industrial technologies and whose technological breakthroughs often come from breakthroughs of basic research; third is knowledge spillovers and absorptive capacity approach, which affirms the spillover effects of basic research, but denies the opinion that basic research should be "free", and believes that one important thing of basic research is to improve the ability of enterprises to absorb and utilize innovations from outside and thus improve the efficiency and effectiveness of technology transfer. Most of these studies suggest that basic research is the potential fundamental variable of industrial technological innovation. Such effects are inflicted on them by influencing industrial learning ability. On empirical research methods, traditionally, we mainly use methods of econometrics, such as surveys and case studies, to examine the impact of academic scientific research on industrial innovation and the ways it impacts. In recent years, people are paying increasing attention to research methods based on patents. For example, Kodama & Suzuki (2007) have extracted citation data of four fields, biotechnology, nanotechnology, IT and environment from the Japanese patent database, and found that different industries use university research results differently. Technology-based industries are inclined to informal technical specialty channels and such kinds of industries are active recipients.

4.2.1 Linear model

The linear process model comes from the thought of Bush (1945), who proposes it in *Science: Endless Frontier*. The model understands the process from science to innovation in this way: scientific investment → basic research → applied research → economic output. It believes that basic research is the source of national competitiveness, so it lifts basic research as national strategy and has repeatedly stressed that ultimately basic research will end in the improvement of national economic interests.

This hypothesis has two key points. First, from a static point of view, basic research does not take into consideration practical results when it is implemented. If basic research caters to practical application goals, its creativity will be ruined. Therefore, there is inherent contradiction between basic research and applied research, which are inherently separated. Second, from a dynamic point of view, basic research is a pioneer of technological development. Basic research should not prioritize practical applications. Basic research can be transformed to technological innovation if combined with applied research. Both views have profound impacts on the S&T policies of the United States, having directly contributed to the institutionalization of its federal government's funding for basic research.

After World War II, many countries have begun to follow the United States to pay more attention to the potential economic effects of basic research. In respect of S&T policies, the government is concerned about how to finance S&T and how to build basic research infrastructure and develop a scientific community (Zhou, 2008). Universities and public research institutes, as main force of basic research, have also received sustained and stable support from the government in terms of funds and talent cultivation. Influenced by the thoughts of Bush (1945), most countries, in practice, have set funds for basic research or applied research in public research institutes and universities. As a result, basic research institutes and foundations began to emerge, such as France's National Scientific Research Center (1941), Germany's Max Planck Society (1948), Chinese Academy of Sciences (1949), US National Science Foundation (1950), and National Natural Science Foundation of China (1986).

However, the view of linear process from basic research to economic innovation has encountered enormous challenges because the basic research driven by curiosity is unable to ensure national victory in the midst of world economic competition. In the Science and Technology Policy White Paper released by the British government in 1993, it has pointed out that it is wrong to simply believe that applied results of basic research will turn out automatically. The core of industrial technological innovation is to develop relevant knowledge and skills to solve practical problems of users, which involve a lot of issues of industrialization of research results.

Beesley (2003) has pointed out that because scientists lack entrepreneurial skills that can commercialize academic research, curiosity-driven basic research has a low level of commercial awareness. The "linear model" of industrial technological innovation from basic research, applied research to product development advocated by Bush (1945) is not enough to guide practice effectively. In fact, some scholars such as Bush see basic research and applied research separately. However, basic research is separated from the actual needs of industrial innovation, so it is very important to strengthen the collaborative relationship between researchers and industries (for example, we can establish joint R&D centers to eliminate the cultural gap between universities and industries) (Figure 4-3).

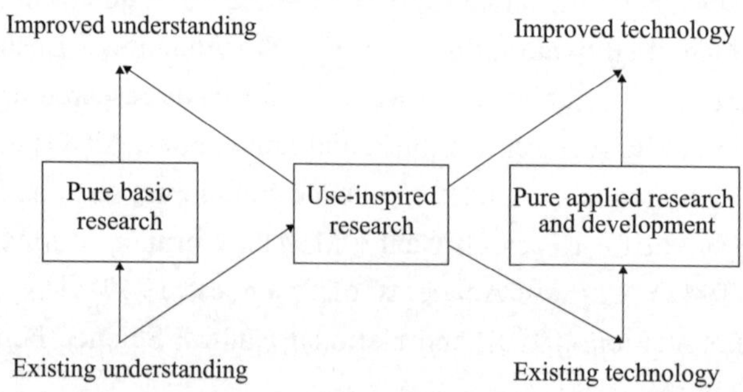

Figure 4-3 Dual-track dynamic model of S&T development

Data source: Stokes, 1997.

In practice, the ways basic research contributes to industrial development are complicated: They may be direct or indirect, implicit or explicit. In addition, the contribution varies in different industries and at different periods (Pavitt, 2001). Mansfield (1980) has conducted empirical study on the relationship between the structure and productivity of industry for the first time, and analyzed the composition of R&D funds, especially the investment ratio of basic research, of 119 American enterprises during the two periods from 1967 to 1977 and from 1977 to 1980, as well as the changes of total factor productivity (TFP) during these periods. Mansfield has an econometric model at both the industrial level and the enterprise level. The model at the industrial level is $\rho_{ti}=\lambda+\Phi_0\frac{dR_{tbi}/dt}{Q_{ti}}+\Phi_1\frac{dR_{tai}/dt}{Q_{ti}}+\Phi_2 u_{ti}+z_i$; the model at the enterprise level is $\rho_{ti}=\lambda+\Phi_0\frac{dR_{tbi}/dt}{Q_{ti}}+\Phi_1\frac{dR_{tai}/dt}{Q_{ti}}+z_i$, wherein, Q_{ti} stands for industrial added value of industry i at the period of t; R_{tbi} means the basic research investment stock of industry i; R_{tai} is applied research investment stock of industry i (Mansfield, 1980). Mansfield (1980) has concluded that statistically, there is obvious direct relationship between an industry's or an enterprise's investment in basic research and the growth rate of its TFP. Jaffe (1989) studied the relationship between cross-field R&D institutions and enterprises' R&D and innovations of 29 states in the United States for 8 years. It shows that in a specific technology field, basic research has significant direct impact on enterprises' innovation in this field (this is reflected in the patents applied by enterprises). Additionally, basic research also has indirect effects on enterprises' innovation by influencing enterprises' R&D.

But with regression analysis based on references, Adams (1990) found that even though we cannot deny that academic research is important for manufacturing industries, it will take 30 years for academic research of basic science to get conspicuous effects. Stokes (1997) has proposed an S&T development mode different from that of Bush (1945). Toole (1999) has also found that although medical instrument in medical basic research can be converted to commercial technology, it often takes a rather long time, probably 17 to 19 years. Taking into consideration the differences of enterprises within the industry and between different industries, Marsili (1999) believes that the relationship between basic research and industrial innovation is complicated, and the ways enterprises obtain knowledge from public basic research are also different.

4.2.2 Radical innovation

Numerous studies have shown that basic research ability is the foundation for enterprises to implement radical innovation. Historically, a great many important innovations are derived from basic research of enterprises. Freeman et al. have proposed the concept of "science-based technology" in related works of innovation economics in order to explain the evolution of industries which have started from the end of the 19th century and rapidly developed after World War II. As Nelson (1995) pointed out, "as extremely basic science, chemistry, physics and biology rose in the 19th century and have brought a lot of improvements and thus changed the nature of the industrial innovation". These so-called new technologies developed since the 1970s, especially information technology and biotechnology, have stimulated people's interest in re-examining the relationships between science, technology, and economic organizations.

Mansfield is one of the scholars who pioneered the study of basic research's impact on industrial radical innovations. Mansfield et al. (1977) and Mansfield (1991) have conducted empirical analysis by interviewing senior R&D directors of 76 enterprises in seven industries in the United States. The analysis shows that without basic research, about 11% of new products and 9% of new technologies would not be produced, accounting for 3% and 1% of corporate sales respectively. They have also studied new products and new technologies that have been supported by basic research to some extent in the past 15 years.

They have also found that without basic research, new products and new technologies, accounting for 2.1% and 1.6% of sales respectively, would not be produced. In the 1970s, both the investment intensity of basic research and the support for high-risk projects have decreased in most American industries. Mansfield (1980) was worried and pointed out that basic research has a positive impact on the TFP of an enterprise or industry. Historically, basic research's impact on TFP has exceeded that of applied research. Mansfield (1998) further clarified the contribution of basic research to industrial innovation in his later extension study (with data from 1986 to 1994). He found that, without basic research, about 15% of new products and 11% of new technologies would not have been developed. These innovative products or services account for 5% of corporate sales, and the lag period from basic research to industrial application has been shortened to 6 years from 7 years.

Studies of Narin et al. (1997) have shown that industrial patents have increasingly referred to papers in science journals. Taking biotechnology industry as an example, Deeds et al. (1997) have established a model to study the impact of enterprises' scientific capabilities on their fund raising at initial public offering (IPD), and proposed five indicator variables for scientific capabilities. They have also taken the paper publishing record of top corporate researchers as a supplementary factor and proven that there is significant positive correlation between the two by a lot of empirical research about US enterprises. Beise & Stahl (1999) have also conducted a similar study of German companies and found that about 5% of new products' output value is closely related to basic research, and basic research has a greater impact on product innovation than on process innovation. However, their study has not taken into consideration the impact of industrial difference on the relationship of "basic research and industrial innovation". Tijssen (2002) has discussed science dependence of technologies, and believes science dependence can be divided into two categories: information source and research environment. Enterprises' technology improvement is indispensable from progress of related scientific research and accumulation of corporate scientific knowledge and related resources. With USPTO's patent data of 10 industries in eight European countries, Leten et al. (2007) found that, in the four fields of biotechnology, pharmaceuticals, organic fine chemicals and semiconductor, the intensity of basic research has a significant positive impact on technological productivity. So these four fields should strengthen their foundation of science and the relationship between science and technologies. However, for fields such as agriculture and food chemistry, optics, analytical measurement and control technology, information technology, basic materials chemistry and telecom, there is no significant impact; so innovations of these six fields are mostly based on evolution of technologies, instead of frontier of science.

After World War II, the United States has had outstanding achievements in basic research; this can be seen from the number of Nobel Prizes, scientific papers and corresponding citations. The astounding growth of American enterprises and industries is also directly dependent on the basic scientific research (such as biotechnology, information and communications technology) (Pavitt, 2001). Other examples are MP3's core components and technologies, including MP3 memory, lithium-ion battery, LCD, DRAM memory and signal compression technique, of which MP3 memory comes from the discovery of giant magneto-resistance effect in 1988. This discovery is an achievement of basic research on multilayer metal films supported by the US Department of Energy. Lithium-ion battery comes from the LCD basic research funded by the US National Institutes of Health (NIH), the National Science Foundation and the US Department of Defense (DOD). DRAM memory is an achievement of basic research on ultra-large-scale integrated system and circuit design co-supported by IBM and DOD in the 1960s and 1970s. Signal compression is the application of Fourier Transform—a basic research achievement of signal processing project supported by DOD in 1965 (Zhou, 2008).

Enterprises especially need to invest in basic research if they would like to have radical innovations because basic research is a necessary part of enterprises to accumulate knowledge and develop ability. Radical innovations become increasingly dependent on enterprises' ability to absorb and reutilize external knowledge, while enterprises with strong basic research ability have a stronger "absorptive capacity" (Cohen & Levinthal, 1990); this provides a sounder knowledge base for learning and is the key to helping enterprises develop radical innovation ability in the end. Although in the short term applied research can bring larger market share and stronger product competitiveness, this kind of competitive advantage is not sustainable because increase of such performance is at the expense of resources that enterprises could use to conduct basic research and create basic knowledge. This is even harmful to the entire industry (Calderini & Garrone, 2001). MP3, as a successful product of radical innovation, is the result of basic research in five fields carried out by the United States. [①] However, how can we manage basic research and achieve radical innovation? This still remains an important issue. If we pay too much attention to basic technologies, it might be too technology-oriented and just like the case of Xerox: Xerox was accumulating a large number of original technologies, but its competitiveness was declining because radical innovation is not just technological breakthrough. However, if we ignore basic research, enterprises will lose their core competencies. In theory, it is a perplexing strategic problem for us to determine the direction and emphasis of basic research. For enterprises, basic research needs to be highly consistent with corporate strategy and becomes an important source of enterprises' radical innovative.

① In January 2006, the Domestic Policy Council of the White House Office of Science and Technology Policy (OSTP) issued "American Competitiveness Initiative" (ACI), which proposed what the United States should do in the next 10 years in order to further improve its national competitiveness; this includes promoting research and development, innovation and educational action in important directions for future basic research, including chemistry, biology, optics, electronic materials, new materials and energy.

4.2.3 Knowledge spillovers

Quite a number of studies have pointed out that only when a user has the ability to understand some information will that information become useful knowledge. Lacking long-term investment in such accumulative scientific knowledge, enterprises will be unable to apply the scientific knowledge they have obtained, especially tacit knowledge (Rosenberg, 1990; Pavitt, 1998). Basic research is a "ticket" for a country's industry to gain access to the global knowledge base. It enables its enterprises to integrate into the global knowledge network effectively, absorbing and utilizing its knowledge and skills (Salter & Martin, 2001). If a country relies on others for basic scientific knowledge, its pace of industrial development will be slowed down (Bush, 1945). So, acquiring external knowledge is the key of absorptive capacity, but it is based on internal knowledge and exchanges. Cohen & Levinthal (1990), with internal R&D costs as an explanatory variable of absorptive capacity, have emphasized the key role of basic research in enterprise learning, and believed that it is the basic research that has created the ability of enterprises to digest and utilize external new knowledge, that absorptive capacity is based on existing knowledge capacity of enterprises, and that the whole process is a cycle.

With knowledge becoming resources and capital, industries are increasingly applying results of basic research to specific fields of production and creating a series of applied knowledge. Adams (1990) believes that the "knowledge base" effects of basic research will have a great impact on corporate innovation. Public basic research can provide opportunities for industries to access specialized knowledge. Studies found that high-tech companies are able to maintain close contact with basic research with two motives: First, expect to obtain the most cutting-edge knowledge in the specialized field in order to ensure its own innovative ability; second, expect to solve some specific technological problems within itself by basic research (Zhou, 2008).

With the relationship between academic research and industrial practice becoming increasingly close, basic research begins to change from pure academic research to applied research and short-term projects, and the interaction between universities/research institutes and industries has been increasingly strengthened. In the empirical research of biomedical industry, McMillan & Hamiton (2003) pointed out that biomedical research grants for students in universities can help produce more innovative scientific literature and patents, and train more qualified talents. In addition, such basic research in universities has also played a very important role in the emergence of a large number of new biotechnology companies in the United States. Salter & Martin (2001) pointed out that basic research mainly has six kinds of contribution to economic growth: increase the stock of useful knowledge; train skilled students; provide new equipment, instruments and methods; establish network of professional knowledge; solve problems of industrial technology; and create new businesses. Calderini & Garrone (2001) believe that internal basic research can help enterprises absorb external knowledge; and the contribution of basic research to enterprises is having improved the quality of their downstream innovation activities. Beise & Stahl (1999) pointed out that basic research can help enterprises absorb achievements of public research and apply them to product/process innovation. Enterprises with high R&D intensity are more frequently using public R&D results than those with low R&D intensity.

Maurseth & Verspagen (2002) found that having a higher level of production-related knowledge, enterprises will benefit more easily from external knowledge spillover effects. According to the absorptive capacity approach, production and transfer of knowledge has strong self-cumulative and path-dependent features. Therefore, more knowledge stock means having a strong research and development capability to develop more new knowledge and to enhance enterprises' R&D investment to absorb, learn and imitate external knowledge, so that enterprises should have a stronger technological ability to absorb external knowledge spillovers.

Besides, Abramovitz's (1986) "social ability" theory, similarly with the "absorptive capacity", stressed the important role of business initiative to absorbing spillovers. Harabi (1997), by surveying 358 R&D managers of 127 Swiss manufacturing companies, found that the primary way of basic research spillover is the development of enterprises' independent R&D, including the implementation of reverse engineering in product innovation, the use of information from the scientific and technical meetings and publications in process innovation.

Research shows that only with high capacity in basic research can a country absorb and effectively utilize scientific knowledge created in other countries. For a country with low innovation capacity, the strategy is to strengthen its basic research, to digest scientific knowledge in other countries. Therefore, the issue of industrial technological policy for catching-up countries is to consider three key questions: independent research and development, the introduction of advanced technology and international technical cooperation (Lim & Song, 1996).

4.3 Rationality of the government to support industrial basic research

For most developed countries in the world, the government is the major force supporting basic research. The thought that the government should support public R&D in order to improve the country's basic research capabilities, promote significant S&T innovation and thus economic benefits can, in theory, be traced back to Bush's views on basic research: Basic research creates new knowledge and thus results in practical applications. Bush pointed out that as the knowledge of scientific output has the nature of "public goods", that is, all people or institutions can easily use it, and enterprises and private investments are reluctant to invest. In terms of its nature, basic research basically reports of market failure. Therefore, it cannot get necessary support from industries. As a result, the government is obligated to fund basic research. The United States has pioneered in the institutionalization of government-funded basic research by establishing the

National Science Foundation (NSF).

Neoclassical economists have demonstrated the rationality of government-funded basic research from the perspective of economics. Nelson (1959) and Arrow (1962) have systematically analyzed the public goods attribute of basic research and believed that with market failure effects of basic research, enterprises are unable to fully invest in basic research and so this responsibility must be borne by the government. They believe that it is rational for the government to intervene in basic research because there is "market failure" for basic research.

In order to deal with the issue of market failure, neoclassical scholars have advocated that we should use the "visible hand" of the government to balance the allocation differences of innovation resources and innovative elements caused by the "invisible hand" of the market. Generally, it is believed that compared to applied research and technological improvements, basic research has greater economical externality. Enterprises have invested less than the amount required for optimal investment because private benefits are lower than social benefits (Guo, 2007). Since the knowledge has the attribute of public goods and effects of spillover (Arrow, 1962), people generally believe that private enterprises are not motivated to invest in R&D. As the output of basic research does not have a high appropriable level, large enterprises are unable to invest a large amount of funds in research with highly uncertain results in order to avoid risk and get short-term profits. In addition, due to information asymmetry of capital market, it is difficult for enterprises to invest in or finance R&D plans, while small businesses that are engaging in the niche market are usually unable to build large R&D labs. In summary, due to the effectiveness of technology and capital markets, inseparability of R&D activities, scale economies effects, enterprises are reluctant to invest in R&D. Therefore, the government must invest in public R&D so as to provide S&T achievements for private enterprises to carry out technological innovations (Beise & Stahl, 1999).

However, there are also opposing views about the dominant role of the government in basic research investment. Rosenberg & Birdzell (1990) have advocated that free market economy is conducive to industrial innovation after their study of "western miracle". In *The Economic Laws of Scientific Research* (1996), British scholar Kealey proposed a complete opposite view against Bush's, believing that investment of basic research should apply a highly-free market-based mechanism and the government-led investment in basic research is "inefficient" and even "invalid". Kealey believes that it has been proven that basic research has great commercial benefits, so with more investment in basic research, enterprises will gain more profits.

However, government-funded basic research is far away from industries. It will result in inefficiency and replacing and crowding-out effects on scientific research of enterprises and private investments. Kealey (1996) has opposed government-funded basic research and believed that market should take charge of basic research. His opinions are mainly based on views that S&T are endogenous activities of market, that there is an alternative relationship between government S&T investment and corporate S&T investment, that government-funded basic research will result in low efficiency of economy and so on (Ma & Zeng, 2007). Kealey (1996) has put forward "two kinds of commercial interests" from an economic perspective and believed that basic research can produce "first mover advantage", while the follower can have "second mover advantage" by imitation. Generally, scientists are interested in the former one, while companies are interested in the latter one. However, the former is the foundation of the latter. Kealey (1996) pointed out that, historically, it is the market that has guided technologies and the technologies that have guided science. Industrial competitiveness is indispensable from innovation. We must carry out basic research if we want innovations. Enterprises with strong basic research capacity will eventually benefit from economy. However, government-funded basic research focuses too much on unconstrained exploration or national strategic plans; they are mostly deviated from the market, thus resulting in inefficiency of economic activities. Government funding of basic research will increase the total investment cost and weaken enterprises' investment capacity. The competition between countries is not simply competition of scientific theories; instead, it is economic competition based on scientific research. For example, different attitudes of the United Kingdom and Japan towards investment in basic research have resulted in different status of development.

Joglekar & Hamburg (1983) warned that not to support industrial basic research is an ill-advised suggestion. They also pointed out that it is not advisable for governments to adopt seed money to encourage industrial basic research because this kind of investment will reduce investment in basic research from industries. To stimulate corporate investment in basic research, governments should increase R&D cooperation willingness between enterprises, reduce competition among enterprises but increase competition between industries, increase industrial concentration, give tax relief to R&D investment, and pay less attention to the heterogeneity of industries.

Overall, it is reasonable for the government to fund basic research in respect of social externalities. However, this S&T development based on linear mode is unable to fully explain new phenomenon: with increasing role of S&T and social progress, development of technologies needs science, while development of science also needs technologies. Basic research is driven by curiosity of scientists and enriched by issues explored by technological and commercial development (Ma & Zeng, 2007). Thus, Nelson (1995) believes that technological policies of a country need to solve three basic questions: First, should investment in enterprises' applied research be invested in individual enterprises or industry, and how can one manage such plan of applied research? Second, how can we organize and manage industry-focused basic research plans? Third, for the achievements of such public research projects, which of them should be public and which should be proprietary?

4.4 Motivation and influencing factors of enterprises' investment in basic research

Because of basic research's externality effects, many people wonder whether enterprises can play an important role in basic research. Actually, profit-oriented organizations are reluctant to invest in such research, unless they can expect significant commercial returns from basic research within a relatively reasonable time-frame or reasonable range of uncertainty. Neoclassical economists believe that because there is market failure in supporting basic science, the government must support basic research. But if basic research is purely research-oriented, the enthusiasm of people who support it will be snubbed out (Stokes, 1997).

Because of the public goods attribute of knowledge, people generally believe that private enterprises have no motivation in R&D investment. According to traditional views, if universities and public research institutes can provide necessary basic research results for industries, why should so many enterprises invest in basic science? With an empirical study, Mansfield (1980) has drawn an important conclusion: The impact of basic research on an enterprise or industry is not only positive, but also higher than that of applied research and experimental development with the same input. Although in the short term, applied research and development can bring larger market share and stronger product competitiveness, this kind of competitive advantage is not sustainable, because improvement of such performance is at the expense of resources that enterprises can use to carry out basic research and create basic knowledge. This is even harmful to the entire industry (Calderini & Garrone, 2001).

In fact, since the 1980s, there has been technological revolution every 10 years in the world, whose major force is always enterprises. With the rise of new S&T revolution, basic research becomes increasingly integrated with applied research. Basic research has become a field with commercial benefits. With increasingly fierce corporate competition, in order to improve their technological innovation capability and market development capacity, a great many large enterprises begin to directly engage in or fund basic research (Ma & Zeng, 2007). The technological revolution based on basic research has enabled enterprises to obtain high returns and increased their interest in exploring basic science.

A large number of empirical studies have shown that the use of external knowledge cannot substitute the knowledge created by enterprises themselves. The more enterprises use knowledge generated from basic research, the more they will be motivated to carry out their own basic research so as to better develop products, obtain and utilize external knowledge from manufacturers (Beise & Stahl, 1999). The more enterprises invest in their own R&D activities, the more external knowledge they can obtain. Berman (1990) has found that innovation-leading enterprises still maintain a high level of internal

R&D investment even after cooperating with universities in R&D. Cohen & Levinthal (1990) believe that enterprises must cultivate their own absorptive capacity by internal R&D, so as to effectively integrate external knowledge into new products/process development. Therefore, enterprises with a high level of internal investment in basic research can be more willing to and more effectively benefit from external knowledge spillover (Beise & Stahl, 1999).

Gibbons & Johnston (1974) believe that there are two reasons why enterprises engage in basic research. First, they themselves need to acquire knowledge from basic research. Second, it helps enterprises to clearly recognize the trend of industrial technologies. Callon (1994) and Pavitt (2001) believe that the reason why enterprises engage in basic research is not to develop specific products; instead, it is intended to acquire relevant knowledge and offer general background knowledge. Cheng & Sun (2000) believe that it is for the need of long-term development and strategic competition that enterprises conduct a large amount of basic research while carrying out technology-oriented research. Hu (2004) believes that many enterprises are actively engaging in basic research, not waiting for the effects of external scientific "public goods" because the achievements of basic research can solve the bottleneck problems of their technical production and bring a large number of patented products. Although the scientific knowledge generated by enterprises' basic research cannot be protected and it will also be used by competitors in R&D, it will take less time for enterprises who engage in basic research to apply their basic scientific theories to production than their competitors. Therefore, they have greater competitive advantages in terms of patent application and technological innovation.

Basic research has enormous social returns, but it cannot create innovations by itself. Without strong R&D capabilities of enterprises as support, economic benefits of basic research can never be achieved. Therefore, it is necessary for enterprises to carry out basic research.

Econometric studies have also shown that the growth of applied technologies is restricted by the reserve of basic scientific knowledge. Without necessary academic research as basis, there will be no substantial long-term economic growth (Yang & Pan, 2003).

In the 1980s and early 1990s, many large enterprises in Japan such as Hitachi, Toshiba, Canon and NEC had a strong interest in basic research. They imitated practices of some large American enterprises ①, and attached great importance to funding research that couldn't generate applications immediately (Stokes, 1997). Such basic research is mostly about Pasteur's quadrant, instead of Bohr's quadrant. However, since it is a long-term process for basic research investment to get returns and it has high risks, internally conducted basic researches in enterprises are mainly for applied research projects, instead of creation and invention of scientific knowledge. For example, Japanese enterprises began to carry out a large amount of basic research since the mid-1960s. However, these basic R&D activities are mainly focused on technological innovation of enterprises. They are secondary technological innovation based on new technological introduction. This model resulted in the lack of ability to industrialize basic research in later time and formed barriers for its development of high technologies, such as the failure of developing the fifth-generation super computer in the 1990s (Ma & Zeng, 2007).

Of course, not all enterprises can obtain the same output from basic research. After taking into account the difference of enterprises within the industry and among different industries, Marsili (1999) believes that the relationship between basic research and industrial innovation is complicated, and the ways enterprises obtain knowledge from public basic research are also different. For example, it is still an unsolved problem that to what extent external environment, especially

① American enterprises do not carry out basic research aimlessly. R&D laboratories established by GE, DuPont, AT&T and other enterprises explored scientific phenomena also associated with their general technology strengths.

industrial environment and market demand, influences the direction and resources allocation of basic research. Calderini & Garrone (2001) found that the market structure affects enterprises' investment decisions on basic research and applied research. In market turmoil, enterprises, with a strong short-term tendency, will invest more resources in applied research and experimental development. For enterprises of different industries, they also have different motives in basic research investment. Joglekar & Hamburg (1983) have studied the influence of industrial characteristics on industrial R&D resources allocation, and found that there are four basic features of industries with high R&D investment: large amount of resources for investment, close cooperation between enterprises, high tendency of risk aversion, high-level investment of enterprises in basic research and low-level dependence on government support. Jaffe (1989) and McMillan & Hamilton (2003) have found that scientific research of fields such as the pharmaceutical industry, biotechnology industry, electronics industry, optics industry and nuclear energy have significant impact on industrial innovation, and these industries have more interest and motivation to invest in basic research. Lim (2004) believes that in the pharmaceutical industry, innovation is closely related to basic research and applied research, while innovation in the semiconductor industry mainly depends on applied research. Toole (1999) believes that both market size and basic research have positive effects on technological research of drugs at an early stage in the biological industry.

It has been a hot topic in economics for many years that whether large enterprises are more innovative than small ones. Many scholars believe that as large enterprises can establish large-scale internal R&D laboratories, they can acquire more innovative knowledge, while most small enterprises tend to imitate the innovative knowledge widespread in their industry. However, Acs & Audretsch (1990) have challenged this traditional view and believed that sometimes small enterprises are more innovative than large ones. In addition, the reason why enterprises choose to carry out basic research may have a relationship with the features of enterprises' management. Buderi (2000) analyzes the history and strategic changes of some famous enterprises in the world (such as AT&T, Lucent, IBM, Siemens, NEC, GE, Xerox, HP, Intel, Microsoft, etc.) in basic research, and finds that these enterprises have a common feature: They attach great importance to basic research. However, these enterprises have different attitudes towards basic research: For example, enterprises such as Microsoft and Xerox adopt basic research out of free exploration spirit, while Intel and AP focus more on user-inspired basic research.

Although basic research has huge potential for economic growth, it has troubled policy makers and researchers for a long time to decide how to encourage enterprises to invest or participate in basic research. This has been a huge problem in both developed and developing countries. Corporate investment in basic research is often considered conflicting with expenses of applied research, experimental development, sales, marketing and fundamental management (Calderini & Garrone, 2001).

4.5 Summary

It has been supported by data in the world that basic research has potential contribution to industrial innovation and economic growth. No matter who they are—researchers, policy-makers or even entrepreneurs, all agree that basic research has strategic value. According to existing literature, there are mainly three aspects people repeatedly debate about.

First, what is basic research? It is difficult to find a universal definition for basic research because the word "basic" itself has a high level of uncertainty. With accelerated integration of S&T and production, basic science and applied science now have increasingly blurred boundaries. In addition, if we completely exclude practical goals from basic research, it is difficult to explain the widespread Pasteur research activities. Is basic research just "from books to books"? Is it just a one-way relationship for "basic research → technological innovation"? These issues still have not been solved in academics.

Second, who should invest in basic research? Basic research is a cause of scientific exploration. It is hard to determine its output in advance. This involves the roles of the government and market in optimizing basic research investment and resources allocation efficiency. The theoretical basis for the government to support basic research is S&T innovation information model (Bush, 1945) and S&T resources allocation "market failure" theory (Arrow, 1962). However, Rosenberg & Birdzell (1990) and Kealey (1996) believe that a free market economy is conducive to the absorption and utilization of scientific knowledge by industrial innovation. The key problem is how government plans and market mechanisms can complement each other in terms of resources allocation.

Third, who will benefit from basic research? This involves two issues. Firstly, Bush (1945) believes that since basic research results of universities and research institutes can provide knowledge support for corporate innovation, should profit-maximizing oriented enterprises focus on obtaining transfer and spillover of these research results, rather than carrying out their own basic research? Second, since basic research results (knowledge and information) are public goods and can be transmitted globally, does it mean that developing countries should be introducing and absorbing such research results from other countries, instead of carrying out basic research at the same level as that of developed countries? Debates and different opinions on these issues have led to different S&T policies in different countries. For example, the United States (as well as Korea) has been very supportive of enterprises to engage in basic research. Japan has also attached great importance to basic research at the catch-up stage. However, at the catch-up phase, China focuses on technological introduction. It is until the independent innovation stage that China begins to strengthen its basic research.

How can we allocate the internal R&D resources of enterprises and how can we strengthen basic research, while at the same time not weakening enterprises' capacity of applied research and experimental development? This has been a major issue that has troubled enterprises for a long time. Although early studies of Nelson (1959) showed that public input in basic research is reasonable in an economic sense, his study at a later time emphasized that the ways for basic research to transform to industrial application are complicated: They may be direct or indirect, implicit or explicit. These complicated effects are reflected in the knowledge field of basic research, as well as the change of time. It has posed a severe challenge for internal management of enterprises in terms of conflicts between short-term benefits and long-term benefits, and between profit-oriented and innovation-oriented options.

References

Abramovitz, M. Catching up, forging ahead, and falling behind. *The Journal of Economic History*, 1986, 46 (2): 385-406.

Acs, Z. & Audretsch, D. *Innovation and Small Firms*. Cambridge: MIT Press, 1990.

Adams, J. Fundamental stocks of knowledge and productivity growth. *Journal of Political Economy*, 1990, 98 (4): 673-702.

Arrow, K. J. Economic welfare and the allocation of resources for invention//Nelson, R. R., et al. *The Rate and Direction of Inventive Activity*. New Jersey: Princeton University Press, 1962: 609-626.

Beesley, L. Science policy in changing times: Are governments poised to take full advantage of an institution in transition?. *Research Policy*, 2003, 32 (8) : 1519-1531.

Beise, M. & Stahl, H. Public research and industrial innovations in Germany. *Research Policy*, 1999, 28 (4): 397-422.

Berman, E. M. The economic impact of industry-funded university R&D. *Research Policy*, 1990, 19 (4): 349-355.

Buderi, R. *Engines of Tomorrow: How the World's Best Companies Are Using Their Research Labs to Win the Future*. New York: Simon & Schuster, 2000.

Bush, V. *Science: The Endless Frontier*. Washington, DC: US Government Printing Office, 1945.

Calderini, M. & Garrone, P. Liberalisation, industry turmoil and the balance of R&D activities. *Information Economics and Policy*, 2001, 13 (2): 199-230.

Callon, M. Is science a public good?. *Science, Technology, & Human Values*, 1994 (19): 395-424.

Calvert, J. *Farewell to the Blue Sky: Basic Research Concepts and Their Roles*. Wuhan: Wuhan University of Technology Press, 2007. [卡尔维特．告别蓝色天空：基础概念及其角色演变．武汉：武汉理工大学出版社，2007.]

Calvert, J. & Martin, B. R. Changing conceptions of basic

research?. SPRU Science and Technology Policy Research, University of Sussex, 2001.

Cheng, S. & Sun, L. How to understand basic research and applied research. *Journal of Dialectics and Nature*, 2000, 22 (4): 56-60. [成素梅，孙林叶．如何理解基础研究和应用研究．自然辩证法通讯，2000，22（4）：56-60．]

Cohen, W. M. & Levinthal, D. A. Absorptive capacity: A new perspective on learning and innovation. *Administrative Science Quarterly*, 1990, 35 (1): 128-152.

Conant, J. B. *On Understanding Science: A History Approach*. New York: Mentor, 1951.

De Marchi, M. & Rocchi, M. Basic research in Italian industry. *R&D Management*, 2000, 30 (1): 79-88.

Dearborn, D. C., Kneznek, R. W. & Anthony, R. N. Spending for industrial research, 1951–1952. Harvard University Graduate School of Business Report, Cambridge, MA, 1953.

Deeds, D. L., Decarolis, D. & Coombs, J. E. The impact of firm specific capabilities on the amount of capital raised in an initial public offering: Evidence from the biotechnology industry. *Journal of Business Venturing*, 1997, 12 (1): 31-46.

Gibbons, M. & Johnston, R. The roles of science in technological innovation. *Research Policy*, 1974, 3 (3): 220-242.

Guo, B. *Modes, Mechanism and Performance Evaluation of Industry-University Cooperation in the Knowledge-based Economy*. Beijing: Science Press, 2007. [郭斌 . 知识经济下产学合作的模式、机制与绩效评价 . 北京 : 科学出版社, 2007.]

Hao, F. The evolution of R&D funded modes of OECD members and its implication to China. *Science Research Management*, 2006, 27 (2): 57-64. [郝凤霞．OECD 研究与开发资助模式演化及其启示．科研管理，2006，27（2）：57-64．]

Harabi, N. Channels of R&D spillovers: An empirical investigation of Swiss. *Technovation*, 1997, 17(11/12): 627-635.

Hu, W. Ideas of R&D and functional orientation for public S&T finance in China. *Collected Essays on Finance and Economics*, 2004 (2):

41-45. [胡卫 .R&D 知识品与我国科技财政的职能定位 . 财经论丛，2004（2）: 41-45 .]

Irvine, J. & Martin, B. *Foresight in Science: Picking the Winners*. London: Pinter, 1984.

Irvine, J., et al. Assessing basic research: Reappraisal and update of an evaluation of four radio astronomy observatories. *Research Policy*, 1987, 16 (2-4): 213-227.

Jaffe, A. B. Real effects of academic research. *American Economic Review*, 1989, 79 (5): 957-970.

Joglekar, P. & Hamburg, M. An evaluation of federal policy instruments to stimulate basic research in industry. *Management Science*, 1983, 29 (9): 997-1015.

Kealey, T. *The Economic Laws of Scientific Research*. Basingstoke: Macmillan, 1996.

Kodama, F. & Suzuki, J. How Japanese companies have used scientific advances to restructure their businesses: The receiver-active national system of innovation. *World Development*, 2007, 35 (6): 976-990.

Leten, B., Belderbos, R. & Looy, B. V. Technological diversification, coherence, and performance of firms. *Journal of Product Innovation Management*, 2007, 24 (6): 567–579.

Levin, et al. Approaching the returns from industrial R&D. Cowles Foundation for Research in Economics at Yale University, 1988.

Lim, K. The relationship between research and innovation in the semiconductor and pharmaceutical industries (1981–1997). *Research Policy*, 2004, 33 (2) : 287-321.

Lim, Y. T. & Song, C. H. An international comparative study of basic scientific research capacity. *Technological Forecasting and Social Change*, 1996, 52 (1): 75-94.

Ma, B. & Zeng, G. "Kealey's problem" and the reason of government-funded basic research. *Science & Culture Review*, 2007, 4 (5): 90-101. [马佰莲，曾国屏 . "基莱问题" 与政府资助基础研究的理性 . 科学文化评论，2007，4（5）: 90-101 .]

Mansfield, E. Basic research and productivity increase in

manufacturing. *The American Economic Review*, 1980, 70 (5): 863-873.

Mansfield, E. Academic research and industrial innovation. *Research Policy*, 1991, 20 (1): 1-2.

Mansfield, E. Academic research and industrial innovation: An update of empirical things. *Research Policy*, 1998, 26: 773-776.

Mansfield, E., Rapoport, J., Romeo, A., et al. Social and private rates of return from industrial innovations. *The Quarterly Journal of Economics*, 1977, 91 (2): 221-240.

Marsili, O. The anatomy and evolution of industries: Technical change and industrial dynamics. University of Sussex, 1999.

Maurseth, P. B. & Verspagen, B. Knowledge spillovers in Europe: A patent citations analysis. *Scandinavian Journal of Economics*, 2002, 104 (4): 531-545.

McMillan, G. S. & Hamilton, R. D. The impact of publicly funded basic research: An integrative extension of Martin and Salter. *IEEE Transactions on Engineering Management*, 2003, 50 (2): 184-191.

Narin, F., Hamilton, K. S. & Olivastro, D. The increasing linkages between U.S. technology and public science. *Research Policy*, 1997, 26 (3): 317-330.

Nelson, R. R. The simple economics of basic scientific research. *Journal of Political Economy*, 1959, 67 (3): 297-306.

Nelson, R. R. Why should managers be thinking about technology policy?. *Strategic Management Journal*, 1995, 16(8): 581-588.

OECD. Frascati Manual: Proposed standard practice for surveys on research and experimental development, 2002.

OECD. Governance of public research: Toward better practices, 2003.

Pavitt, K. The social shaping of the national science base. *Research Policy*, 1998, 27 (8): 793-805.

Pavitt, K. Public policies to support basic research: What can the rest of the world learn from US theory and practice? (And what they should learn). *Industrial and Corporate Change*, 2001, 10 (3): 761-779.

Rosenberg, N. Why do firms do basic research (with their own money?). *Research Policy*, 1990, 19 (2): 165-174.

Rosenberg, N. & Birdzell, L. E. Science, technology and the western miracle. *Scientific American*, 1990, 263 (5): 42-54.

Rosenberg, N. & Nelson, R. American universities and technical advance in industry. *Research Policy*, 1994, 23(3): 323-348.

Salter, A. J. & Martin, B. R. The economic benefits of publicly funded basic research: A critical review. *Research Policy*, 2001, 30 (3): 509-532.

Solow, R. M. Technical change and the aggregate production function. *The Review of Economics and Statistics*, 1957, 39 (3): 312-320.

Stokes, D. *Pasteur's Quadrant: Basic Science and Technological Innovation*. Washington, DC: Brookings Institution Press, 1997.

Tijssen, R. J. W. Science dependence of technologies: Evidence from inventions and their inventors. *Research Policy*, 2002, 31 (4): 509-526.

Toole, A. The impact of federally funded basic research on industrial innovation: Evidence from the pharmaceutical industry. Discussion paper for Stanford Institute for Economic Policy Research, 1999.

Waterman, A. The changing environment of science. *Science*, 1965, 147 (3653):13-18.

Yang, L. & Pan, H. Human capital, basic scientific research and economic growth. *Economic Research Journal*, 2003 (4): 72-78. [杨立岩，潘慧峰．人力资本、基础研究与经济增长．经济研究，2003（4）:72-78．]

Zhou, J. *Foundation and Source of Innovation: Investment, Evaluation and Coordination of Basic Research*. Beijing: Science Press, 2008. [周寄中．创新的基础和源泉：基础研究的投入、评估和协调．北京：科学出版社，2008．]

Chapter 5

Industry-Driven Basic Research

Both theoretical and practical circles agree that basic research will potentially contribute substantially to industrial innovation and economic development (Freeman, 1987). However, the path is highly complicated. Very often, we need to go through the lengthy and painful process from the theory to the realization of lab technology, and then to the manufacture of industrialized products. In terms of contribution of basic research to industrial technological innovation, the national strategic goal does not necessarily coincide with that of the enterprises, so it is unrealistic to pin hope on directly converting the basic scientific results given by universities and public research institutes to industrialization.

Since industry-driven basic research relates to the need of industrial innovation, it is one of the effective methods of exploring and solving major social and economic hurdles and has important realistic values. In Europe (such as Britain), many governments have begun to pay attention to the issue of how to consider factors of demand side when distributing the resources of basic research (Pavitt, 2001).

5.1 The features and necessity of industry-driven basic research

Industry-driven basic research is similar to Pasteur's quadrant (utility-driven basic research), proposed by Donald Stokes (1997). Unlike Bohr's quadrant (pure basic research) and Edison's quadrant (pure applied research), Pasteur's quadrant is neither application-oriented nor pure theory-targeted. According to this quadrant, the research is carried out in the way of basic research but under the guidance of a practical target. A comparison among industry-driven basic research, scientists' curiosity-driven basic research, and government's program-driven basic research is demonstrated in Table 5-1.

Table 5-1 A comparison among the three types of basic research

Item	Driving force	Subject of investment	Subject of implementation	Subject of benefit	Evaluation standard
Scientists' curiosity-driven basic research (CDBR)	Scientists' curiosity about the unknown world and the sense of mission about research	Relying mainly on government's S&T projects, with a few non-governmental organizations	Scientists in universities and scientific research institutes and a few amateur scientists	Scientists' possession of the results	The innovation of knowledge
Government's program-driven basic research (GDBR)	Government's attention, anticipation and arrangement of major economic and social development issues	Government's S&T projects	Scientists in universities and scientific research institutes and a few corporate researchers	National project departments, scientists	The degree of reaching the national profit expectation
Industry demand-driven basic research (IDBR)	Industry demand generated in industry competition, including product development, technological improvement and service innovation	Relying mainly on corporate investment in R&D, supported by government's S&T projects	Corporate R&D labs, a few scientists in universities and scientific research institutes	Enterprises, some scientists in universities and scientific research institutes	The core industrial technological capability

It is traditionally believed that scientific research will lead to the application of knowledge in technology. However, scholars of industrial innovation history do not agree with this idea. Kline & Rosenberg (1986) point out that science expands to the process of innovation and that scientific research and technological progress are connected with each other. Stokes (1997) is doubtful about Bush's opinion that national investment in basic science can be rewarded with technology. He discovers that when America's basic research lagged behind Europe, its technology was reversely in the lead. In R&D practice, there was not only the flow from science to technology but also the counter flow from technology to science. Technological development has become a more significant source of scientific research.

Quite often, the emphasis on application will define the directions of basic research. The study of Narin et al. (1997) shows that industrial patent is increasingly referring to academic articles in science magazines. Industry-driven basic research needs interactions from various organizations. Leydesdorff & Etzkowitz (1996) suggest that the triple helix model of "university/research institutes—industry—government" is regarded as the organization binding mechanism of basic research and that the country needs the interactive connections among different organizations and institutions to conduct basic research. Scientists who work in application-induced basic research areas play a role in the process of technological development. Once basic research is influenced by the idea of potential application, the uncertainty is reduced as to who will gain the technological profits from new scientific knowledge (Stokes, 1997).

Many people believe that since basic scientific knowledge is a public product, developing countries do not need to undertake basic research themselves. Instead, they can facilitate innovation by absorbing the research results of developed countries. If different countries have the same ability of converting information according to their context, then it is not necessary for developing countries to contribute in basic research. They should invest more scientific research funds in applied research and development research. In other words, developing countries can "take a free ride of the developed countries". Therefore, it is more important for developing countries to transform basic research results into applied and development research, in order to enhance rapid economic development (Zhou, 2008).

However, basic research is by no means free of charge (Pavitt, 2001). Without basic research, if enterprises want to understand and use the basic research results of other countries, they have to be equipped with related background knowledge and technical resources, and these both depend on research. The "late-mover advantage" based on introduction and imitation can be realized only when the enterprise has its own internal R&D ability. China's enormous R&D investment has not effectively led to a parallel growth of industrial innovation capacity. One important reason is that Chinese enterprises often neglect basic research, and the public basic research sponsored by the government lacks attention to the needs of the industry.

Basic research, led by technological advances and market demands, is becoming a trend (Stokes, 1997). Technological progress in dyeing, electricity and public health clearly depends on the progress of chemistry, physics and bioscience ever since the Second Industrial Revolution. In fact, since the late 19th century, practical application has become one of the driving forces of basic research. For instance, Pasteur's application-triggered basic research facilitated the development of public health. To some extent, Japan's post-war experience in several decades abolishes Bush's theory about the relationship between basic research, industrial development and international trade. Japan's success is more commonly attributed to its learning and improvement of world advanced technology than to its construction of a science engine, which drove its industrial development internally (Stokes, 1997). In order to upgrade properties of products and cost advantage, Japan laid too much emphasis on the improvement of production technology, but its poor basic research slowed the momentum of radical innovation. Although enterprises like DuPont and IBM, which once enjoyed the benefits of basic research, suspended massive investment in basic research in a given time period after the 1990s, they are aware of the fact that enterprises should carry out a wider range of basic research to maintain their lead in the industry and to build more intellectual foundations for applied research.

In the long run, most basic research will be induced by the consideration of application. The advantage of this idea is that it can meet the goals of politicians and the public's support for policy organizations. Breakthroughs in pure research can stir up further application-triggered basic research; likewise, breakthroughs in application-triggered basic research can stimulate further pure research (Stokes, 1997). Even the United States, which values basic research over a long period of time, once got trapped in the difficulty of declining competitiveness because of the lack of basic research. On June 15th, 2009, *Business Week* had an article on its cover, stating that the United States lacked innovation during the past decade and that the commercialization process of most early record-breaking scientific and technological progress was far below people's expectation. For example, genetic therapy did not receive recognition in the United States. People in the outer suburbs still could not use the satellite Internet, which was successfully developed 10 years ago. Economy that substituted for energy did not make any substantial progress. Biotechnology was on the rapid progress; however, medical treatment and healthcare did not provide satisfactory results compared with the huge investment and enormous efforts in the early stage.

In order to turn the tide of radical innovation, the United States encouraged innovation in all industrial areas in *American Competitiveness Initiative* in 2006, which set the strategic target of becoming the world's leader in basic research and planned to double their investments in basic research, especially focusing on developing basic science based on industrial needs. Science of science policy (SoSP) should think about the new challenge of how to keep a relatively stable and moderately balanced relationship among three kinds of basic science and how to adopt an indirect way to encourage privately owned enterprises to make long-term investment in basic research.

5.2 Why enterprises do basic research

The ability to do research, especially that of basic research, shapes enterprises' market position, and is thus the decisive factor of enterprises' survival and downfall (Buderi, 2000). In the 1870s, the idea was gradually rising that successful and systematic research could bring huge advantages to enterprises. Many people believe that well-planned basic research helps enterprises not only to protect their own core business market, but also to exploit new business areas. For example, in the early 20th century, GE's scientific exploration in electricity and AT&T's theoretical research in telephone helped the enterprises to develop new businesses, such as wireless, wireless communication and television, which helped enterprises to become leaders in these areas (Buderi, 2000). In the 1960s, large-scale corporate R&D became the new trend of that time. The efficiency and benefits of corporate intervention or participation in basic research depend on the answers to the two questions: First, in terms of the relationship between input and output, how much can enterprises' participation in basic research impact the competitiveness? Second, since basic research is a kind of universal knowledge, can enterprises easily profit from others' basic research results? It is typical for late developing countries to gain core knowledge through technological import because the enterprises generally lack knowledge accumulation. The enterprises' basic research usually serves the purpose of product development in practical terms and is normally used to fill the gap between general basic research and the knowledge needed for product development. The basic research in the business circle is distinctly different from that in higher education institutions. It does not try to answer the most fundamental scientific questions, but tries to obtain the knowledge needed for further technological development (Zhou, 2008).

In countries with leading innovation abilities, many enterprises are involved in basic research and then are able to form long-lasting competitiveness. Many radical technological innovations that played a decisive role in international economic development originated in enterprises in the recent 100 years. For example, modern communications technology was derived from Bell Laboratory; Ford Laboratory first realized mass production of automobiles; Voestalpine and Nippon Steel Corporation Laboratories were the founders of iron and steel modernization; the development of machine tool technology should be attributed to Ingeer and Siemens Laboratories in the first place. Other examples also demonstrated the importance of basic research in enterprises, such as electrical technology research in GE Laboratory ①, chemical science innovation in DuPont and Bayer Laboratories, modern aircraft technology R&D in Boeing and Airbus Laboratories, as well as IBM, Intel and Microsoft Laboratories' enormous contribution to the modernization of computers.

---

① General Motors boasts of a powerful R&D laboratory, which invented four-wheel brake systems and the refrigerant Freon.

In developed countries, entrepreneurs often attach great importance to internal R&D because of the intense competition and pursuit of profit. Innovation in the United States relies to a great extent on the large quantities of enterprises that pay much attention to R&D. Laboratories in enterprises have exerted overwhelming influences on industrial development in the United States. They played a decisive role in its industrial development in the 20th century, and more importantly, their achievement made its government and its industry, commerce and financial circles realize the importance of industrial basic research.

Enterprises' basic research, unlike academic research, is very practical. For example, when researchers are asked to investigate the abnormal behaviors of metal-oxide semiconductors at low voltages, it is a kind of practical basic research. Former Bell Laboratory Chief Arno Penzias believed that applied research and basic research belonged to different dimensions. In fact, the divisions of corporate research into either academic and applied research or short-term and long-term research both include basic research (as is shown in Figure 5-1). For instance, in recent years, the volume of the semiconductors is decreasing, while their speed is increasing and functions are upgrading, which urges enterprises' researchers to devote themselves to exploring the possibility of chips of atomic size. This also impels enterprises such as HP, IBM and NEC to allocate more resources into the research area of basic quantum physics (Buderi, 2000).

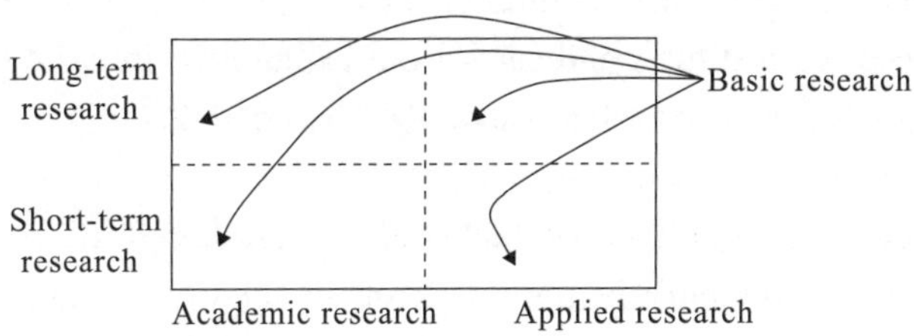

Figure 5-1 Basic research in enterprises

Data source: Buderi, 2000.

Therefore, corporate R&D will not follow the linear way of "research—development—production—marketing" and adopt instead an interactive method among different modules to improve the quality and utility of basic research. Of course, enterprises cannot invest in basic research aimlessly, and we cannot judge enterprises' R&D just according to investment. In the 1980s, following the trend of "university-style research", basic research in many enterprises of the United States deviated from their commercial purposes, which provided opportunities for Japanese enterprises to raise their competitiveness in the market. One case in point is IBM, whose R&D expenditures were once higher than the total of 10 other enterprises including Hitachi and HP. However, IBM did not get much of an absolute advantage in the market for this reason. In the early1990's, even Bell Laboratory reduced R&D expenditures. Hence, in addition to resource input, allocation of resources is also an important aspect. Furthermore, it is very important to keep balance between short-term and long-term research, as well as between basic research or strategic research and applied research.

Unlike Microsoft, which encourages a free and exploratory method of doing basic research, Intel adopts a relatively cautious way, which is closer to engineering, with more development activities than research ones. The research, under clear directions, must be closely connected with the enterprise's business scope. But it does not mean that it lacks innovation or basic research. In 1995, Central Laboratory was established in Intel and its aim was to undertake basic research. This kind of directed research made good use of intellectual resources in universities and public research institutes and carried out exploration in basic scientific theories, outsourcing or collaborative R&D (Buderi, 2000).

In business practice, the target of enterprises' R&D is often to track realistic problems. During this process, a lot of meaningful basic research may be generated. In the 1980s, Xerox could not make good use of the research results from personal computer initiated in Xerox Palo Alto Research Center, such as graphical user interface, local area network and laser printing technology. Thus the value of its basic research was much questioned. Later, after Xerox adjusted its R&D strategies, solving the dilemma of modest profits in basic research mode and avoiding repeating the same mistake, it became a world giant of S&T innovation again.

Enterprises' basic research is not only related to the leader's strategic vision and organizational culture but also connected with its stage of development. Many enterprises such as Xerox and NEC set up their own central laboratories when their business took the leading position in the market and the enterprise was developing rapidly. Some enterprises, on the other hand, started to develop and pay attention to basic research after obtaining external scientific technology. For example, HP became outstanding in the industry of laser printers by utilizing Canon's authorized laser technology and its own innovative products of printing technology. For a long period, HP has placed emphasis on R&D between product development and basic research (Figure 5-2).

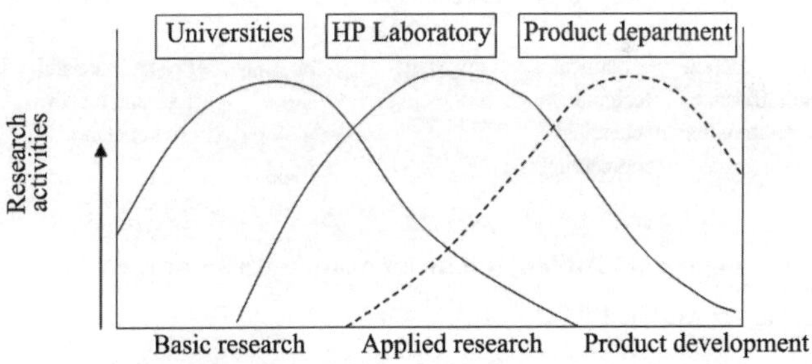

Figure 5-2 Role positioning of R&D in HP Laboratory

Data source: Buderi, 2000.

DuPont managed its research strategy very well. Since 1927, DuPont has experienced four innovation waves or cycles (Figure 5-3), each starting with discoveries in basic research, and then going through the period of new technological invention and product creation. After that, the demand of product creation and market change stimulates enterprises to undertake the next round of basic research. The whole process is repeated over and over again. Almost every 18 years, basic research reaches the predominant position in corporate research for a period of time, lasting for several years, and then giving way to applied research. DuPont continuously achieves breakthroughs in innovation through its own basic research, mergers and acquisitions, and various authorizations and patent transactions, including fiber-viscose, damp-proof cellophane, Teflon, neoprene and nylon.

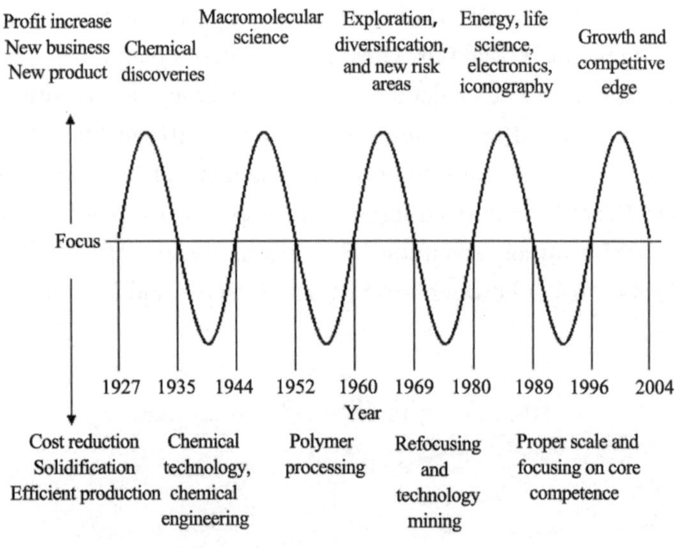

Figure 5-3 DuPont's four innovation waves or cycles

Data source: Miller, 1997.

However, no matter how attractive basic research is, many people are still doubtful about whether it is necessary for enterprises to conduct basic research themselves. Research cost is on the increase, and some enterprises achieved success even without conducting any basic or applied research. For example, in the late 1990s, the three dominant enterprises in the world's PC markets, namely, Compaq, Dell and Gateway, did not do any research activities of their own. Their success lies not in first mover innovation, but in the rapid commercialization of production, sales and delivery and in their competitiveness in time and space. Furthermore, the increasing diversification and professionalization of technology makes it impossible for enterprises to do everything by themselves, so strengthening joint basic research has become the trend. As a reflection of this trend, research-oriented top enterprises in the world starts to establish research laboratories all over the world, to make full use of global intellectual resources, to better understand the demand of local customers and local market, and to better utilize the research results of local enterprises and universities. Nowadays, enterprises attach more importance to potential business reward when carrying out basic research. But in the past, research laboratories in enterprises with huge R&D investment like Xerox placed more emphasis on the quantity and significance of academic articles. In fact, only large-scale laboratories like IBM and Bell Laboratory can afford the free and exploratory method of "ivory-tower style" to do basic research, but this kind of R&D expenditure never surpasses 1% of total R&B expenditure in enterprises.

5.3 Resources investment mechanism of industrial basic research: The complementary relationship between the government and the market

How can we support industrial basic research? According to Nelson's study (1995), to support industry-driven basic research, national S&T policy needs to answer three basic questions: First, does the investment go to one single enterprise or to the whole industry? Second, how can we organize and deal with this type of S&T plan? Third, considering the results of publicly supported research plan, which part is public and which part is proprietary?

The idea of facilitating major S&T innovation to gain economic benefits via government-supported basic research can be traced back to Bush's argument for the rationality of government's investment in basic research and to Nelson's and Arrow's opinions that the output of basic research has the nature of public goods. The linear model of "science–technology–innovation" that Bush (1945) put forward is regarded as the theoretical foundation for the United States

Federal Government to support basic research. Nelson and Arrow believe that the free market system cannot optimize the resources allocation of R&D input from the perspective of the public goods nature of basic research or knowledge spillover effect, for market failure makes it impossible for enterprises to fully invest in basic research. Since the output of basic research is a kind of information that has the nature of public goods (Arrow, 1962), enterprises cannot obtain all the profits from their own basic research by means of technological secret or through intellectual property rights (Nelson, 1959). Because of the high risk of R&D, most enterprises are not willing to invest in basic research, so it is the government that should make investment in public R&D (Beise & Stahl, 1999). Under the influence of these scholars, it is generally believed that enterprises lack the driving force to invest in basic research. For large enterprises, it is impossible for them to put massive expenditure on R&D of highly uncertain results, under the idea of risk aversion and the pursuit of short-term profits. Besides, restricted by the asymmetric information of the capital market, it is difficult for enterprises to do investment and financing of R&D plans. As for small enterprises that contend with each other in the niche market, they do not have the ability to build large-scale research laboratories themselves (Joglekar & Hamburg, 1983). From the practice both in developed countries (such as the United States, Japan and France) and some emerging developing countries (such as Korea), in most cases, the government is the main fund provider for basic research. In these countries, up to 12% to 30% of R&D expenditures is invested in basic research, which lays a solid foundation for industrial innovation. For example, the emergence of IT and bio industry in the United States mainly benefits from government support in basic research in related areas. In Japan and other countries and regions, which are in the phase of catching up, although basic research input is not very high, the proportion of basic research in overall R&D input is continuously on the increase, especially in Japan, where the investment in industrial basic research is strongly emphasized in some competitive industries (such as chemical industry).

How can government support and market mechanism better coordinate with each other to facilitate optimized resources allocation of industrial basic research? According to the model of the rate of return of the project (Figure 5-4), the government should make investment if the private rate of return of a certain basic research project is lower than the lowest private acceptable rate of return. If the project makes some breakthroughs after obtaining national support but still cannot introduce private sectors to enter, then public input should continue. Generally, we suppose the estimated lowest social rate of return of a project is at least equal to its private rate of return ①. In Figure 5-4, the points on the dotted line signify that the lowest social rate of return equals to the lowest private rate of return. If a project is run by a private sector and achieves success, its research results may be utilized by other institutions and organizations, then leading to the knowledge spillover effect to some extent. The different degrees are marked by the lengths of vertical arrows. As to Projects A, B and C in the four projects, since the private rate of return is lower than the lowest private rate of return, enterprises are usually unwilling to get involved. So these three projects are alternatives for government sponsorship. Both Projects A and B have a relatively low private rate of return but demonstrate a considerably high social rate of return. These projects require huge investment, but usually have strong universality, and thus should be sponsored by the government to obtain high social rate of return. Project D has a higher private rate of return than the lowest private rate of return, so enterprises will run Project D independently. Project C is under the corporate lowest rate of return, so usually enterprises will not invest in C without government sponsorship. However, since Project C has a relatively high social rate of return, the government can lower the corporate lowest rate of return through R&D policies including tax deduction and exemption or tax credits, in order to make the private rate of return higher than the lowest private rate of return. In this way, enterprises will also participate

① It is commonly believed that the lowest private rate of return is greater than the lowest social rate of return. This is mainly because of the risk aversion attitude of private sectors and of other related issues such as access to capital and cost.

in Project C. Of course, the overall continuity of the project needs to be taken into consideration in terms of specific basic research projects. With the development of the projects, private and social rates of return keep changing. Therefore, the government needs to consider whether further sponsorship is needed or enterprises should get involved. In terms of application-oriented basic research, if the private rate of return is lower than the lowest private rate of return, and meanwhile its social rate of return is very high, then government sponsorship will achieve great efficiency. If a certain project has not reached the stage of corporate involvement, then the government should continue to support the project.

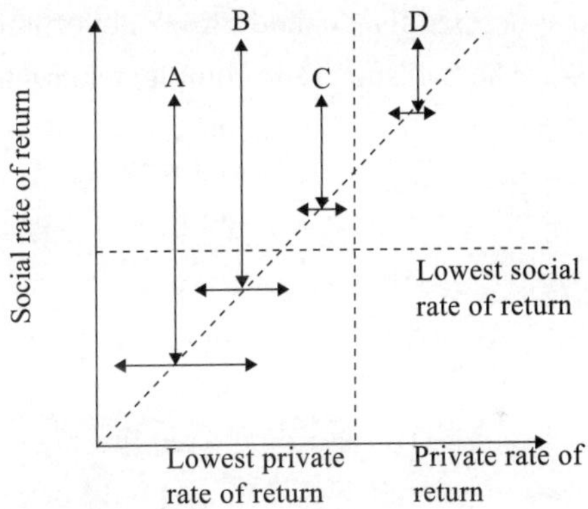

Figure 5-4 Selection of publicly-funded R&D projects: Standard model of rate of return

Data source: Tassey, 1997.

In China, it is commonly believed that the enterprises' ability for basic research is low, which results in the fact that the government needs to make investment and solve the problem of industry-related core technological supply. Government-led major S&T projects often lack the interactive mechanism of industrial demands, and thus cannot effectively support industrial independent innovation. In fact, this is one of the major reasons for growing R&D expenditures but slow growth of innovation ability of industry. Market-oriented innovation does not need too much input from the government in all actuality. The government-led R&D pattern in China is mainly to balance enterprises' extremely weak capacities of development, which leads to the reversed roles of the government and the enterprises. With the heritage of socialist economy, the government loves big science and large-scale R&D projects. However, since enterprises do not have much R&D capability, the government's fund mainly flows to universities and scientific research institutes. In a research report on China's innovation policy, OECD points out that China needs more policy-making from bottom up, giving a more important role to private sectors and promoting innovation via more coordination among organizations (OECD, 2008). Because industrial basic research is market-oriented, it is not enough to allocate basic research resources only depending on government planning or market mechanism. The two should be complementary to each other.

5.4 Organizational models of industrial basic research

Models for industry-driven basic research can be divided into three organizational models: endogenous model (guided by enterprises), coordinated model (industry-education-research institute cooperation), and planned model (national investment).

The endogenous model is affected by many factors including the educational level, the development phase of enterprises, organizational features and cultures, resources and capability, so it is only a small number of countries such as the United States, Germany, the United Kingdom, Japan and Korea that can afford to do that. The planned model does not provide a high efficiency of resources allocation, because under the guidance of national strategies, sometimes, basic research cannot meet the practical needs of industrial innovation. In the future, the coordinated model, guided by the government, will become the mainstream.

5.4.1 The endogenous model: Corporate R&D laboratories

Throughout history, internal R&D laboratories are often a symbol of basic research in enterprises. For instance, pharmaceutical companies usually invest a huge sum of money in basic research hoping to discover new medicine or sometimes to enhance internal R&D abilities by conducting basic research so that they can better understand new conclusions in frontiers of science (Cockbrun & Henderson, 1998).

The US enterprises always maintain the tradition of emphasizing basic research. Many federal laboratories are devoted to industrial basic research as well. At the end of the 19th century, the emergence of science-based US industry witnessed the evolution of the laboratory system, which is regarded as the origin of modern R&D laboratories in enterprises. These laboratories changed the enterprises' characteristics and introduced well-planned R&D laboratories, which aimed at forming science-based technological system or products. Electric commercialization propelled by Edison, Brush, Thomson and Westinghouse, as well as the commercialization of communications application advanced by Bell, required the creation of new technology through research of natural phenomenon (Crow & Bozeman, 1987). Nineteen US scientists who worked in enterprises or industrial laboratories won the Nobel Prizes from 1901 to 1999, accounting for 9.41% of the overall number of award-winning people of the United States (Zeng et al., 2002). In 1998, Lucent decided to invest 11% of its total sales revenue in R&D, while Bell Laboratory's annual budget for R&D reached 3.7 billion USD. Lucent also established an internal risk capital to sponsor the creative ideas that were proposed by researchers but did not match its current business. In this way, any scientist had the opportunity to win financial support, provided they had good ideas[①]. Microsoft went against the tradition of product development in the IT industry and made huge investment in basic research. Microsoft

①An ivory tower that spins pure gold. *Business Week*, April 19, 1999.

Research, founded in 1991, has become one of the largest basic research institutions globally and attracts more than 850 doctoral researchers to undertake various R&D work there. These research projects, though seemingly "basic" and far away from merchandising, brought many intellectual properties and new technologies to the enterprise (Microsoft Research produced around 20% of basic patents in Microsoft). Meanwhile, it solved some key theoretical problems in product design and development through basic research. Microsoft Research adopted a "university-style" management mode, which did not exert any pressure on researchers, in order to ensure an open and creative atmosphere for real future technologies.

Japanese enterprises are also typical examples that lay great emphasis on basic research. For instance, Shimadzu always follows its corporate philosophy of "Contributing to Society Through Science and Technology" ever since its foundation in 1875. With outstanding product innovation capabilities, it keeps endeavoring to learn the S&T knowledge that goes ahead of the time and meets social demands in bioscience and environmental protection. Based on its long-term investment in basic research, Shimadzu has successively developed equipment including photoelectric spectrophotometer, gas chromatograph and X-ray analyzer. Especially in areas such as analytical and testing instruments, medical instruments and aviation industry machinery, it keeps releasing creative products that cope with the market demands based on three core technologies, namely, lighting technology, X-ray technology and image-processing technology. So the company enjoys high reputation among high-tech industries in the world.

5.4.2 The coordinated model: Basic research under industry-university-research institute cooperation

This model can be generally divided into two subtypes.

First of all, enterprises have the need to carry out basic research during the process of development, but they lack related human resources or capability. So they choose to outsource the R&D function to universities or to conduct research with universities. Sometimes, the basic research under industry-university-research institute cooperation provides a way for large enterprises without knowledge accumulation in the early stage to enter a new technological field. More often, it is a way for enterprises with little R&D abilities to enter a certain product field through the cooperation with universities or scientific research institutes. The advantage of this pattern is that enterprises can avoid blind investment and save research funds and that external excellent talents can help to improve the R&D efficiency. But the disadvantage is that enterprises cannot fully comprehend the scientific principles because of the tacit knowledge and the poor communication.

Second, enterprises choose to apply for nation-funded research projects together with universities or scientific research institutes. Thus, enterprises can build the system of knowledge sharing with the academic community. Under this circumstance, enterprises are generally the project leader or main undertaker. In China, 50% of National Science and Technology Major Projects, 35% of National High-tech R&D Program (Program 863) are all led by enterprises. In 2011, enterprises take the lead or participate in 68% of the National Awards for Science and Technology Progress (though such research focuses more on applied research and product development) (Wan, 2011). In actual practice of basic research under industry-university-research institute cooperation, Chinese enterprises, because of their weak internal R&D capabilities and the lack of abilities to apply for basic R&D projects, rarely undertake basic research under industry-university-research institute cooperation via national S&T projects. And then it is difficult for them to cooperate with universities and scientific research institutes in the field of basic research.

5.4.3 The planned model: Basic research guided by national investment

The planned model is the current practice throughout the world. The United States is just one example. Ever since the 1980s, many countries, including China, have enhanced the top-level design of basic research resources input in succession and formed a growing supporting system influenced by Bush's (1945) theory, and the successful experience of the US National Science Foundation. China has designed comparatively complete scientific and technological plans based on basic research. These plans concentrate on guiding scientists toward free exploration and satisfying the national strategic demands. The government's leading role is quite obvious in fields such as space technology, bioscience and microphysics. For example, China's successful manufacture of artemisinin is the product of governmental promotion from its beginning.

5.5 Strategy for industrial basic research in an open environment for innovation

Globalization has greatly changed the corporate innovation mode. It has become a new trend to seek sources of radical innovation by opening up to the external world. Open innovation creates opportunities for enterprises to improve the ability of innovation. However, too high a level of opening will exert negative influences on enterprises' ability of basic research. More and more enterprises turn their attention to external innovation resources. Consequently, they stress on the need for rapid obtainment and utilization, but gradually lose the driving force to do internal R&D with great concentration. However, if an enterprise lacks input in basic research, it usually cannot make effective use of external scientific knowledge resources. Although the key to absorptive capacity is focusing on external knowledge, open innovation cannot bring any long-lasting effect if the enterprise lacks internal knowledge foundation. Cohen &

Levinthal (1990) consider internal R&D expenditures as an explanatory variable of absorptive capacity. They emphasize that basic research plays an essential role in corporate learning. They believe that basic research generates the enterprise's ability to absorb and utilize new external knowledge. And its absorptive capacity is based on its past ability of knowledge. This is a cycle. Research also shows that the more an enterprise invests in basic research activities, the more external knowledge it can obtain. Enterprises with substantial input in basic research refer more frequently to the results of the public research that are far from themselves than those with low density of input. Besides, they are willing to and can more effectively benefit from external knowledge spillover (Beise & Stahl, 1999). Many leading enterprises in innovation still keep a high level of internal R&D investment even after research cooperation with universities (Berman, 1990).

In the era of globalization, market competition is growing increasingly fierce and enterprises are becoming fickle. So, enterprises like to focus on short-term market share instead of long-term advantages. This is the case even for US enterprises that have been emphasizing basic research for ages. Now, the US industrial competitiveness is also affected by the neglect of industrial basic research and too much emphasis on open innovation. Therefore, enterprises need to use the mode of open innovation in a proper way. The utilization of external knowledge cannot substitute the enterprise's own creation of knowledge. The more an enterprise invests in basic research, the better it can absorb the results of public R&D and put it into product or process innovation.

5.6 Summary

Industry-driven basic research is important for innovation. For China, industrial basic research helps enterprises to understand not only the "how" but also the "why" in applied research and experimental development, because many scientific principles in applied research need to be solved via basic research.

In history, some enterprises, though they had the foundation in basic research, were faced with difficulties in development and even temporary failure because of various reasons. But relying on their remarkable abilities of technological innovation, they quickly rose from the ashes. For example, in the 1980s, Xerox was confronted with tough challenges from Japanese companies because it focused too much on developing high-end products. Its market competitiveness once dropped to the lowest point. The academia criticized Xerox for putting too much emphasis on basic research that deviated from the market, which led to the declining of innovation performance. However, the massive knowledge of scientific knowledge finally helped Xerox to become the world-leading manufacturer of office printers again. Of course, the resurgence of Xerox is the comprehensive reflection of various factors. But it is the technological advantages constructed by the "preference" and "persistence" in basic research that helped Xerox to rapidly enhance its competitiveness and accelerate its revival. The *Fortune* Global 500 companies, including GE, Canon, DuPont and IBM, which are familiar to us all remain at the forefront of world competition stage via the construction of sound knowledge foundations.

Without a proper level of basic research in enterprises, domestic enterprises can only import foreign technology. In fact, China's current investment and distribution of basic research both prefer the planned model. But planned basic research lacks attention to the industrial needs, and the S&T evaluation system is not reasonable. As a result, many research results, shelved and forgotten, cannot be fully transferred into core industrial competitiveness. The S&T policies in China should consider more how to undertake basic research for industrial technological innovation.

References

Arrow, K. J. Economic welfare and the allocation of resources for invention//Nelson, R. R., et al. *The Rate and Direction of Inventive Activity*. New Jersey: Princeton University Press, 1962: 609-626.

Beise, M. & Stahl, H. Public research and industrial innovations in Germany. *Research Policy*, 1999, 28 (4): 397-422.

Berman, E. M. The economic impact of industry-funded university R&D. *Research Policy*, 1990, 19 (4): 349-355.

Buderi, R. *Engines of Tomorrow: How the World's Best Companies Are Using Their Research Labs to Win the Future*. New York: Simon & Schuster, 2000.

Bush, V. *Science: The Endless Frontier*. Washington, DC: US Government Printing Office, 1945.

Cockburn, I. M. & Henderson, R. M. Absorptive capacity, coauthoring behavior, and the organization of research in drug discovery. *The Journal of Industrial Economics*, 1998, 46 (2): 157-182.

Cohen, W. M. & Levinthal, D. A. Absorptive capacity: A new perspective on learning and innovation. *Administrative Science Quarterly*, 1990, 35 (1): 128-152.

Crow, M. & Bozeman, B. R&D laboratory classification and public policy: The effects of environmental context on laboratory behavior. *Research Policy*, 1987, 16 (5): 229-258.

Freeman, C. *Technology Policy and Economic Performance*. London: Printer, 1987.

Joglekar, P. & Hamburg, M. An evaluation of federal policy instruments to stimulate basic research in industry. *Management Science*, 1983, 29 (9): 997-1015.

Kline, S. & Rosenberg, N. An overview of innovation//*The Positive Sum Strategy: Harnessing Technology for Economic Growth*. Washington, DC: National Academy of Sciences, 1986: 275-306.

Leydesdorff, L. & Etzkowitz, H. Emergence of a triple helix of university-industry-government relations. *Science and Public Policy*,

1996, 23: 279-286.

Miller, J. Basic research at Dupont. *Chemtech*,1997, 4: 29-33.

Narin, F., Hamilton, K. S. & Olivastro, D. The increasing linkages between U.S. technology and public science. *Research Policy*, 1997, 26 (3): 317-330.

Nelson, R. R. The simple economics of basic scientific research. *Journal of Political Economy*, 1959, 67 (3): 297-306.

Nelson, R. R. Why should managers be thinking about technology policy?. *Strategic Management Journal*, 1995, 16(8): 581-588.

OECD. Reviews of innovation policy: China, 2008.

Pavitt, K. Public policies to support basic research: What can the rest of the world learn from US theory and practice? (And what they should learn). *Industrial and Corporate Change*, 2001, 10 (3): 761-779.

Stokes, D. *Pasteur's Quadrant: Basic Science and Technological Innovation*. Washington, DC: Brookings Institution Press, 1997.

Tassey, G. *The Economics of R&D Policy*. London: Quorum Books, 1997.

Wan, G. The financial investment of central government in science and technology has grown over 20% annually ever since the 11th FYP. Central People's Government of the P.R.C., 2011-04-02. [万钢．“十一五”以来中央财政对科技投入每年增长 20% 以上．中国政府网，2011-04-02．]

Zeng, G., Li, Z. & To, C. Thinking about contemporary differentiation of scientific community and academic internationalization. *Journal of Dialectics Nature*, 2002, 24 (6): 32-37, 43. [曾国屏，李正风，杜祖贻．当代科学共同体的分化与学术国际化问题的思考．自然辩证法通讯，2002，24（6）: 32-37，43．]

Zhou, J. *Foundation and Source of Innovation: Investment, Evaluation and Coordination of Basic Research*. Beijing: Science Press, 2008. [周寄中．创新的基础和源泉：基础研究的投入、评估和协调．北京：科学出版社，2008．]

Chapter 6

International Experience on Basic Research Investment and Industrial Innovation Promotion

The purpose of this chapter is to understand the practice of some international countries in doing basic research and promotion of industrial innovation.

There is a trend in the United States and other industrialized countries to separate basic research and applied research in the public R&D since the post-war period. Following this trend, and recognizing the fact that basic research plays a fundamental role in technological development, as well as in the industrial innovation since 1970s, developed countries had accelerated the basic research input by means of enacting policy initiatives, competition guidance and optimal configuration mechanisms sequentially. An obvious example is that some emerging technologies (such as information, biological medicine, new materials, etc.), which heavily rely on the advances of basic research, have been given long-term support in the United States and Japan, expecting that these industries could introduce disruptive innovations, and thereby enhance strong competitive advantages.

Different countries have their own characteristics in terms of the scale, proportion and structure of basic research investment. In the United States, universities, large enterprises, government research institutes are the three pillars of basic research system. Enterprises-based basic research is industrial innovation oriented; meanwhile, basic research (including universities, federal laboratories) supported by the government is more concentrated in the strategic areas of national defense, health, environment and so on. In Japan, enterprises play an extremely important role in basic research, partly due to learning and imitation of the US enterprises, so they establish internal laboratories, to compete with the US enterprises. In Korea, there is a similar trend of enterprise-dominant basic research, but here, these resources are basically mastered in a few large enterprises (such as Samsung, LG). But Korea is seen as vulnerable with the weak basic research capacity of universities, which leads to insufficient support to industrial innovation from academia. In contrast, although companies in European countries (such as Germany, France, the United Kingdom, Sweden) also attach great importance to R&D, their basic research investment is relatively small.

6.1 The United States: Excellent in basic research with good industry-university-research institute cooperation

The United States is the strongest nation worldwide in scientific research, industrial innovation and economic development. Long-term attention to basic research is a fine tradition of the United States (Figure 6-1). In the R&D investment system of the United States, federal government, enterprises and non-profit organizations have been increasing their investment in basic research in the past half century, which makes the basic research investment share of total R&D stable at 13%–20%. The ratio of basic research, applied research and experimental development was relatively stable at 1 : 1.4 : 3.9 during 1981 to 2009 (Figure 6-2).

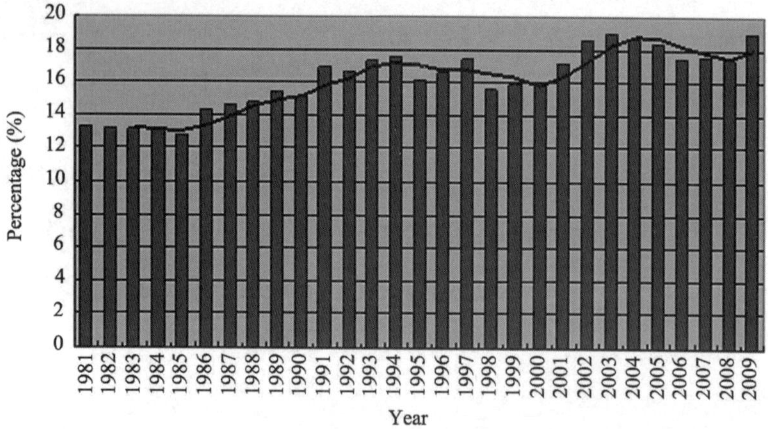

Figure 6-1 Basic research investment share of R&D expenditure in the United States

Data source: NSB, 2012.

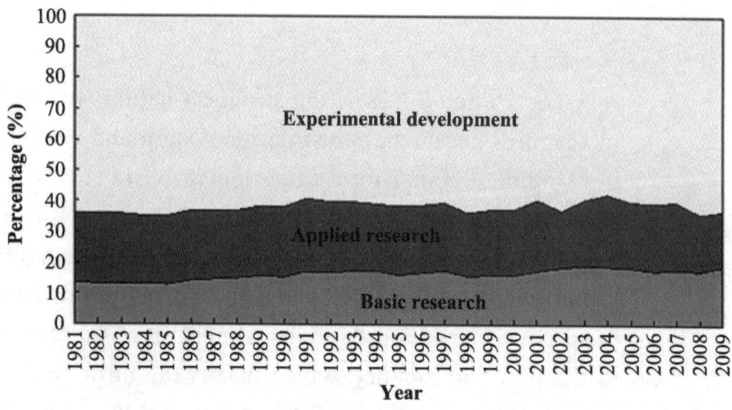

Figure 6-2 Structural changes in the US R&D investment system

Data source: OECD, 2011.

In the United States, government funding for basic research mainly goes to universities and federal laboratories (Li et al., 2008). Universities are the core of the national R&D system, as well as the most appropriate institutional locus for basic research. Universities now combine the functions of education and research, and as a consequence, its basic research funding accounts for around 60% of the national funding for basic research stably since the 1970s (the funding was 40.544 billion USD in 2009, accounting for 53.37% of national funding). For the universities, about 64% of the funds obtained from government R&D investment were used to carry out basic research (Figure 6-3), including scientific research and training students. US Federal Laboratory is a vital basic research department (its funding was 11.373 billion USD, accounting for 14.97% of the national funding in 2009), whereas the expenditure only accounts for 15%–17% of acquired R&D funding (Figure 6-4, Figure 6-5).

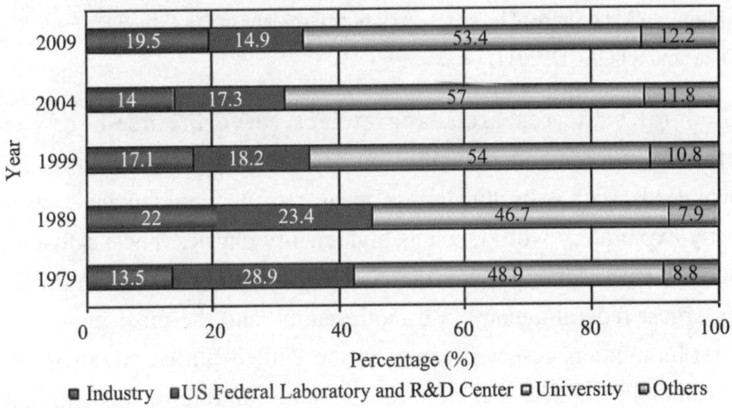

Figure 6-3 The performance of basic research in R&D subjects of the United States

Data source: OECD, 2011.

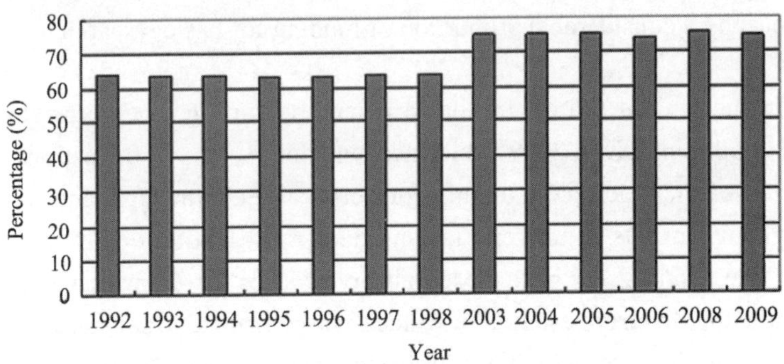

Figure 6-4 The share of basic research in universities of the United States

Data source: NSB, 2012.

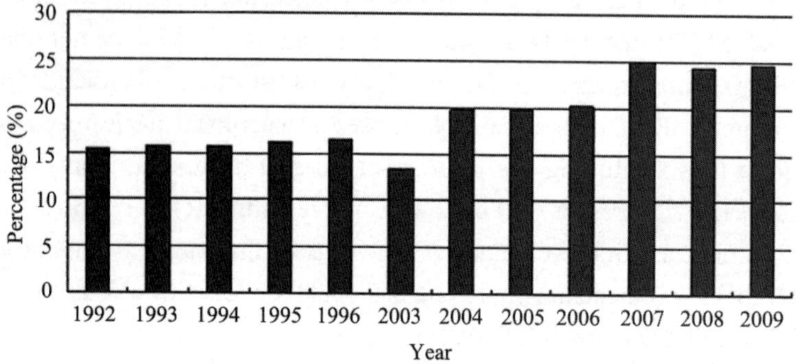

Figure 6-5 The share of basic research in federal labs of the United States

Data source: OECD, 2011.

Federal basic research laboratories serve the needs of the government, which focuses more on research projects meeting the national needs, such as health, disease, measurement, environment, as well as costly exploratory work (such as high energy physics, nuclear fusion, astronomy, etc.). Among them, the National Institutes of Health (NIH) is the largest federal agency in basic research, and the most important national biomedical research center in the United States. Meanwhile, NIH is also an important agency for the state funding of medical and biological research. In addition, the Department of Defense (DOD), the Department of Energy (DOE), the National Aeronautics and Space Administration (NASA) and the National Science Foundation (NSF) also occupy a considerable proportion of funding for basic research.

In the United States, the basic research funding of enterprises (14.784 billion USD in 2009, which accounts for 19.46% of the national basic research funding) is mainly for corporate laboratories, and the rest mainly invests in universities and non-profit institutions (Figure 6-6). The funding for basic research invested by enterprises varies by industry differences. For instance, bio-pharmaceutical industry and other industries, which are highly dependent on long-term basic research, have relatively larger basic research investment than other industries. A growing number of enterprises have realized the benefit of funding basic research of universities, and as a consequence, their investment in universities accounts for an increasing share of their total basic research investment since the 1970s. The United States enacted the Federal Tax Law in 1936 allowing enterprises' R&D expenditure in scope of tax deductions. The Bayh-Dole Act, adopted in 1980, which also greatly stimulated the interest of enterprises entrusting basic research to universities (Zhou, 2008). The majority of the increased funds in biomedicine field help the quick conversion from basic research achievements into products in respect to therapeutic, diagnostic and medical equipment.

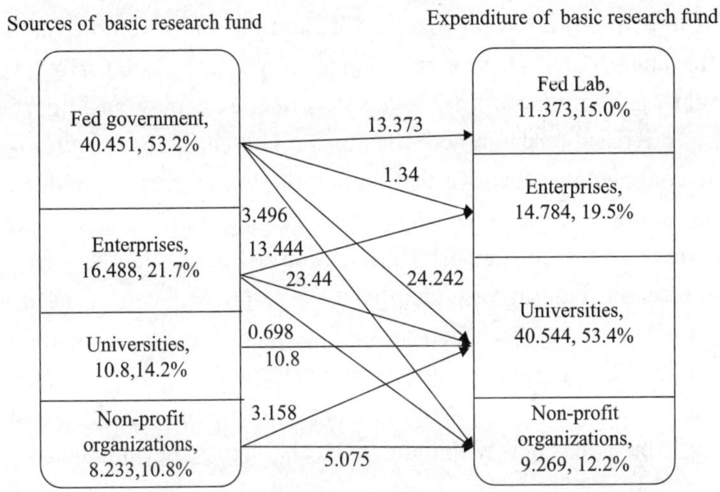

Figure 6-6 The source and performance of basic research in the United States (billion USD)(2009)

Data source: NSB, 2010.

The rise of various types of laboratories is the source of the US innovation; meanwhile, the corporate lab is a symbol of the US innovation. Bell Labs, now called Lucent Technologies, established in 1925, boasts an unrivaled 11 Nobel Prizes winners, which had not only led the enrichment of the US communications industry in the first half of the 20th century, but also laid the foundation of the enterprise's innovation (AT & T formed Bell Labs at early times, and now Bell Labs belongs to Lucent Technologies) in quite a long time.

Although Bell Labs and Xerox PARC are often addicted to basic research, therefore the enterprises' development suffered severe challenges, superior R&D capabilities rejuvenate these giants after a brief setback, which shows that competitiveness can in many cases be traced back to the consistent emphasis on basic research and long-term knowledge accumulation.

The US enterprises' investment in basic research had two substantial growth periods in the 1990s. In the 1980s, only 4% of the R&D expenses were spent by US enterprises for basic research. By the late 1980s, the proportion rose to 7%, and the same value remained till the late 1990s, whereas it dropped to 6% since 2000. However, in light of the situation in 2009, enterprises still play an important role in the basic research activities of the United States (Figure 6-7). Some commentators believe that among the US enterprises, reduction in the funding for basic research would influence the supporting role of basic research on industrial innovation. Although the majority of basic research funding provided by the enterprises has been used for their own basic research activities, they fund the basic research of universities and non-profit organizations as well. A large number of enterprises are confident of the prospect for basic research, and link scientific breakthroughs with industrial technological development.

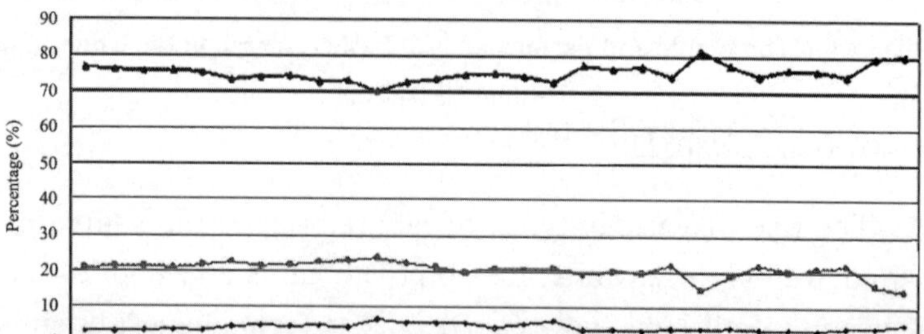

	1981	1982	1983	1984	1985	1986	1987	1988	1989	1990	1991	1992	1993	1994	1995
Basic Res.	2.9	3.1	3.3	3.4	3.3	4.6	4.6	4.4	4.8	4.3	6.4	5.6	5.6	5.5	4.3
Applied Res.	21	21	21	21	21	22	21	21	22	23	24	22	21	20	21
Experiment Dev.	76	76	75	76	75	73	74	74	73	73	70	72	73	75	75

	1996	1997	1998	1999	2000	2001	2002	2003	2004	2005	2006	2007	2008	2009
Basic Res.	5.3	6.3	3.5	3.6	3.5	4	3.9	4.1	3.8	3.8	3.4	4.2	4.3	5.2
Applied Res.	20	21	19	20	20	22	15	19	22	20	21	21	16	15
Experiment Dev.	74	73	77	76	77	74	81	77	74	76	76	74	80	80

Figure 6-7 The share of different research in R&D expenditure of US enterprises

Data source: OECD, 2011.

The most important organization in the US S&T system is the National Science Foundation (NSF). The NSF is the major source of federal government support for basic research, and also the major source of federal government support for universities' basic research except for the US Department of Health and Human Services (HHS). The NSF invested 6.9 billion USD in 2009 to carry out R&D development, of which the basic research project accounted for 92%. The NSF attaches great importance to industrial basic research, and its engineering research center focuses more on cooperating with enterprises in research, which differs considerably with basic astronomical laboratories affiliated with universities throughout the country. For example, in the mid-1980s, the NSF collaborated with several important enterprises to establish an industrial design center in Carnegie Mellon University, with the purpose of developing technology and knowledge of computer-aided design used in manufacturing systems. Researchers emphasize the applications in the middle stages and targeted basic research. This kind of research promoted the US enterprises to develop new competitiveness and brought a fresh paradigm for engineering schools (Crow & Bozeman, 1987).

The US government has been focusing on strengthening industry-driven basic research. According to the NSF statistics, in recent years, 73% of the scientific basis for the US enterprises' patent comes from basic research supported by the government, and every new technological patent relates to an average of two basic research papers since 1997. Over the past 25 years, the economic value analysis for basic research conducted by the US economists revealed that 50% of the US economic growth is attributed to R&D, which was driven by basic research (Chen, 2005). Except for national defense, the federal government allocated the rest 41% of technological funding to basic research in 2002. Also in 2002, the US Congress approved the bill of doubling the budget of National Science Foundation in the next 5 years. In order to maintain its leadership in innovation, the US put forward the American Competitiveness Initiative (ACI) in January 2006, promising to double the funding for important federal research institutes in the next 10 years for their basic research in physics and engineering. In addition, the role of non-profit organizations in the US research system is significant, of which the basic research expenditure has exceeded federal laboratories; by this it became the third largest sector to perform basic research. Non-profit organizations are characterized by small-scale investments but stable financial support and relaxed environment, which attracted numerous high-level scientists (Zhou, 2008).

Historically, the dominant ideology of the US basic research in the early stage is Bush's innovation model (1945). This helps the US to be a global superpower until now. But in the US, there was a big problem of a low conversion rate of technological achievements (less than 5%), which further resulted in a mismatch between scientific capacity and industrial technological innovation capacity. In the 1980s, the rapid rise of developed European countries and emerging Asian countries (such as Japan and Korea) brought tremendous competitive pressures to the United States. On reflection, the US government, academia and industries believe that the patent system needed to be reformed—basic research achievements and technology must be transferred to industry somehow. Bayh-Dole Act and Stevenson-Wydler Technology Innovation Act of 1980 were just a manifestation of this idea, which changed the US S&T policy, and provided a powerful policy support for the movement of basic research from ivory tower to industry. In 1994, the Executive Office of President and Science and Technology Office published a document representing the development idea of scientific policy in new era—"Science in the National Interest", which pointed out that basic research was an important prerequisite to enable the national goals to be met; however, the linear model between basic research and economic development was questionable. An interdependent relationship exists among basic research, applied research and technological development. This idea had been embodied in "Unlocking Our Future: Toward a New National Science Policy", which was published by the House Science Committee of the United States in 1998.

It is worth mentioning that the US public reaction on federal government supporting basic research was overwhelmingly favorable. Since 1985, the NSF carried out investigations every year about questions like "Do you think whether the federal government should support basic research so as to promote the United States in the forefront of innovation or not, although basic research can't bring benefit directly?". About 82% of respondents strongly supported the investment for basic research in 2010 (the lowest approval rate was 76% in 1992, and the highest was 87% of 2006) (NSB, 2010).

However, in the United States, the emphasis on industry-driven basic research has declined in recent years. In the last 10 years of the 20th century, there has been a significant transformation in the field of scientific research. Overadherence to open innovation led to too much emphasis on monitoring, identification and access to external resources; therefore, basic research of large corporate research labs was in a lower position than before. High-tech enterprises that have been keen on basic research, such as Lucent, IBM, Xerox, AT&T, Exxon and DuPont, also began to refocus on profitable short-term applied research. Numerous enterprises shifted their focus from "blue sky" (basic research) down to the "earth" (experimental development) [①], and the US industrial basic research and development research investment once again dropped to less than 20%. Bell Labs stopped most of the long-term research in physics, and focused on meeting the current needs of users; meanwhile, HP, IBM and other enterprises had asked the researchers to focus on the pragmatic ideas and plans that might generate financial revenue.

However, Japanese industries have shown their emphasis on long-term research projects during this period (almost 50% of R&D funding was invested in basic research and development research). Kodak's failure of the film camera is partly because it focused too much on incremental innovation, while its rival Sony successfully developed a filmless camera as a result of up to 15 years' basic research and became

① Blue sky research comes down to earth. *Business Week*, July 3, 1995.

the new industry leader. Although many emerging US S&T enterprises tried to make up for the deficiency by strengthening basic research, scholars still worried about how the weakening basic research would have a negative effect on the US innovative capability (Rosenbloom & Spencer, 1996; Hicks et al, 2001). Payson & Jankowski (2000) found that an important trend of the US economy is that, the industrical R&D level shows unprecedented growth since the 1990s (the average annual growth rate of national R&D expenditure is 6%, of which the industrial average annual growth rate is 7.4%). However, the industrial R&D level presented a "go down" trend: During the period from 1994 to 2000, the average annual growth rate of the US industrial R&D funding which was used for experimental development was more than 6.1%, while the annual growth rate of the funding used for basic research and development research was between 5.7% and 5.8%. A recent survey from the US Census Bureau found that there is a trade surplus of 30 billion USD for advanced high-tech products in 1998, but a trade deficit of 53 billion USD in 2007 [①]. The most surprising fact is that the US trade deficit unexpectedly appeared on the world's leading life sciences industry. Intel CEO Paul Otellini worried about the status of the United States, and believed that the ignorance of education, scientific research and the construction of digital domain have led the US economic and technological competitiveness to fall behind the world's standards. In fact, in the first decade of the 21st century, the government spending on basic research showed a downward trend, which has led to the weakening of the competitiveness of the US innovation capacity.

① http://finance.sina.com.cn/money/20090609/10176323577.

6.2 Germany: Centering on corporate innovation

Germany has a tradition of emphasizing on basic research, and basic research investment expenditure has accounted for 20% of R&D investment chronically since the 1980s (Zhou, 2008).

German basic research is based on three pillars: universities, German Research Federation, and research institutes outside universities (represented by the Max Planck Society and Helmholtz Association of German Research Center). In addition to the scientific research, universities also undertake a wide range of other important tasks, such as education and medical research; because of this, the basic research topics of German universities are generally limited (Wang, 2006). In terms of structural layout, universities have an advantage in subject distribution, which is also an important facet of scientific research in Germany. Max Planck Society, founded in

1948, is an institution specialized in basic research, which is powerful in inter-discipline, and has been recognized as the strongest research institute to perform basic research in Germany. A fine tradition of the national research centers is their collaboration with universities, research institutes outside the universities (such as the Max Planck Society and Fraunhofer Institute) and enterprises in doing basic research.

Basic research intensity of enterprises has been maintained at a stable level in Germany (Figure 6-8). Consequently, strengthening the industrial demand-oriented basic research is a prominent trend of German S&T policy in recent years. For R&D expenditure in Germany, a stable share of 60% is financed by the business sector for a long time; the enterprises' R&D investment mainly focuses on experimental development, and the major investor of basic research is the government (Zhou, 2008). According to statistics, in 2010, the total research investment of German enterprises was 49 billion euros, accounting for 70% of the total national research investment. In the R&D investment rankings of EU, 11 German enterprises are in the top 25. Large German enterprises generally have research departments and development departments, thereby conducting research activities at different levels. Research departments focus on the exploration of strategic technology; meanwhile, the development departments are responsible for realizing indigenous innovation and new product development by exploiting existing knowledge and technology.

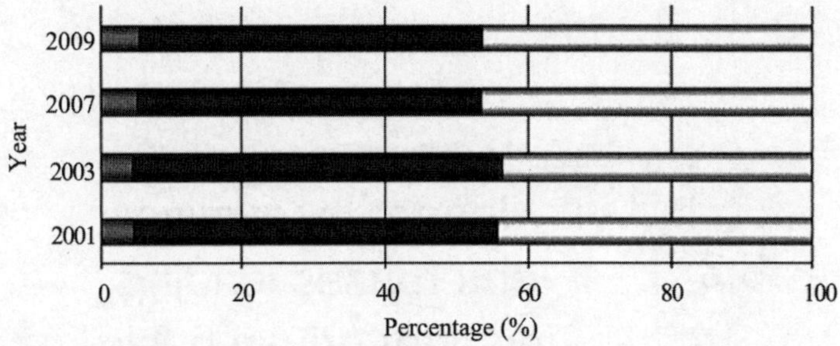

	2001	2003	2007	2009
Experiment Dev.	44.3%	43.8%	46.5%	46.2%
Applied Res.	51%	51.8%	48.5%	48.4%
Basic Res.	4.7%	4.47%	4.963%	5.42%

Figure 6-8 German enterprises' R&D input and the variation of basic research intensity

Data source: OECD, 2011.

In contrast, there is only 8 billion euros of R&D investment for each year from SMEs, but the German government at all levels has helped them with a variety of financing or joint R&D projects. Additionally, SMEs, universities and research institutes have carried out extensive and in-depth cooperation. The German government has always attached importance to providing financial support to enterprises and research entities. The annual funding for scientific research invested by the German government is on an average of around 20 billion euros in recent years. Promoting technological standardization is also the focal point of the government, the purpose of which is to facilitate German enterprises holding the initiative in the competition. In Germany, research cooperation between enterprises and universities is very common. Official figures reveal that about 14% of German universities' research funding is financed by enterprises (Han, 2012).

6.3 Japan: Innovative enterprises leading industry-driven basic research

The feature of Japanese R&D expenditure usage is "who pays, who uses", showing strong independence. Japanese innovation system is characterized by a large enterprises-centered innovation model, whereas government influence is also obvious. The government intervenes and supports R&D activities; meanwhile, it integrates national universities, research institutes and R&D departments of large enterprises, consequently to establish a market-targeted R&D system. Government-led R&D only accounts for 20% of Japanese technological innovation and the rest is all completed by enterprises. Enterprises' R&D expenditure is mainly used for internal R&D activities (Zhou, 2008). The Ministry of Education, Culture, Sports, Science and Technology (MEXT) is the key department for Japanese government to support basic research, of which one important role is to invest in national universities for their regular funding. Moreover, competitive funding has been significantly increased since 2000, and reached a percentage of 25% of total investment in 2006, and the proportion of the two changed from 6 : 1 in the early stage to 1 : 1 at present. With respect to the structure of various agencies' R&D expenditure (Table 6-1), the internal structure of enterprises' R&D expenditure is remaining at a steady level of 1 : 3 : 12 (Figure 6-9); meanwhile, the structure of Japan as a whole remains at around 1 : 2 : 4.

Table 6-1 R&D performance of enterprises, research institutes and universities from 1990 to 2009

Year	Enterprises (%)			Research institutes (%)			Universities (%)			National scale (%)		
	Basic	Applied	Experimental	Basic	Applied	Experimental	Basic	Applied	Experimental	Basic	Applied	Experimental
1990	6.4	21.8	71.8	14.3	28.6	57.1	54.6	37.1	8.3	13.0	24.2	57.3
1991	6.8	22.2	71.1	14.6	29.3	56.1	54.5	37.1	8.4	12.3	22.7	56.8
1992	6.9	22.1	71.1	16.6	27.6	55.8	54.3	37.3	8.3	12.8	22.4	55.9
1993	6.7	21.4	71.9	18.7	26.9	54.5	54.0	37.4	8.6	13.5	22.5	55.3
1994	6.8	22.2	71.1	18.5	27.8	53.6	54.2	37.1	8.7	13.7	22.8	55.0
1995	6.6	22.0	71.3	20.6	27.7	51.7	54.6	36.5	8.9	14.2	22.3	60.0
1996	6.2	22.1	71.8	19.8	26.9	53.3	54.7	36.4	9.0	12.2	22.5	60.0
1997	6.2	21.6	72.2	21.3	31.8	46.8	54.3	36.7	9.0	12.0	22.6	60.0
1998	5.6	21.9	72.6	24.4	29.9	45.6	54.8	36.3	8.9	12.0	21.6	60.1
1999	5.8	20.5	73.7	25.0	27.9	47.1	54.0	36.9	9.1	12.3	22.0	60.0
2000	5.8	21.3	73.0	27.6	26.8	45.7	53.6	37.3	9.1	12.4	21.3	60.0
2001	5.8	20.4	73.9	30.5	24.1	45.3	53.5	37.5	9.0	12.2	21.0	60.0
2002	5.9	19.5	74.6	31.0	26.9	42.1	54.0	36.4	9.6	12.6	21.2	61.0

Continued

Year	Enterprises (%)			Research institutes (%)			Universities (%)			National scale (%)		
	Basic	Applied	Experimental	Basic	Applied	Experimental	Basic	Applied	Experimental	Basic	Applied	Experimental
2003	6.0	19.4	74.7	30.9	30.5	38.6	55.0	36.5	8.5	12.6	23.0	60.9
2004	6.0	19.3	74.5	30.7	31.2	38.1	54.0	36.7	9.3	12.0	21.2	61.6
2005	6.3	19.6	74.0	29.8	30.2	40.0	55.0	35.7	9.3	12.0	21.0	61.8
2006	6.5	18.6	74.7	31.3	29.2	39.5	54.6	38.0	7.4	11.6	20.5	62.9
2007	6.4	20.1	73.3	32.0	30.1	37.9	55.2	35.2	9.6	11.6	21.5	62.0
2008	6.4	19.9	73.4	30.1	29.9	40.0	54.5	37.1	8.4	11.4	21.7	62.6
2009	6.7	20.5	72.6	29.8	30.3	39.9	53.1	36.4	10.5	15.0	21.3	58.9

Data source: OECD, 2011;Japanese White Paper on Science and Technology (2005–2010); Japan Science and Technology Indicators (2009–2012).

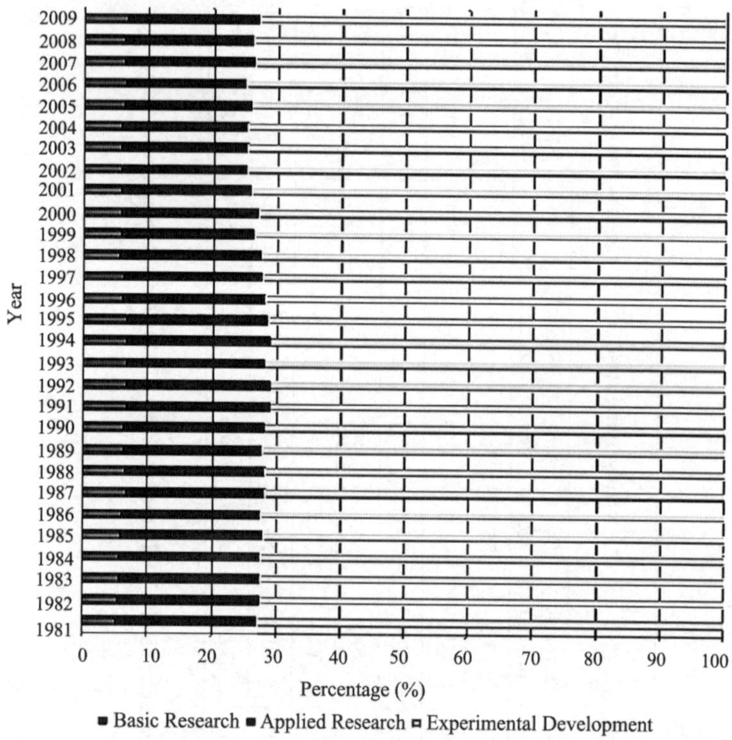

Figure 6-9 Structural changes in basic research, applied research and experimental development of Japanese enterprises

Data source: OECD, 2011; Japanese White Paper on Science and Technology (2005); Japan Science and Technology Indicators (2009, 2010, 2011).

Japan is also a model of attaching great importance to industrial basic research. Japanese companies are very good at learning. The effective utilization of global basic research resources after World War II, coupled with Japan's formidable technological application capability, promote Japan to be the second-largest economy in the world; a good example is Canon, which pioneered in successfully developing LCD based on papers of physicists.

In a traditional sense, universities have been the frontier of basic research after World War II, whereas enterprises have increased their investment in basic research, and hence, universities and enterprises have been two major camps of Japanese basic research after World War II (Zhou, 2008). In 1965, Japan's basic research intensity was up to 30.3%, which was closely related to Japan's technological catch-up strategy. From the beginning of the Iwato boom in 1959, there was a dramatic increase in the number of research institutes belonging to Japanese enterprises, and nearly all the larger enterprises had their own R&D unit; an era so called "prosperity in central research centers" was thus formed (i.e. a large number of central research institutions were established within enterprises) (Zhou, 2008). However, the intensity of basic research in Japan declined sharply after 1965. In addition, Japanese enterprises shifted their focus from basic research to technological development, combined with the Oil Crisis in 1973, contributing to the decline in the proportion of basic research.

After the 1980s, Japanese companies again stepped up investment in basic research, because of which basic research intensity was increasingly stable, and the enterprises anchored their hope on basic research to maintain strong competitive edge (Jie, 2004). More than three-quarters of Japanese companies are engaged in basic research, which are on the stand of national and international development, but still more for their own survival and development.

However, after the collapse of the bubble economy in the 1990s, in consideration of R&D cost, quality and efficiency, numerous enterprises began to entrust basic research to universities; meanwhile, Japanese universities employed researchers from enterprises as teachers due to their better understanding of industrial needs, and so they were found to be more suitable for carrying out basic research in line with industrial development[①]. In 2011, more than one-fifth of the world's

① Japan's national innovation system with the participation of government and business sector, Xinhuanet, June 8, 2012.

patent applications came from Japan, which was eight times that of the United Kingdom. Since the government formulated a clear preferential national policy, together with the close linkage between universities and enterprises, Japan is rated as the best country of the world to commercialize basic research. ②

② Japan is worth learning in technological innovation, etc. *Reference News*, http://finance.cankaoxiaoxi.com/2012/0424/31794.shtml.

6.4 Korea: "Bottom-up" industrial basic research investment system

Korea is one of the most successful countries in catch-up innovation. Just like Japan, the R&D funding also maintains at a high level. R&D funding financed by enterprises is mainly used for their internal R&D activities; only a small part will be invested in universities and national research institutes. Meanwhile, government funding mainly flows into the national research institutes and universities (Li & Zeng, 2008). Korean S&T innovation system is characterized by "bottom-up" capacity-building rather than the traditional "top-down". In Korea, if enterprises lack independent intellectual property rights, they will receive neither government support nor public welcome; therefore, enterprises tend to have a strong impetus to conduct R&D and innovation.

During the period from the 1960s to the 1970s, Korea's R&D investment is dominated by the government; after the 1980s, the proportion of government R&D declined gradually and enterprises became the main actor for R&D (Figure 6-10).

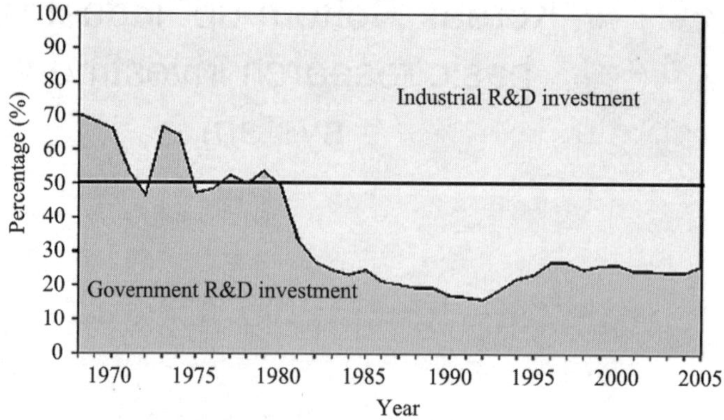

Figure 6-10 Variation of Korean R&D investors

Data source: OECD, 2011.

After the Cold War, the Korean government found it was increasingly difficult to build international competitiveness through simple technological import and imitative innovation; therefore, it began to pay attention to strengthening basic research. In May 1994, Korea's National Science and Technology Council developed "Comprehensive Plan to Revitalize Basic Research". This five-year comprehensive plan was targeted at enabling Korean basic science to be the world's top seven in 2010, and further strengthening the comprehensive support for basic research of universities. In November 1994, Korean Institute of Science and Technology Policy Studies proposed "2010 Long-term Science and Technology Development Plan" to the government; in the "Basic Research" section of this plan, they put forward improving university research systems, strengthening the support for university basic research, in addition to establishing basic research cooperation mechanisms and cooperation funds between universities and industries.

Basic research of Korea is responsible for a majority of the changes that have happened for Korea to have now grown into a significant scientific center of Asia Pacific and one of the most important high-tech industry gathering places in the world and this transition has been from a backward agricultural country that it was in the 1960s. The most striking aspect of Korea is that enterprises are the core strength of basic research. At present, in Korea, basic research funding performed by enterprises accounts for about 50% of total basic research funding (55.73% in 2003), followed by universities (25.22% in 2003) and national research institutes (19.05% in 2003). The share of funding for basic research in total R&D expenditure of enterprises in 2002 and 2003 was 9.10% and 10.60% respectively (Figure 6-11).

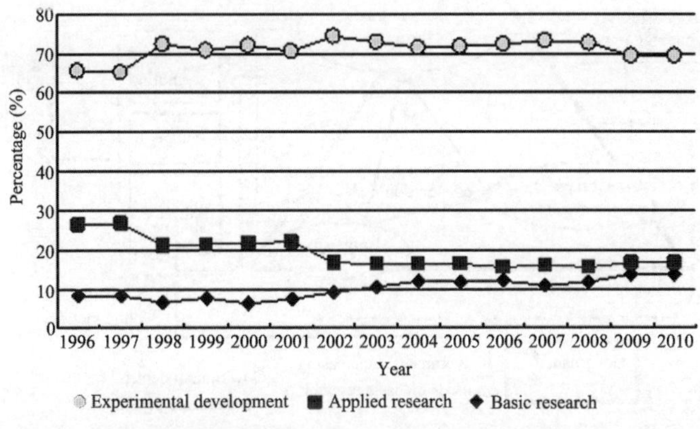

Figure 6-11 R&D investment structure change of Korean enterprises

Data source: OECD, 2011.

Korea adopts a government-led model for S&T innovation. Government funding for basic research is mainly invested in universities, of which the majority is for purposive basic research. The government used to actively support government research institutes and university research centers, which includes helping them to set up laboratories overseas (Figure 6-12). For the purpose of promptly transforming basic research achievements into productive forces, the Korean government attaches great importance to establishing research collaboration mechanisms among industry, universities and research institutes, and sets up specialized industry-university-research institute consortia funding as well as cooperative parks, so as to enable the enterprises to fully take advantage of research strength, information and equipment of universities, and enhance the transform of research achievements from universities. By way of providing matching funds, promotion funds and specific R&D expenses, the Korean government has fulfilled a series of tax concession policies for research institutes which are affiliated with enterprises set up in special industry-university-research institute cooperative parks.

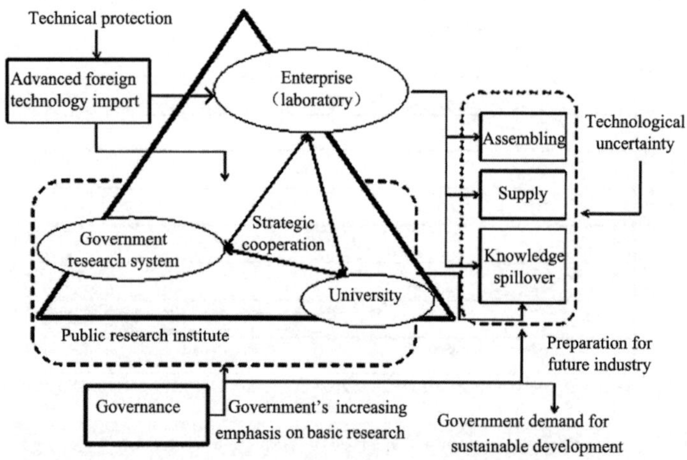

Figure 6-12 Technology and innovation framework of Korea in the 21st century

Data source: Shi, 2011.

A distinctive feature of Korean S&T innovation is "R&D for practical use". They implement a package of S&T support system for the entire process from R&D to R&D achievements transformation, with the development and implementation of a series of S&T policies centered on basic research. In 1967, Korea promulgated the National Science and Technology Promotion Law, which formalized the strategy of promoting national economic development through S&T innovation.

In late 1970s, Korea's S&T policy shifted from being export-driven to technology-driven and the "Science Foundation" was established in 1977 as a national supporting organization for basic research in order to get rid of dependence on technological import and imitation. Since 1978, the Korean government has made Basic Scientific Research Implementation Plan every year, and established Basic Scientific Research Supporting Center in 1988. Korea started to increase investment in basic research, and support and guide private enterprises to conduct R&D activities through a variety of tax and finance policies, which led to enterprises' R&D funding more than that of the government's in the 1980s; meanwhile, enterprises set out to build research institutes and paid more attention to basic research (Zhou, 2008).

Since the 1990s, Korea initiated to strengthen policy control for the linkages between technological suppliers and demanders. Instead of following the trends of advanced countries, Korean R&D policy objective during this period was committed to competing with the technological innovation of developed countries through R&D, particularly in terms of basic research of enterprises. In December 1989, Korea formulated the Basic Scientific Research Revitalization Act, specifically to accelerate the investment of enterprises' basic research through the implementation of R&D tax incentives, so as to boost the basic research collaboration and communication among universities, government-owned research institutes and corporate labs. However, the strength of enterprises' basic research declined during this period, the reasons for which are similar to those of Japan's. Increased technological innovations and technological improvement, combined with the shift towards experimental development, directly led to the decline of basic research and applied research intensity. In December 1990, the presidential decree Basic Scientific Research Revitalization Act Enforcement Order was enacted to support basic research, which further stipulated an important role of the Ministry of Education, Science and Technology, which is committed to improving and strengthening the research collaboration among academia, industry and research institutes. Since then, the Basic Scientific Research Revitalization

Act and Basic Scientific Research Revitalization Act Enforcement Order went through revision for 8 and 14 times respectively. Korea explicitly made "to expand investment in S&T, to advance cutting-edge technology industrialization" a central goal of national S&T policy in 1993. In December 1997, the Science and Technology Promotion Act was amended to Science and Technology Innovation Special Law (Technology Policy and Regulation Department, 2010). In the 21st century, Korea's investment in basic research once again began to rise, which was partly due to the government's surged R&D budget and partly due to enterprises' increasing investment in basic research (Zhou, 2008). In the last 20 years, the core of Korean S&T policy has been to strengthen the international competitiveness of local enterprises, and construct competitive advantages of Korea in high-tech sectors.

An important feature of Korean basic research is to make the enterprises the first performer for investment as well as the usage of basic research funding. Although research funding only accounts for 10% of enterprises' R&D expenditure, the share of enterprises' funding in basic research is more than 50% of the national basic research funding. Basic research of Korean enterprises is typically industry-driven.

Currently, Korea's S&T system has changed over from being government-led to enterprise-led; a large number of high-tech and industrial technologies are developed by the enterprises, while the government's priority has shifted to supporting innovation environment and facilitating collaborative research among industries, universities and research institutes. Meanwhile, national research institutes are mainly responsible for national technological development, and universities are primarily in charge of carrying out basic research.

6.5 The implications for China

Strengthening industry-driven basic research is a consistent choice for both advanced countries and emerging industrialized countries. However, in theory, industry-driven basic research has not been given sufficient attention, and many scholars and policy-makers who insisted on strengthening basic research are heavily influenced by Bush, Nelson, Arrow and other scholars about the rationality of government investment in basic research; they have neglected the necessity and possibility of enterprises involved in basic research. The experience of insisting on strengthening industry-driven basic research of innovation-leading countries, such as the United States, Japan, Korea and European countries, is worthy of being used for reference by China.

In China, the government plays a key role in the S&T system, from support for S&T funds, evaluation of S&T achievements, the promotion of S&T achievements, and even to the award of S&T personnel. The business side relies on imitation strategy, which allocates more resources to technological absorption and experimental development. Experience from the United States, Japan, Germany, Korea and other countries shows that basic research should maintain a proper, stable proportion in R&D investment (roughly between 10% and 20%), so as to sustain current industrial competition and future industrial development. Therefore, these countries attach great importance to improving the government funding in basic research, meanwhile encouraging enterprises to strengthen their investment in long-term research, from which co-development of national scientific capability and innovation capacity can be realized. Such experience has significantly inspired industrial innovation in China.

References

Chen, J. Basic research: The foundation of independent innovation. *Anhui Science and Technology*, 2005 (6): 4-6. [陈佳洱．基础研究：自主创新的基石．安徽科技，2005（6）: 4-6．]

Crow, M. & Bozeman, B. R&D laboratory classification and public policy: The effects of environmental context on laboratory behavior. *Research Policy*, 1987, 16 (5): 229-258.

Han, M. German multi-level scientific research system ensures the quality and vitality of innovation. Xinhuanet, 2012-06-08. [韩墨．德国多层次科研体系保证质量创新与活力．新华网，2012-06-08．]

Hicks, D., Breitzman, T., Olivastro, D., et al. The changing composition of innovative activity in the US: A portrait based on patent analysis. *Research Policy*, 2001, 30 (4): 681-703.

Jie, Y. The impact of Japanese university's scientific research institution on the development of basic research. *Impact of Science on Society*, 2004 (2):14-17.[节艳丽．大学科研体制对于日本基础研究发展的影响．科学对社会的影响，2004（2）: 14-17．]

Li, H., Sun, X. & Zeng, G. On the basic research investment and its coordination mechanism in typical countries. *Studies in Science of Science*, 2008, 26 (6): 1174-1180. [李红林，孙喜杰，曾国屏．典型国家基础研究投入及其协调机制研究．科学学研究，2008，26（6）: 1174-1180．]

Li, H. & Zeng, G. On the evolvement and coordination mechanism of basic research investment in Japan and Korea. *Scientific Management Research*, 2008, 26 (5): 89-93. [李红林，曾国屏．基础研究的投入演变及其协调机制：以日本和韩国为例．科学管理研究，2008，26（5）: 89-93．]

NSB (National Science Board of the US). *Science and Engineering Indicators 2010.* Arlington, VA: National Science Foundation, 2010.

OECD. Science, technology and R&D statistics, 2011.

Payson, S. & Jankowski, J. *Sixth Year of Unprecedented R&D Growth Expected in 2000*. Arlington, VA: National Science Foundation,

2000.

Rosenbloom, R. S. & Spencer, W. *Engines of Innovation.* Boston, MA: Harvard Business School Press, 1996.

Shi, T. Korea's strategy for development of STI capability: A historical perspective. China-OECD Roundtable on Innovation Policies, October 18-19, 2011, Beijing.

Wang, D. Overview of German basic research and management system. *Global Science, Technology and Economy Outlook*, 2006 (1): 47-50. [王大川．德国基础研究概况和管理体系．全球科技经济瞭望，2006（1）: 47-50．]

Zhou, J. *Foundation and Source of Innovation: Investment, Evaluation and Coordination of Basic Research*. Beijing: Science Press, 2008. [周寄中．创新的基础和源泉：基础研究的投入、评估和协调．北京：科学出版社，2008．]

Chapter 7

Analysis of Basic Research Investment Mechanism of China

This chapter studies the relationship between the country's economic development stage and basic research. A lot of empirical research finds that the investment in basic research follows a model in terms of economic development: increasing in the process of the country's growth and then declining when the country reaches a certain stage of development. But this is not true in China: In the stage of rapid economic development since the 1980s until now, basic research's proportion in total R&D investment is constantly at a low level, especially for industrial basic research; does this cause the low capability of industrial innovation in China?

In this chapter, we analyze how China invests in basic research by comparing other catching-up and developed countries, to see what the key factors are in shaping the investment in a country, stage development and national strategy. Does the Chinese government make a good policy or is it contrary to what most of the other countries do? How different are the roles of universities, enterprises and public institutes in basic research in China when compared to their counterparts in leading countries?

7.1 The evolution of basic research investment in China

Through consistent efforts, the Chinese government has made rapid growth in scientific investments, which has altered the former layout of global scientific research resources and charted a new global map of science and research (Table 7-1).

Table 7-1 Global ranking of major scientific indicators in R&D investment and expenditure of China

Year	R&D expenditure		Number of inventions and patents granted		S&T papers indexed by SCI, EI and ISTP
	Percentage (%)	Rank	Percentage (%)	Rank	Percentage (%)
2003	2.5	6	6.0	4	5.1
2004	2.9	6	7.9	3	6.3
2005	3.4	6	8.4	4	6.9
2006	4.0	6	7.7	4	8.4

Continued

Year	R&D expenditure		Number of inventions and patents granted		S&T papers indexed by SCI, EI and ISTP
	Percentage (%)	Rank	Percentage (%)	Rank	Percentage (%)
2007	4.6	6	8.8	4	9.8
2008	5.7	4	12.1	3	11.5
2009	7.1	4	16.9	3	11.6
2010	8.6	3	—	3	12.3
2011	10.1	3	—	3	14.1
2012	11.1	3	—	3	16.5
2013	13.7	2	—	3	17.3

Data source: *National Science and Technology Funds Statistical Bullwtin 2014.*

The number of R&D staff in China has tripled in two decades. But the proportion of basic research staff is quite low, which is under 10% (Table 7-2).

Table 7-2 Changes in the full-time equivalent of R&D and basic research staff in China from 1991 to 2013

Year	Full-time equivalent of R&D staff (1,000 person-year)	Full-time equivalent of basic research staff (1,000 person-year)	Share of basic research (%)
1991	670.5	61.3	91.4
1992	674.3	58.4	86.6
1993	697.8	63.3	90.7
1994	783.2	76.4	97.6
1995	751.7	66.6	88.7
1996	804.0	69.6	86.5
1997	831.2	71.7	86.3
1998	755.2	78.7	104.2
1999	821.7	76.0	92.5

Continued

Year	Full-time equivalent of R&D staff (1,000 person-year)	Full-time equivalent of basic research staff (1,000 person-year)	Share of basic research (%)
2000	922.1	79.6	86.3
2001	956.5	78.8	82.4
2002	1,035.1	84.0	81.2
2003	1,094.8	89.7	81.9
2004	1,152.6	110.7	96.1
2005	1,364.8	115.4	85.0
2006	1,502.5	131.3	87.4
2007	1,736.2	138.1	79.5
2008	1,965.4	154.0	78.3
2009	2,291.3	164.6	71.8
2010	2,553.8	173.7	68.0
2011	2,883.0	193.2	67.0
2012	3,247.0	211.1	65.0
2013	3,533.0	222.6	63.0

Data source: The series of *China Statistical Yearbook on Science and Technology* (1992–2014).

What is not in line with the trend of increased R&D investment is that the proportion of basic research investment is comparatively low. In China, more expenditure is on experimental development and even the growth in applied research expenditure is slow (Figure 7-1, Figure 7-2). This shows that the support for basic research in China is relatively limited. Although China's GDP ranks second in the world, basic research investment has not been improved accordingly. In the past decade, the proportion of basic research in GDP hovered at around 0.05% (exceeding 0.08% after 2009), nearly one tenth of that of other innovative leading countries (Table 7-3).

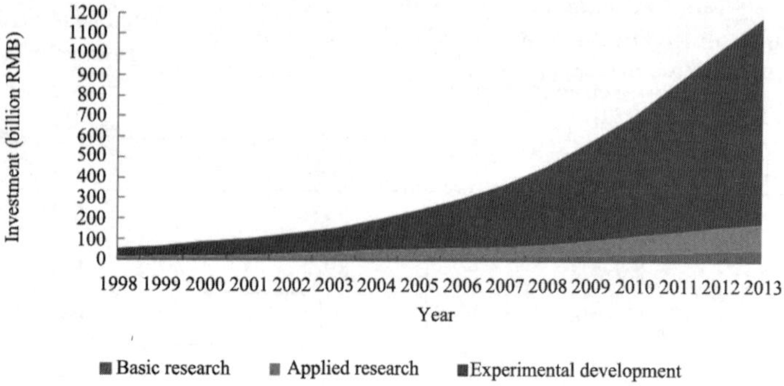

Figure 7-1 Changes of R&D investment and structure in China

Data source: The series of *China Statistical Yearbook on Science and Technology* (1999–2014).

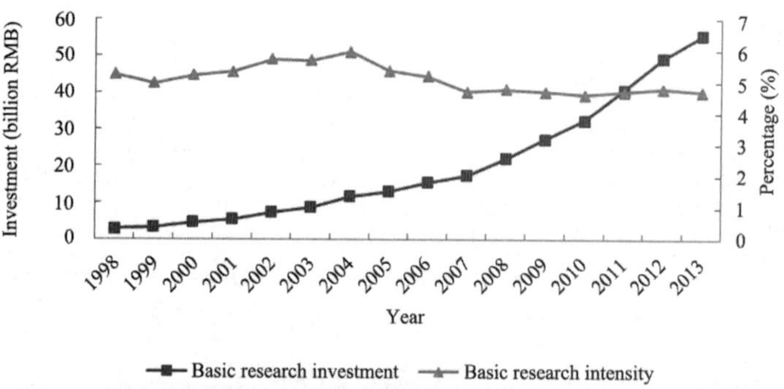

Figure 7-2 Trend of basic research investment and intensity intensity in China

Data source: The series of *China Statistical Yearbook on Science and Technology* (1999–2014).

Table 7-3 Comparison of the proportion of basic research investment in GDP in major countries (%)

Country	2000	2005	2006	2007	2008	2009	2010	2011	2012	2013
United States	0.43	0.48	0.46	0.47	0.49	0.55	0.50	0.47	0.45	0.48
Japan	0.38	0.40	0.40	0.40	0.39	0.42	0.39	0.41	0.42	0.44
France	0.51	0.50	0.50	0.52	0.54	0.59	0.55	0.53	0.54	—
United Kingdom	—	—	—	0.19	0.19	0.21	0.18	0.16	0.18	—

Continued

Country	2000	2005	2006	2007	2008	2009	2010	2011	2012	2013
Korea	0.29	0.43	0.46	0.50	0.54	0.64	0.63	0.68	0.74	0.75
China	0.05	0.05	0.05	0.05	0.05	0.08	0.08	0.09	0.10	0.10

Data source: OECD, 2014; the series of *China Statistical Yearbook on Science and Technology* (2001–2014).

7.2 The relationship of the stage of economic development and basic research

One key question that often confuses scholars of S&T policy and policy-makers in China is: What is the appropriate proportion of basic research in the total R&D investment? (Xiao, 2005) Should the investment in basic research be maintained at a stable level or the more the better?

In order to answer these questions, we take 20 years' data of basic research investment intensity in China from 1991 to 2011 to check whether the change with time of this variable has any nonlinear characteristics.

After using curve fitting tests-based SPSS software, we find that only the *S* curve and the quadratic curve (*Q* curve) pass the significance test. According to the fitting condition of the fitting curves and the actual values, the fitting degree of the quadratic curve is more ideal and its goodness of fit is better than that of the *S* curve. Therefore, we choose the quadratic curve, the result of which is shown in Table 7-4 and Figure 7-3.

Table 7-4 Results of parametric test of the *S* curve fitting of the basic research intensity in China

Coefficient	Regression coefficient value	Significance (*p*-value)
Constant term	4.338	0.010
b_1	0.208	0.000
b_2	–0.010	0.000
R^2	0.601	—
F-statistic	13.546	0.000

Dependent variable: Intensity of basic research in China (%), $N = 21$.

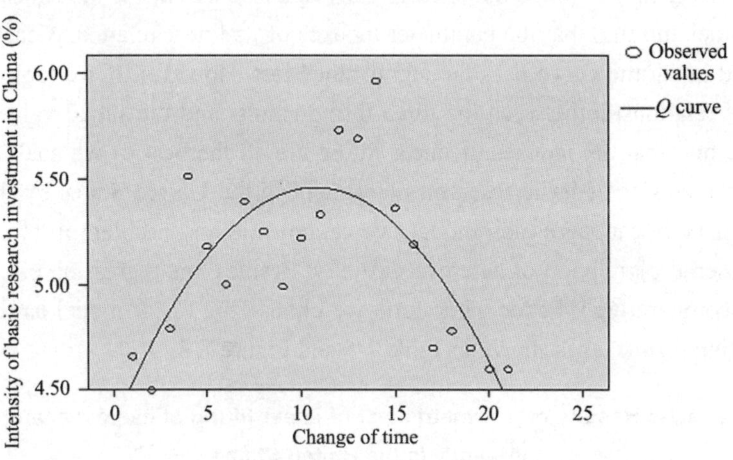

Figure 7-3 Results of *Q* curve fitting of basic research intensity in China (1991–2010)

Note: Intensity of basic research (%) = Expenditures on basic research/ Expenditures on R&D. The scales of the horizontal axis are reset, with the year 1991 as the starting point 1.

The following fitting model can be made according to the statistical result:

$$GEBR/R\&D=4.338+0.208\times t-0.0.1\times t^2,\ t = 1991, 1992, ..., 2011$$

This means that in the recent 20 years, the intensity of basic research investment in China presents an "inverted U-shaped" developmental track. It was continuously on the rise since 1991 and began to fall from the inflection point (10.4, 5.42). In other words, the intensity of basic research investment reached 5.42% in 2000 before its overall trend started to decrease. That could be because enterprises became the subject of R&D after 2000 but the investment in basic research did not persistently increase as the overall R&D investment did.

For purposes of comparison, we also look at the historical changes of basic research intensity in the United States and in Japan the same way. When doing the curve fitting of the historical data of basic research intensity in the United States from 1981 to 2009, we choose the "linear" model and find that the parameter indexes of its linear relation, *S* curve and quadratic curve pass the significance test. However, if we look at the relationship between the three fitting results and the actual values, we find that the indexes of linear fitting are all the best. If we analyze the intensity of basic research investment in the United States by the unitary linear regression model, we cannot find any problem in terms of serial correlation or heteroscedasticity. Besides, the regression effect is comparatively better. Therefore, we choose the linear model as the fitting results, as is shown in Table 7-5 and Figure 7-4.

Table 7-5 Results of parametric test of linear fitting of basic research intensity in the United States

Coefficient	Standard regression value	Significance (*p*-value)
Constant term	13.241	0.000
B	0.196	0.000
R^2	0.772	—
F-statistic	91.284	0.000

Dependent variable: Intensity of basic research in the United States (%), $N = 29$.

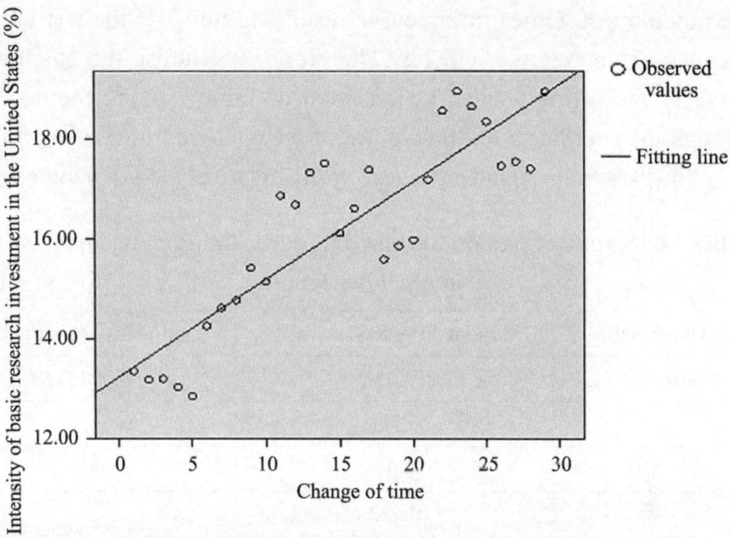

Figure 7-4 Results of linear fitting of basic research intensity in the United States (1981–2009)

Note: Intensity of basic research (%) = Expenditures on basic research/ Expenditures on R&D . The scales of the horizontal axis are reset, with the year 1981 as the starting point 1.

The following fitting model can be made according to the statistical result:

$$GEBR/R\&D=13.241+0.196\times t,\ t = 1981, 1982, ..., 2009$$

This means that the intensity of basic research in the United States shows an overall rising development track during the past 30 years since 1981, which benefits not only from the long-term emphasis on basic research of the US Federal Government (though there was slight decline during the 1980s), but also from the huge investment in basic research by innovative enterprises in the United States.

When doing the curve fitting of the historical data of basic research intensity in Japan, we find that only the parameter index of quadratic curve passes the significance test, while that of the *S* curve does not pass the module test. Other fitting curve results do not pass the test either because of one or two indexes. Therefore, we choose the quadratic curve as the fitting results, as is shown in Table 7-6 and Figure 7-5 (though the coefficient of module test of the *Q* curve fitting is not high, only this passes the significance test, with a relatively low *F* value).

Table 7-6 Results of parametric test of S curve fitting of basic research intensity in Japan

Coefficient	Standard regression value	Significance (*p*-value)
Constant term	12.031	0.000
b_1	0.111	0.039
b_2	–0.004	0.014
R^2	0.522	—
F-statistic	4.864	0.016

Dependent variable: Intensity of basic research in Japan (%), $N = 29$.

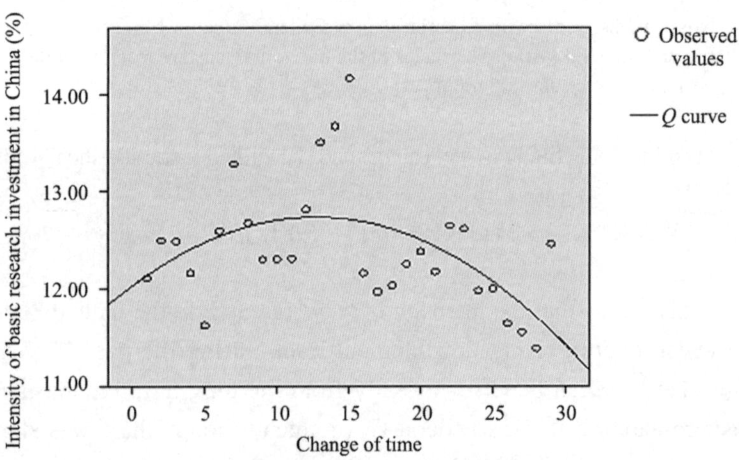

Figure 7-5 Results of *Q* curve fitting of basic research intensity in Japan (1981–2009)

Note: Intensity of basic research (%) = Expenditures on basic research/ Expenditures on R&D. The scales of the horizontal axis are reset, with the year 1981 as the starting point 1.

The following fitting model can be made according to the statistical result:

$$GEBR/R\&D=12.031+0.111*t-0.004*t^2,\ t = 1981, 1992, ..., 2009$$

This means that in the recent 30 years, the intensity of basic research investment in Japan also presents an "inverted U-shaped" development track. It was continuously on the rise since 1981 and began to fall from the inflection point (13.88, 12.80). In other words, the intensity of basic research investment reached 12.80% in 1994 before its overall trend started to decrease.

Next, we do the curve fitting of historical changes from 1991 to 2010 in terms of the basic research investment intensity of Chinese enterprises in a similar way.

We also find that only the parameter index of the quadratic curve passes the significance test, while those of the *S* curve, logarithmic curve or other curves do not pass the module test. Therefore, we choose the quadratic curve as the fitting results, as is shown in Table 7-7 and Figure 7-6 (though the coefficient of module test of the *Q* curve fitting is not high, only this passes the significance test, which could be because the small sample affects the choice of fitting curve module).

Table 7-7 Results of parametric test of *S* curve fitting of basic research intensity of Chinese enterprises

Coefficient	Standard regression value	Significance (*p*-value)
Constant term	0.447	0.012
b_1	0.085	0.025
b_2	–0.005	0.012
R^2	0.685	—
F-statistic	7.502	0.005

Dependent variable: Intensity of basic research of Chinese enterprises (%), $N = 20$.

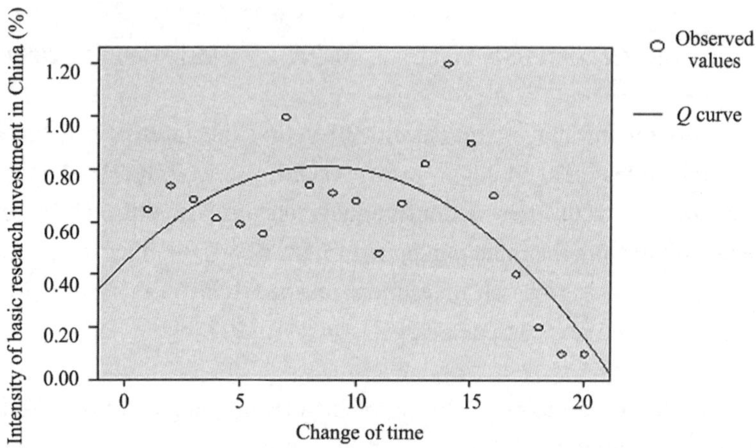

Figure 7-6 Results of *Q* curve fitting of basic research intensity of Chinese enterprises (1991–2010)

Note: Intensity of enterprises' basic research (%) = Enterprises' expenditures on basic research/Enterprises' expenditures on R&D. The scales of the horizontal axis are reset, with the year 1991 as the starting point 1.

The following fitting model can be made according to the statistical result:

$$\text{EGEBR/R\&D}=0.447+0.085\times t-0.005\times t^2,\ t = 1991, 1992, \ldots, 2010$$

This means that in recent 20 years, the intensity of basic research investment of Chinese enterprises presents an "inverted U-shaped" development track. It was continuously on the rise since 1991 and began to fall from the inflection point (8.5, 0.81). In other words, the intensity of basic research investment reached 0.81% in 2001 before its overall trend started to decrease.

Two conclusions can be drawn from the discussions above:

First of all, one of the largest gaps between China and the innovation-leading countries is the weak ability of enterprises to conduct basic research. The proportion of R&D, especially basic research input of Chinese enterprises is relatively low. The massive R&D expenditures are for experimental development, while enterprises still neglect the long-term research.

We believe that the current challenge is to enhance the enterprises' investment in basic research, because China's economic growth pattern has started to change from being resource-dependent to innovation-driven and the innovation mode of industrial technology has changed from simple introduction and imitation to independent innovation. To increase input in R&D, we first need to increase input in basic research. The lack of basic research has been slowing the improvement speed of China's industrial innovation ability. Meanwhile, the problem of long-term weakness in innovation performance cannot be completely solved. Although many successful Chinese enterprises emerge on the high-tech industries and international markets, only less than 1% of them have applied for the patent. Furthermore, only around 2,000 domestic enterprises have their own intellectual properties, accounting for 0.03% of total enterprises (Ran, 2006).

Second, the intensity of basic research investment of the country and that of enterprises both show a certain pattern. The intensity of basic research investment in the United States is steadily rising in the past 30 years, and the turning point has not yet appeared. And both the intensity of basic research investment in Japan and in China present an "inverted U-shaped" development track, first increasing and then decreasing. Japan starts to fall at 12.8%, while the number in China is 5.4%.

We believe that the basic research investment of a country or an enterprise should be maintained at a proper level. Excessive investing will be a heavy burden on the government while too low investing will hurt innovation in economic development, and thus the international competitiveness of the country and the industry. The problem in China is the comparatively low proportion of basic research. Besides, the growth rate of R&D input is relatively low, affecting to some extent the enterprises' input in basic research.

7.3 The imbalanced phenomena among China's basic research, applied research and experimental development

Ever since the 1990s, China started to implement the developmental strategy of "steadily strengthening basic research". As was analyzed in the previous section, the total amount and the proportion of R&D input is in steady increase, benefiting a lot from the support for basic science of S&T projects, including National Natural Science Foundation of China, National Basic Research Program (Program 973), National High-tech R&D Program (Program 863) and Knowledge Innovation Program. However, in the three major modules of R&D in China, massive expenditures are for applied research and experimental development, while the proportion of basic research never goes above 6% and fluctuates sharply. On the whole, China's basic research investment demonstrates three obvious structural "disproportionalities".

The absolute value of R&D expenditures in China is growing quickly, but its relative proportion is growing slowly and its structure is imbalanced, especially the too low proportion of basic research expenditures (as is shown in Figure 7-7). In 2013, the ratio of investment in basic research, applied research and experimental development in China was 1:2.3:18, while in 2012, the ratio in the United States was 1:1.2:3.9, and in Russia 1:1.2:3.9; in 2011, the ratio in France was 1:1.2:1.4:0.2, Italy 1:2:1.1, Japan 1:1.7:5:0.4 and Korea 1:1.1:3.4 (as is shown in Figure 7-8). In history, the proportion of basic research in China also fell far behind that in developed countries (Table 7-8).

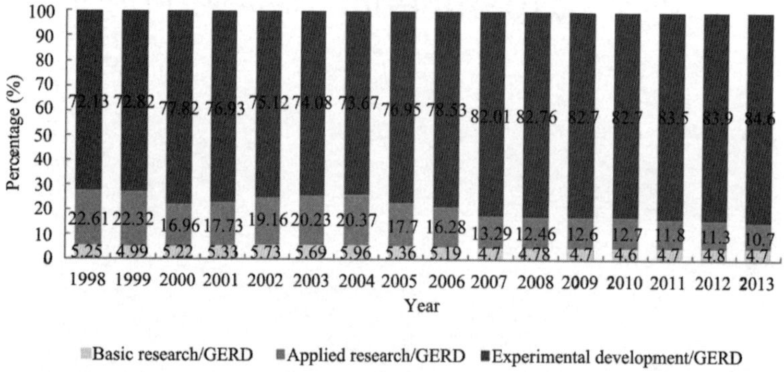

Figure 7-7 The R&D investment structure in China from 1998 to 2013

Data source: The website of China Science and Technology Statistics, *China statistical Yearbook on Science and Technology.*

According to a research report released by OECD in 2007, the total R&D input in China has surpassed that in Japan, ranking second in the world. However, most of the scientific research expenditures are used in product development and adjustment based on the rapidly growing Chinese market rather than basic scientific research work. In recent years, following their global strategies, transnational enterprises have set up sole proprietorship R&D institutions in China one after another and continually improved the scale and intensity of basic research, and this places domestic enterprises under tremendous pressure (Wen & Li, 2007).

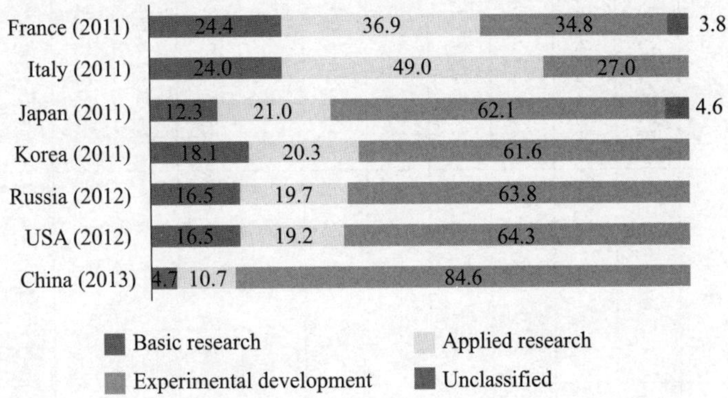

Figure 7-8 Comparison among the structure of R&D expenditures in some countries

Data source: *China Statistical Yearbook on Science and Technology 2014.*

Table 7- 8 Comparison among the structure of R&D expenditures in China and developed countries (%)

Year	China	US	Japan	France	Italy	Korea
1995	5.2:26.4:68.4	16.2:21.1:62.7	14.2:22.4:63.6	—	—	12.5:25.0:62.5
1996	5.0:24.5:70.5	16.6:22.2:61.6	12.2:22.3:66.5	22.0:28.5:49.5		13.2:26.9:59.9
1997	5.4:26.0:68.6	13.8:24.5:61.7	12.0:22.5:65.8	—	22.1:43.8:34.1	13.1:28.5:58.3
1998	5.2:22.6:72.1	15.6:20.5:63.9	12.0:22.6:65.4	—	22.2:43.7:34.1	14.0:25.1:61.0
1999	5.0:22.3:72.7	12.3:21.6:60.2	12.3:21.6:65.4	24.4:27.5:48.1	—	13.6:25.7:60.7
2000	5.2:17.0:77.8	18.1:20.8:61.1	12.4:20.9:66.7	—	—	12.6:24.3:65.1
2001	5.0:16.9:78.1	17.2:23.2:59.5	13.0:22.8:64.2	23.3:33.5:43.2	—	12.6:25.3:62.1
2002	5.7:19.2:75.1	18.6:18.5:63.0	13.4:22.2:64.4	—	—	13.7:21.7:64.5
2003	5.7:20.2:74.1	19.1:23.9:57.1	13.3:22.4:60.0	24.1:36.2:39.7	—	14.5:20.8:64.7
2004	6.0:20.4:73.7	18.7:23.4:58.0	12.2:21.2:66.6	—	—	15.3:21.2:63.4
2005	5.4:17.7:76.9	18.7:21.3:60.0	12.7:22.2:65.2	23.7:39.0:37.3	27.7:44.4:27.9	15.3:20.8:63.8

Continued

Year	China	US	Japan	France	Italy	Korea
2006	5.2:16.8:78.0	18.6:23.1:58.3	12.3:21.5:66.2		28.7:44.7:26.6	15.2:19.9:64.9
2007	4.7:13.3:82.0	17.5:22.1:60.4	12.2:22.6:65.2	25.1:39.2:35.7		15.7:19.8:64.4
2008	4.8:12.5:82.8	17.4:22.3:60.3	8.8:40.6:50.6	25.4:39.0:35.6	27.0:45.6:27.4	16.1:19.6:64.3
2009	4.7:12.6:82.7	17.3:22.4:60.3	12.5:22.3:65.2	—	—	18.1:20.0:61.9
2010	4.6:12.7:82.8	—	—	—	—	18.2:20.0:61.8
2011	4.7:11.8:83.5	—	12.3:21:62.1:4.6	24.4:36.9:34.8:3.8	24.0:49.0:27.0	18.1:20.3:61.6
2012	4.8:11.3:83.9	16.5:19.2:64.3	12.4:20.8:62.1:4.7	24.2:37.4:35.0:3.4	25.3:48.9:25.8	18.3:19.1:62.6
2013	4.7:10.7:84.6	—	12.6:20.9:61.8:4.7	—	—	18.0:19.1:62.9

Data source: The series of *China Statistical Yearbook on Science and Technology* (2007–2014). The ratio in the table is basic research : applied research : experimental development : unclassified .

7.4 Evolution of basic research investment by enterprises

Against the background that the international competition relies more on innovation and the stimulus of independent innovation policies adopted by the government, Chinese enterprises have accelerated their investment in R&D. In 1998, expenditures in R&D by enterprises amounted to 24.7 billion RMB. This figure reached the historical peak of 907.6 billion RMB in 2013, up by 37 times for over one decade (Figure 7-9).

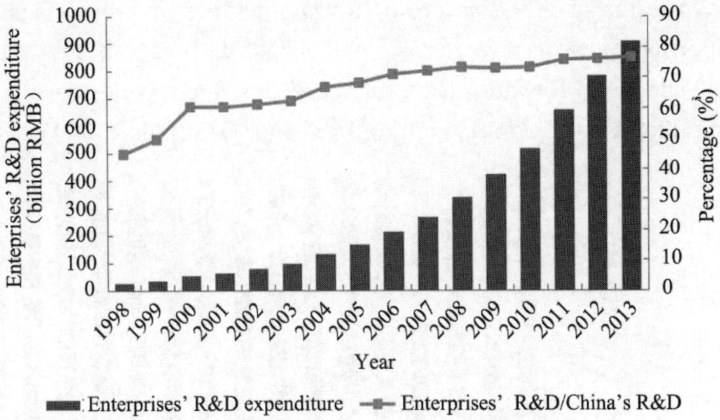

Figure 7-9 Changes of R&D expenditure by Chinese enterprises and its proportion in national R&D spending

Data source: The series of *China Statistical Yearbook on Science and Technology* (1999–2014).

However, a large amount of R&D spending by Chinese enterprises is applied to experimental development. Only a small part goes to applied research. And the spending allocated to basic research is scarce (Figure 7-10).

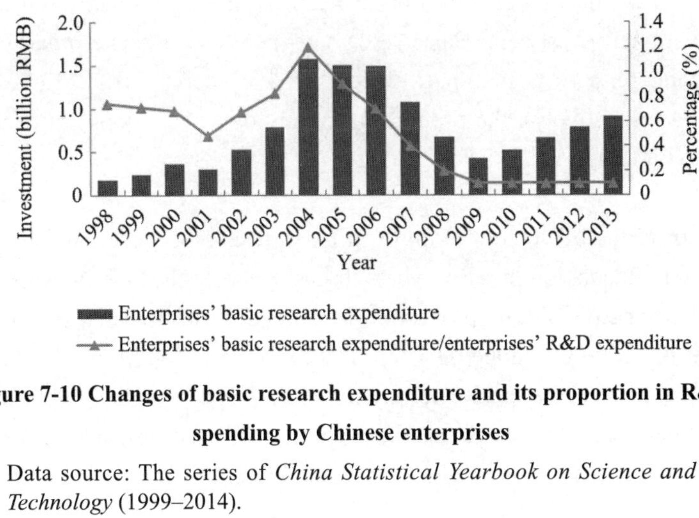

Figure 7-10 Changes of basic research expenditure and its proportion in R&D spending by Chinese enterprises

Data source: The series of *China Statistical Yearbook on Science and Technology* (1999–2014).

In terms of the structure of R&D spending by enterprises, the proportion of basic research is small, which is less than 1% of their R&D spending. Besides, in recent years, this figure is decreasing at a faster rate. It was only 0.1% during 2009 and 2013 (Figure 7-11).

	1998	1999	2000	2001	2002	2003	2004	2005	2006	2007	2008	2009	2010	2011	2012	2013
■Experimental development	88.77	92.21	91.29	92.14	92.1	90.96	89.2	91.6	91.7	96.1	97.6	97.9	97.5	97.0	96.9	97.1
■Applied research	10.51	7.12	8.06	7.40	7.25	8.24	9.6	7.5	7.6	3.5	2.2	2.0	2.4	2.9	3.0	2.7
■Basic research	0.74	0.71	0.68	0.48	0.67	0.82	1.2	0.9	0.7	0.4	0.2	0.1	0.1	0.1	0.1	0.1

Year

Figure 7-11 Structural changes of basic research, applied research and experimental development in R&D investment by Chinese enterprises

Data source: The series of *China Statistical Yearbook on Science and Technology* (1999–2014).

Furthermore, basic research expenditure in China mainly relies on the national investment. Basic research expenditure by enterprises only accounts for a very small part. Except 2004 to 2006 (15.0%, 12.6% and 10.6% respectively), the proportion for any other year does not exceed 10% (Figure 7-12).

In terms of personnel input in basic research, there has been an apparent fluctuation in recent years. It has never reached the threshold of 10 thousand people. Its proportion in national basic research personnel input is within the range of 5% to 8% for a really long time. In recent years, both the total personnel input and the proportion have witnessed rapid decreases. In enterprises, the proportion of basic research personnel input in R&D is also low (Figure 7-13 and Figure 7-14).

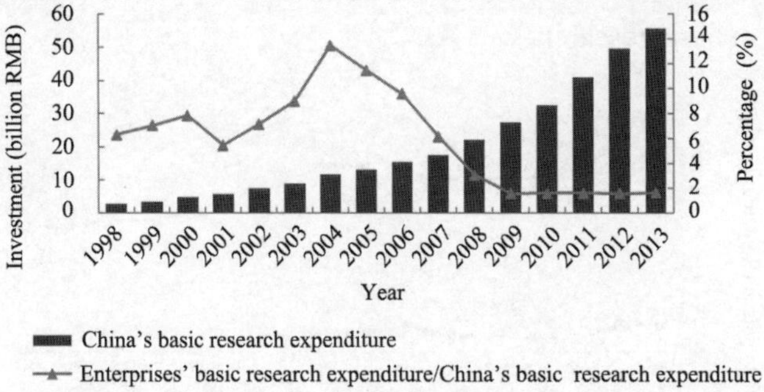

Figure 7-12 National basic research expenditure and the proportion of basic research expenditure by enterprises in China

Data source: The series of *China Statistical Yearbook on Science and Technology* (1999–2014).

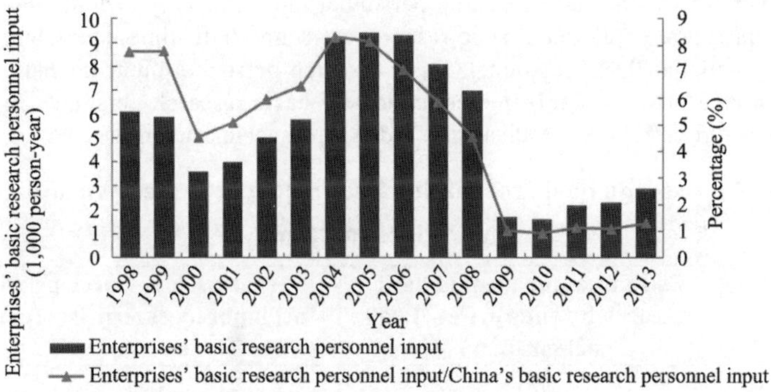

Figure 7-13 Changes of enterprises' basic research personnel input and its proportion in national basic research personnel input

Data source: The series of *China Statistical Yearbook on Science and Technology* (1999–2014).

If we regard enterprises engaged in R&D activities (medium and large enterprises) as analysis objects, we can find that the average basic research personnel input by each enterprise is less than 1 person (Table

7-9) ①. This seems to indicate that most of the Chinese enterprises barely engage in basic research.

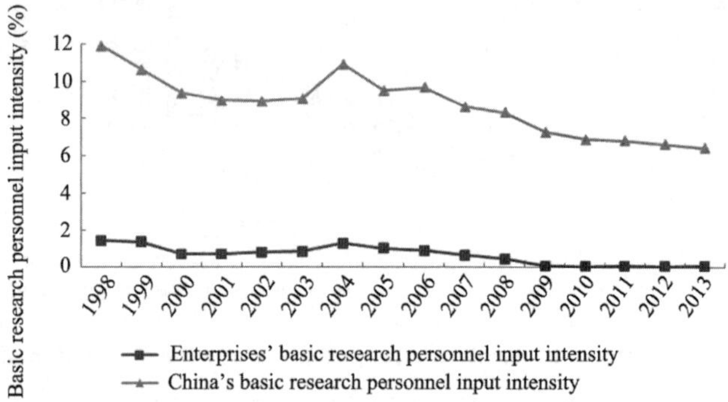

Figure 7-14 Comparison between basic research personnel input intensity by Chinese enterprises and national basic research personnel input intensity

Data source: The series of *China Statistical Yearbook on Science and Technology* (1999–2014).

Note: National basic research personnel input intensity = Full-time equivalent of national basic research personnel/Full-time equivalent of national R&D personnel; Basic research personnel input intensity in enterprises = Full-time equivalent of basic research personnel in enterprises/Full-time equivalent of R&D personnel in enterprises.

Table 7-9 Average basic research personnel input in medium and large enterprises in China

Year	Full-time equivalent of basic research by enterprises (1,000 person-year)	Average basic research personnel input by enterprises (person-year)
1998	6.1	0.84
1999	5.9	0.82

① We should point out that the statistics of the average may result in some misjudgments. Actually, basic research resources of enterprises are often concentrated in several enterprises. It is almost the same case in the United States, Japan, Germany and Korea. In the United States, for example, basic research personnel inputs in IBM, Microsoft, Bell Labs, Xerox, DuPont and other large enterprises occupy a large percentage of the entire country. Basic research spending in many other enterprises is often conducted in the form of R&D outsourcing and R&D cooperation, with really small personnel input. The statistical analysis is just to illustrate the status quo in China. It does not show that there are no basic research personnel in Chinese enterprises.

Continued

Year	Full-time equivalent of basic research by enterprises (1,000 person-year)	Average basic research personnel input by enterprises (person-year)
2000	3.6	0.51
2001	4.0	0.59
2002	5.0	0.70
2003	5.8	0.88
2004	9.2	0.14
2005	9.4	1.37
2006	9.3	1.19
2007	8.1	0.90
2008	7.0	0.69
2009	1.7	0.14
2010	1.6	0.12
2011	2.2	0.18
2012	2.3	0.14
2013	2.9	0.17

Data source: The series of *China Statistical Yearbook on Science and Technology* (1998–2011). Analysis objects of 1998 and 1999 were medium and large enterprises engaged in scientific activities.

We can draw some conclusions based on the aforementioned analyses.

First, Chinese enterprises have become the main force of China's R&D investment, but their basic research investment is insufficient. Prior to the 21st century, the government took the lead in R&D investment. A large amount of R&D expenditure was spent by research institutes and universities. This situation has gone through major changes since 2000. Enterprises have become the main force of China's R&D investment. In the past decade, the R&D expenditure from Chinese enterprises has accounted for over two-thirds of the total spending, which is in line with the global trend. It demonstrates that

Chinese enterprises are taking innovation as their major strategy for the market competition. However, basic research investment only accounts for a small part on the national scale, which hovers at about 10%. This figure has never exceeded 15%. There is still a certain gap compared with the US enterprises (21.7% in 2009).

Second, Chinese enterprises are more interested in experimental development (improving the efficiency of product development) and lack investment in long-term basic research (applied research included). Major trends are the rapid increase in total R&D, a high percentage of experimental development and a low level of basic research. These trends are becoming more apparent. In terms of the spending on R&D activities in enterprises, the ratio of basic research and experimental development is 1 : 120, 1 : 130, 1 : 134, 1 : 192, 1 : 137, 1 : 111, 1 : 74, 1 : 102 and 1 : 131 respectively from 1998 to 2006. From 2007 to 2013, this figure has reached an amazing level, which is 1 : 240, 1:488, 1:979, 1 : 975, 1 : 970, 1 : 969 and 1 : 971 respectively. If we combine basic research with applied research and compare it with the expenditure on experimental development, the ratio from 1998 to 2013 was 1 : 8, 1 : 12, 1:10, 1:12, 1:12, 1:10, 1:8, 1:11, 1:11, 1:25, 1:41, 1:47, 1:39, 1:32, 1 : 31 and 1 : 35. It is also far behind the corresponding ratio in OECD countries (about 3 : 7).

Third, R&D personnel are mainly engineers engaged in product development and experiments, which does not require high expertise. The basic research personnel ratio of Chinese enterprises is low in two aspects: the national basic research personnel investment and R&D personnel input in enterprises. In recent years, this figure is further dropping. It is generally acknowledged that medium and large enterprises have abundant capital and are prone to place emphasis on basic research. But this phenomenon is rarely seen in China according to the statistics.

Our primary conclusions can be supported by a comparison with leading countries.

Although Chinese enterprises are gradually becoming the main actor of R&D (the proportion of enterprises' R&D expenditures in the total amount of national R&D expenditures has grown from 44.8% in 1998 to 76.6% in 2013), the total amount of enterprises' basic research is relatively low and the input structure is imbalanced. The expenditures on research are mainly spent on the improvement of current products and current technology. In 2013, 97.87% of basic research resources in China were concentrated in universities and research institutes, while the research expenditures of enterprises only accounted for 4.33% of the total research expenditures in China. However, from 1991 to 1998, the research expenditures of the US enterprises strikingly accounted for about 25% of the total research expenditures (Payson, 1998). In terms of basic research investments, Chinese enterprises fall far behind the enterprises in leading countries (as is shown in Table 7-10).

Table 7-10 Comparison between the intensity of enterprises, basic research investment in China and that in leading countries (%)

Country	2000	2005	2006	2007	2008	2009	2010
US	2.02	1.80	1.86	1.92	2.02	2.04	—
Japan	2.16	2.54	2.63	2.68	2.70	2.53	—
Germany	1.74	1.72	1.77	1.77	1.86	1.88	1.90
UK	1.34	1.30	1.32	1.31	1.32	1.37	1.38
Britain	1.18	1.06	1.08	1.11	1.10	1.12	1.13
Korea	1.70	2.15	2.32	2.45	2.53	2.64	2.80
27 EU countries	1.11	1.08	1.11	1.11	1.15	1.16	1.16
25 EU countries	—	1.10	1.13	1.13	1.18	1.18	—
15 EU countries	—	1.18	1.20	1.22	1.26	1.27	—
OECD total	1.53	1.51	1.55	1.58	1.60	1.62	—
China	0.54	0.91	0.99	1.01	1.08	1.25	1.29

Data source: OECD, 2011.

7.5 Evolution of basic research investment by universities

Universities play a significant role in formulating independent innovation strategies in China. The basic research level and the status quo of cultivating the scientific capability of students largely decide the scientific competitiveness of China. Since the policy of reform and opening-up was implemented more than 30 years ago, higher education in China has achieved vigorous development. This is not only reflected in the boom in the number of recruits and graduates, but also reflected in the scientific research capacity of universities. Specifically speaking, universities play a leading role in domestic basic research.

(1) In terms of basic research personnel input, the full-time equivalent of basic research personnel in universities has increased steadily for the past decade. The proportion in national basic research personnel input remains at over

60% (reached the peak of 70.8% in 2008). The proportion in R&D personnel input of universities has also been on a steady increase in the last few years (Figure 7-15, Table 7-11).

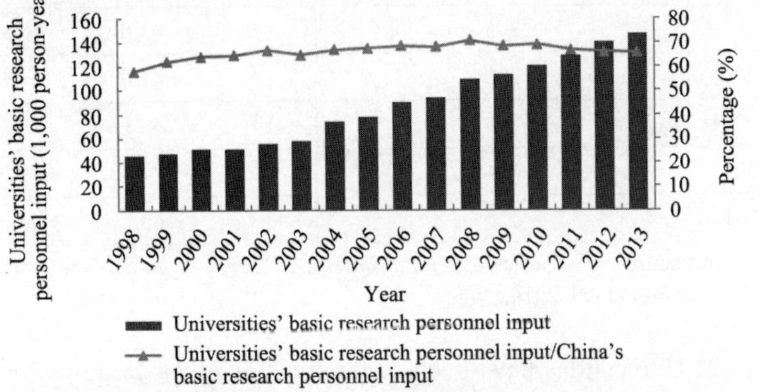

Figure 7-15 Changes of basic research personnel input and its proportion in national basic research personnel input by Chinese universities

Data source: The series of *China Statistical Yearbook on Science and Technology* (1999–2014).

Table 7-11 Changes of full-time equivalent of R&D personnel in Chinese universities and the proportion of basic research personnel

Year	Full-time equivalent of basic research personnel (1,000 person-year)	Proportion of basic research personnel (%)
1998	169	27.0
1999	176	26.8
2000	159	32.0
2001	171	29.8
2002	181	30.9
2003	189	30.7
2004	212	35.1
2005	227	34.4
2006	242	37.2
2007	254	37.0

Continued

Year	Full-time equivalent of basic research personnel (1,000 person-year)	Proportion of basic research personnel (%)
2008	266	41.0
2009	275	41.0
2010	290	41.4
2011	299	43.1
2012	314	44.6
2013	325	45.2

Data source: The series of *China Statistical Yearbook on Science and Technology* (1999–2014).

(2) Universities have become the main actor for basic research. Their expenditure for basic research soared from 4.23 billion RMB in 1995 to 85.67 billion RMB in 2013. The increase is more than 20 times. The proportion in national basic research spending has increased steadily. In 2012, it reached a peak of 55.77%. Universities have undertaken more than half of basic research tasks. In terms of R&D activities in universities, the percentage of basic research investment is within the range of 15% to 35% (Figure 7-16).

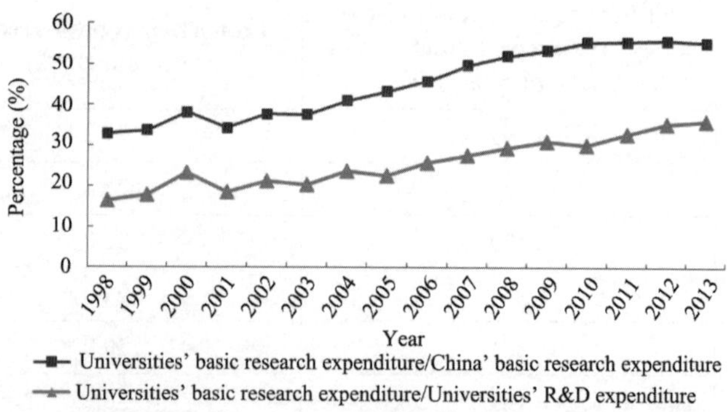

Figure 7-16 Changes of basic research investment intensity of Chinese universities

Data source: The series of *China Statistical Yearbook on Science and Technology* (1999–2014)

Besides, in terms of the sources of R&D for university research, the ratio of the government funding and enterprise-commissioned research remains at 6 : 4 from 2005 to 2013 (Figure 7-17). In China, R&D outsourcing by enterprises to universities is mainly applied research and experimental development. The funding for basic research in universities comes mainly from government funding.

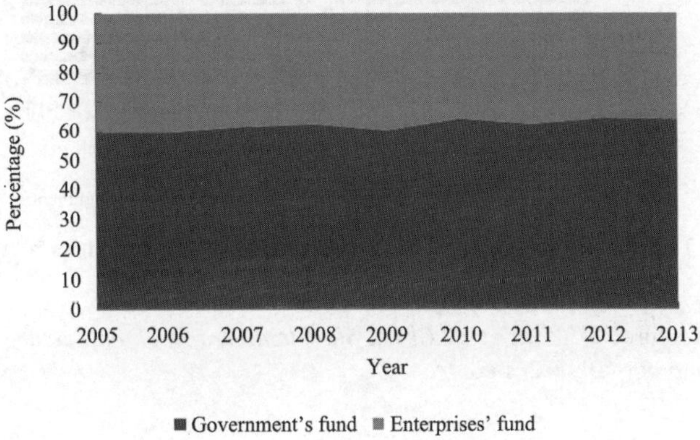

Figure 7-17 Structural changes of the sources of R&D spending in Chinese universities from 2005 to 2013

Data source: The series of *China Statistical Yearbook on Science and Technology* (2006–2014).

(3) In terms of the internal structure of R&D spending by universities, the ratio of basic research : applied research : experimental development changed from 1 : 3 : 2 to 1 : 1.4 : 0.4 from 1998 to 2013 (Figure 7-18). Basic research has received more emphasis, but it is not adequate. The proportion of applied research is too high. It demonstrates that universities have done too much applied research, which hinders its role of disseminating knowledge.

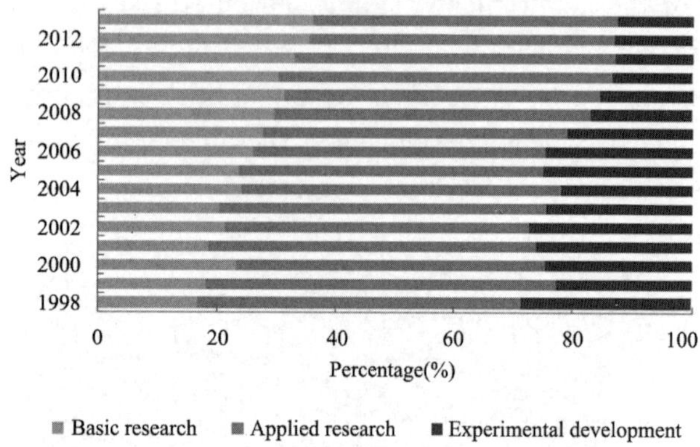

Figure 7-18 Structural changes of expenditure on R&D activities by Chinese universities

Data source: The series of *China Statistical Yearbook on Science and Technology* (1999–2014).

7.6 Evolution of basic research investment by public research institutes

There are 3,696 public research institutes in China; among them are 686 central-level research institutes and 3,010 regional institutes. Establishing public research institutes is one of the effective methods for the government to solve the problem of the market dysfunctions (Salter & Martin, 2001). They are important cradles for knowledge innovation and technological development. After China implemented the scientific institutional reform, related institutions in scientific and disciplinary fields have conducted some integration. Other institutions have turned into enterprise-oriented research institutes by the ownership transformation. These policies vigorously promoted research institutes to conduct R&D and innovation activities oriented towards the industrial needs. From the point of basic research activities, public research institutes play an important role in basic research in China. In areas such as nanotechnology, biology and communications, research institutes represented by the Chinese Academy of Sciences have made remarkable achievements in the international scientific arena. A large number of papers and patents have come from scientific research institutes.

(1) In terms of basic research personnel input, the full-time equivalent of basic research personnel of scientific research institutes remains at 200,000 to 400,000 person-year each year (Figure 7-19). Basic research personnel input remains within the range of 23,000 to 60,000 person-year. In terms of the proportion of basic research personnel, scientific research institutes have remained at 25%–30% of national input in the past decade. Now it stays at a level of 25%. But this proportion is relatively low among R&D personnel in scientific research institutes. It has not exceeded 15% in the past decade (Figure 7-20).

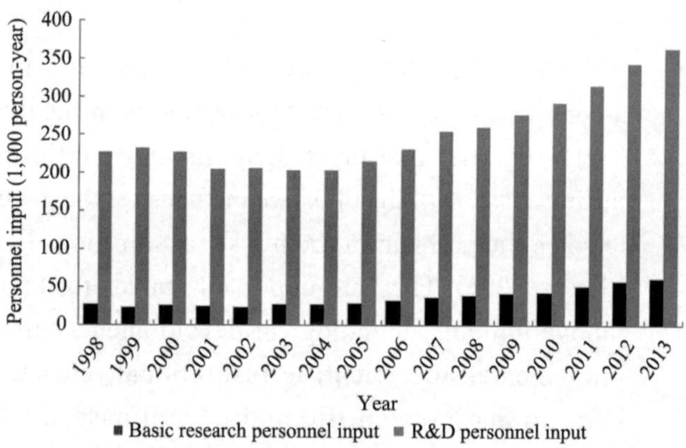

Figure 7-19 Changes of R&D activities and basic research personnel inputs by Chinese public scientific research institutes

Data source: The series of *China Statistical Yearbook on Science and Technology* (1999–2014).

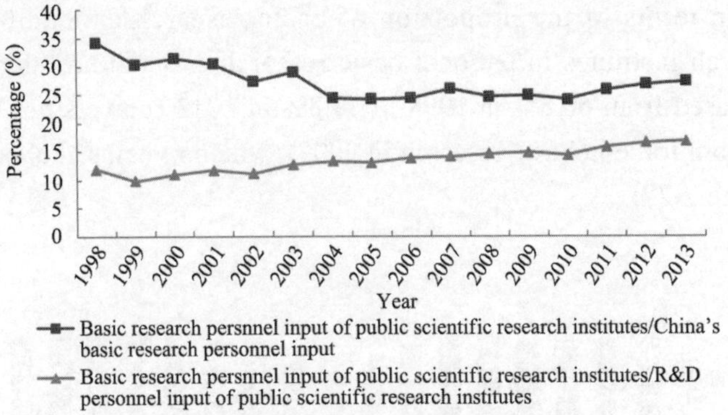

Figure 7-20 Proportion of basic research personnel input by Chinese public scientific research institutes

Data source: The series of *China Statistical Yearbook on Science and Technology* (1999–2014).

(2) Basic research expenditure of research institutes soared from 1.76 billion RMB in 1998 to 22.16 billion RMB in 2013, with an annual increase of about 20%. But this proportion is not high in the R&D expenditure, which has not exceeded 12% for a really long time (Figure 7-21).

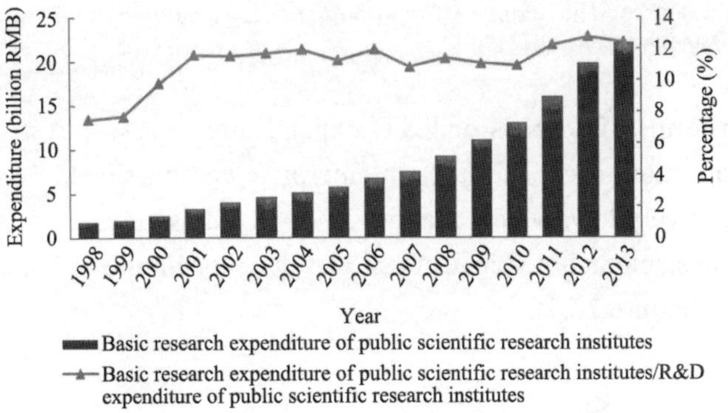

Figure 7-21 Changes of basic research expenditure by public scientific research institutes and its proportion in R&D expenditure

Data source: The series of *China Statistical Yearbook on Science and Technology* (1999–2014).

In terms of the proportion of basic research expenditures by research institutes in national basic research expenditure, this figure decreased from 60.8% in 1998 to 39.8% in 2013 (universities became the main force in basic research in 2007), which remains at about 40%. (Figure 7-22).

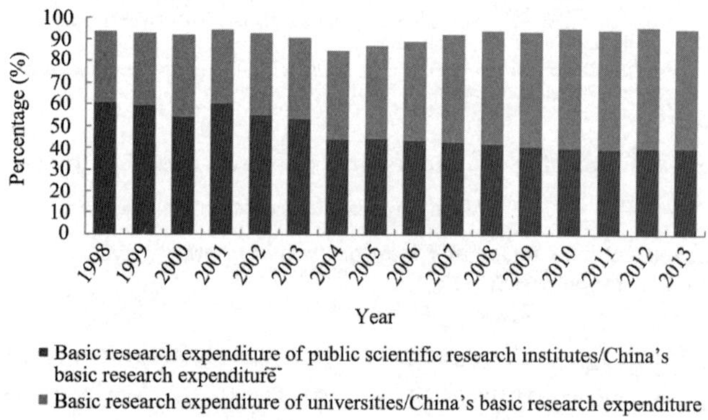

Figure 7-22 Changes of the relation between basic research expenditure by public scientific research institutes and universities in national basic research expenditure in China

Data source: The series of *China Statistical Yearbook on Science and Technology* (1999–2014).

In terms of sources of R&D expenditure in research institutes, the proportion of expenditures in enterprise-commissioned projects is comparatively low, which is about 3%–4%. The sources mainly come from the stable national investment and the expenditure of competitive projects (Figure 7-23).

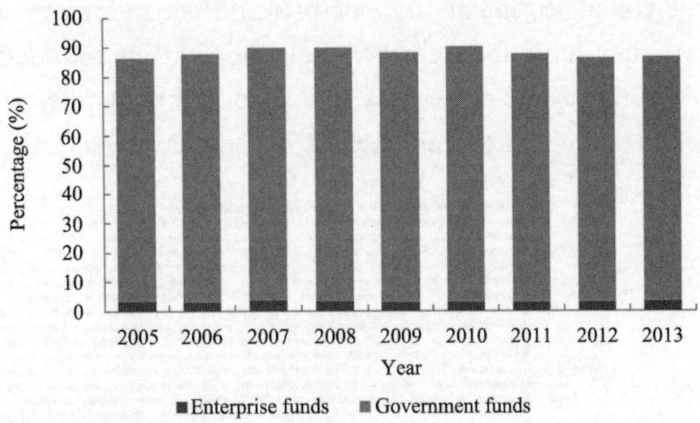

Figure 7-23 Sources of R&D expenditure by Chinese public scientific research institutes (government funding and enterprise funding)

Data source: The series of *China Statistical Yearbook on Science and Technology* (1999–2014).

(3) In terms of the internal structure of R&D expenditure of scientific research institutes, the proportion of basic research, applied research and experimental development changed from 1:4:9 to 1:2:5 from 1998 to 2013 (Figure 7-24). The status of applied research is upgraded. However, investments in experimental development are relatively high. Over half of the expenditure is applied to tests before the product development, which is not closely connected with scientific research. This has undermined the status of scientific institutes as the sources of technological innovation in enterprises.

Based on the high expenditure of experimental development, it seems to show that Chinese scientific research institutes respond strongly to market demand. But the proportion of basic research and applied research is not high. The contribution to innovation of core industrial technologies is limited. But in China, there are still many enterprises that do not have the capacity and momentum to conduct technological innovation. The work undertaken by scientific research institutes helps enterprises improve their product quality and meet the new market demand. But the crux lies in whether scientific research institutes need to invest so much in experimental development, which should be conducted mainly by enterprises. Since the reform of those institutes, they have focused their development on application and the market. But at the same time, this also highlights that there is a mismatch in the role of scientific research institutes in R&D.

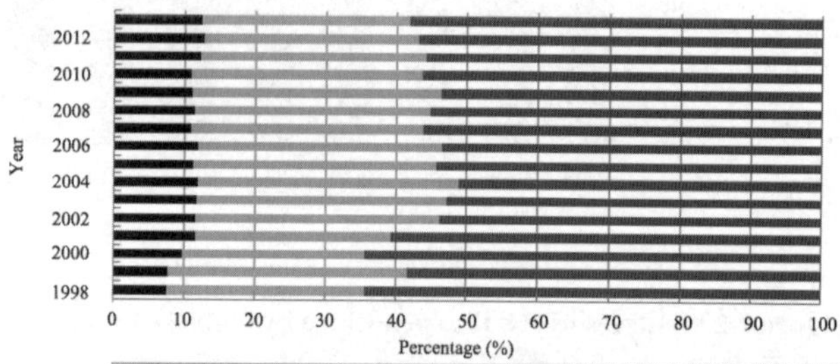

	1998	1999	2000	2001	2002	2003	2004	2005	2006	2007	2008	2009	2010	2011	2012	2013
■Basic resarch/R&D	7.51	7.72	9.81	11.65	11.59	11.75	11.98	11.30	11.97	10.86	11.43	11.10	10.95	12.30	12.80	12.40
■Appiled resarch/R&D	28.13	34.01	25.85	27.73	34.50	35.36	36.85	34.36	34.58	33.01	33.44	35.23	32.67	31.90	30.30	29.50
■Experimental development/R&D	64.36	58.31	64.34	60.62	53.91	52.88	51.19	54.32	53.45	56.13	55.12	53.65	56.38	55.80	56.90	58.10

Figure 7-24 Structural changes of R&D activities expenditure by Chinese public scientific research institutes

Data source: The series of *China Statistical Yearbook on Science and Technology* (1999–2014).

7.7 Summary

In terms of their functions, universities should carry out more basic research and public research institutes should do more applied research. However, from Figure 7-25 and Table 7-12, we can find that Chinese universities are not highly devoted to basic research. The public research institutes have done too many experimental development programs, which should be done more by enterprises. In order to help enterprises to improve their innovation abilities, the government uses enormous S&T resources to carry out product development. But in fact, it is doubtful whether the government's method is efficient, for usually enterprises are more efficient in experimental development.

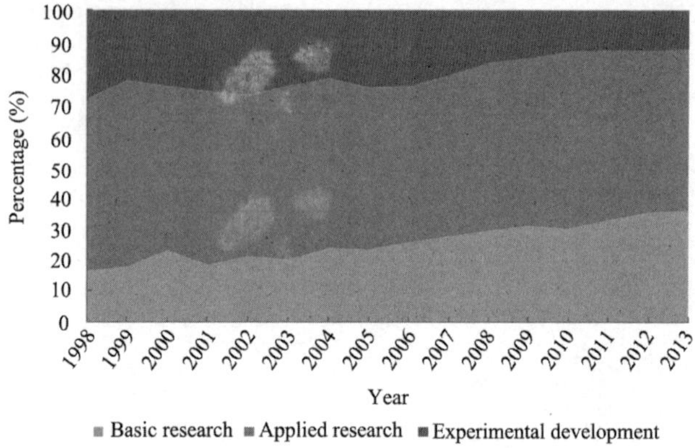

Figure 7-25 Historical changes of the structure of R&D investment in Chinese universities from 1998 to 2013

Data source: The series of *China Statistical Yearbook on Science and Technology* (1999–2014).

Table 7-12 Structural changes of R&D investment of Chinese public scientific research institutes

Year	Expenditures on R&D (billion RMB)	Proportion of expenditures on basic research(%)	Proportion of expenditures on applied research(%)	Proportion of expenditures on experimental development(%)
1995	14.64	6.97	31.08	61.95
1996	17.29	6.55	27.36	66.11
1997	20.64	7.27	33.72	59.01
1998	23.43	7.51	28.13	64.36
1999	26.05	7.72	34.01	58.31
2000	25.80	9.81	25.85	64.34
2001	28.85	11.66	27.73	60.62
2002	35.13	11.59	34.50	53.91
2003	39.90	11.75	35.36	52.88
2004	43.17	11.98	36.85	51.19
2005	51.31	11.30	34.37	54.32
2006	56.73	11.98	34.58	53.45

Continued

Year	Expenditures on R&D (billion RMB)	Proportion of expenditures on basic research(%)	Proportion of expenditures on applied research(%)	Proportion of expenditures on experimental development(%)
2007	68.29	10.94	33.26	56.54
2008	81.13	11.43	33.44	55.12
2009	99.60	11.10	35.23	53.65
2010	118.64	10.95	32.68	56.38
2011	130.67	12.30	31.90	55.80
2012	154.89	12.80	30.30	56.90
2013	178.14	12.40	29.50	58.00

Data source: The series of *China Statistical Yearbook on Science and Technology* (2007–2014).

Before reform and opening-up, the basic research, applied research and experimental development in China belonged to different executive departments. Most of the national basic research was conducted by Chinese Academy of Sciences and other national research institutes. Most applied research and experimental development was done in the research departments in enterprises of different industries and institutions under related ministries. Universities, whose main function was training and education, also undertook limited research; they did not have fixed expenditure support for R&D (Zhu & Gong, 2008)

After the reform of S&T system by the central government in 1985, the situation mentioned above has changed. The country started to make huge investment in S&T. In recent years, enterprises' R&D is also growing rapidly, thanks to the transformation of many public scientific research institutes to enterprise entities. About 1,149 public scientific research institutes transformed into enterprises from 1998 till the end of 2003. Among the 204,000 people who were transferred to enterprises, 111,000 were S&T personnel. Compared to those who stayed in the scientific research institutes, they are, on average, older and their abilities are relatively weaker (OECD, 2008, 2011).

In 2001, China reinforced the transformation into enterprises of research institutes that undertook technological development. However, China's reform of public research institutes brought some negative influences. First of all, some research institutes often leave the results that can be easily commercialized for themselves and transfer the results that cannot be easily commercialized to enterprises. Second, over-commercialized research institutes tend to focus mainly on product technology, rather than take the risks of the research, which to a certain extent leads to the conflict between the innovation activities of the research institutes and those of enterprises in the market and which in some way lowers the efficiency of the national innovation system (OECD, 2008).

In the enterprises-centered national innovation system, the role of public research institutes needs repositioning. We need to emphasize their role in basic research and applied research, because such research will influence the direction of future technological development. Instead of carrying out experimental development for enterprises, universities and research institutes should help private enterprises, whose R&D abilities are still weak at the current stage and improve their technological and innovative abilities. The new national innovation system in China should further emphasize the role of universities in basic research, the comprehensive roles of enterprises in integrating all the innovation factors and in increasing innovation investment, as well as the connection and cooperation among industries, universities, and enterprises (OECD, 2008). Because research expenditures and personnel are highly concentrated in universities and public research institutes in China, the limited mobility of scientists and engineers hinders technological transfer and separates the function of R&D and the process of production in the national innovation system. Policy-makers and enterprises often regard public research results as "free public goods" and focus on how to convert them to productivity.

7.8 Conclusion

From the analysis of China's investment in R&D and the history of development of basic research resources allocation, we can find that the expenditures on R&D in China have been growing rapidly. The structure of expenditure source is reasonable and enterprises have become the main actor of investment in R&D. However, at the same time, the historical problem is still obvious that the proportion of basic research is relatively low (the ratio of basic research investment does not increase along with the increase of R&D input, but fluctuates between 5% and 6% over a long period of time, falling far behind developed countries).

On the one hand, although top leadership in China sharply focuses on the planning and resources investment in basic research, most of the government R&D expenditures are put on experimental development rather than basic research. It is easy to stumble into the low-level equilibrium trap under such circumstances. Although the source structure of R&D expenditure in China is similar to that in the United States, Germany, Switzerland, Japan and Korea, expenditures on experimental development grow faster than those on basic research and applied research in terms of the research types. Massive R&D expenditures are spent on enterprises' experimental development; applied research takes the second place; and basic research is relatively low. Generally speaking, experimental development is the systematic work for the production of new products, new materials and new devices, for the establishment of new technologies, new systems and new services, and for the substantial improvement of the above from the knowledge obtained from basic research, applied research and hands-on experience. As for basic research, it can achieve far more long-term returns on investment than applied research and experimental development. Without investment in basic research, China cannot achieve more breakthroughs in innovation.

On the other hand, basic research for the innovation need of industrial technology still lacks attention. The main enforcement department of basic research in China is universities and research institutes, accounting for up to 90% in all, while the enterprises' basic research is too weak. Besides, the positioning is imbalanced in terms of R&D expenditures in universities and research institutes. Government research institutes emphasize more on applied research and experimental development, while universities and enterprises that should do well in basic research are more devoted to applied research. All of them over-substitute the applied and development programs for enterprises, making the executive body of basic research weaker and producing the economic inefficiency mentioned by Kealey (1996). As a result, although enterprises have become the main subject of R&D in China, they have not become the main subject of technological innovation activities. And technological innovation activities have not become the internal requirement of the survival and development of enterprises either. China has gradually come to realize that the development of enterprises' innovation abilities is the key to the establishment of national innovation system. But in S&T programs, the enterprises' level of involvement is relatively low. Future S&T policies in China should focus on how to attract enterprises to participate in S&T programs and how to integrate national resources to carry out basic research of innovation of core industrial technologies.

References

Kealey, T. *The Economic Laws of Scientific Research*. Basingstoke: Macmillan, 1996.

OECD. Reviews of innovation policy: China, 2008.

OECD. Science, technology and R&D statistics, 2011.

Payson, M. *National Patterns of R&D Resources*. Washington, DC: National Science Foundation, 1998.

Ran, Y. Chinese corporations still do not have strong innovation abilities in terms of the number of patents. *People's Daily*, 2006-12-10. [冉永华．从专利数量看中国企业创新能力还不高．人民日报，2006-12-10．]

Salter, A. J. & Martin, B. R. The economic benefits of publicly funded basic research: A critical review. *Research Policy*, 2001, 30 (3): 509-532.

Wen, K. & Li, L. Analysis of basic research in domestic enterprises in order to enhance indigenous innovation capability. *Science of Science and Management of S.&T.*, 2007 (2): 1-9, 20. [温珂，李乐旋．从提升自主创新能力视角分析国内企业基础研究现状．科学学与科学技术管理，2007（2）：1-9，20．]

Xiao, G. What percentage of basic research investment to R&D investment is advisable. *Studies in Science of Science*, 2005, 23 (2): 197-203. [肖广岭．中国基础研究经费占 R&D 经费的比例多大为宜．科学学研究，2005，23（2）：197-203．]

Zhu, Z. & Gong, X. Basic research: Its impact on China's future. *Technology in Science*, 2008, 30 (3-4): 293-298.

Chapter 8

The Role of Enterprises in Government Research Programs in China

In recent years, R&D investments in China have increased rapidly. But basic research investments still hover at a low level. Basic research oriented towards needs of industrial technological innovation is scarce. In terms of advancing basic research, what China lacks is not a scientific program but the participation of enterprises. This really hinders breakthroughs of core technologies.

In history, many enterprises were mainly manufacturing factories and most of them did not conduct R&D work, only some state-owned enterprises were equipped with independent R&D laboratories, but those laboratories were only engaged in experimental development (OECD, 2008). During this period, public research institutes played the most important role in the innovation system. Enterprises were based on output and had little incentive to improve efficiency and profits, with much less focus on knowledge and intellectual property. So, even though many of China's new industries—automobile, information and communications technology and the iron and steel took off at the same time with Korea, 20 years later, China lagged behind that of Korea in this respect. "Importing–lagging behind–re-importing–lagging behind again" is the rule of this period (OECD, 2008). Until the 1970s, China's basic research and R&D activities have almost been entirely concentrated in the public domain, namely government-funded research institutes established in universities and governments. Although the industrial sector in China is a participant of R&D activities and its importance has improved significantly, it is not the very core of Chinese national innovation. The rapid growth of R&D in the industrial sector is largely attributable to the fact that some public research institutes have been transformed into industrial entities. The number of research institutes in large- and medium-sized enterprises declined from 13,107 in 1995 to 12,568 in 2011. The proportion of large- and medium-sized enterprises with scientific research institutes in all large- and medium-sized enterprises decreased from 39.8% in 1995 to 27.6% in 2011. The proportion of S&T activities of large- and medium-sized enterprises in those of all enterprises also decreased from 56.9% in 1995 to 28.3% in 2011.

This chapter will analyze the mechanism, the model and the issues of basic research by the government in China. It will also analyze related issues concerning basic research oriented towards industrial technological innovation.

8.1 National basic research investment

In the past three decades, in order to meet the needs of different stages of scientific development, China has established diverse channels for the funding of basic research; see Table 8-1 and Figure 8-1.

Table 8-1 Major funding channels of basic research in China

Project	Year	Administration	Strategic positioning and objectives
Major Scientific Program	1983	National Development and Reform Commission (NDRC)	Meeting national strategic needs, funding large-scale modern key equipment at the forefront of international science and technology, including areas such as nuclear physics, astronomy, geosciences, information science, etc.
State Key Laboratory Construction	1984	Ministry of Science and Technology (administered by the former State Planning Commission and related ministries)	Selecting top-notch teams and disciplines in universities and scientific research institutes, providing key equipment, and undertaking numerous major national scientific research tasks
Research Fund for the Doctoral Program of Higher Education of China (RFDP)	Unknown (originally Fund for Scientific Research Projects of Major Institutions of Higher Education)	Ministry of Education	Funding innovative subjects with novel academic thinking and important application prospects
National "Climbing Program"	1991–1998	Ministry of Science and Technology (the former State Science and Technology Commission)	Supporting research teams and implementing more than 100 projects, each project for 3 years, with an average funding intensity of 1 million RMB/year

Continued

Project	Year	Administration	Strategic positioning and objectives
National Program on Key Basic Research Project of China (Program 973)	1998	Ministry of Science and Technology	Conducting research of major key scientific issues geared to major national needs and future science frontiers, covering 8 comprehensive and multidisciplinary fields such as population and health, information, agriculture, resources and environment, energy, and new materials, and funding major frontier basic research
Project 211	1993	Ministry of Education	Founding about 100 institutions of higher education and key disciplinaries as a national priority for the 21st century
Project 985	1998	Ministry of Education	Founding world-class universities of international advanced level in the 21st century
Knowledge Innovation Program	1998	Chinese Academy of Sciences	Aiming at the global frontiers of science and technology, meeting the needs of our economy and society, targeting at basic research and applied research, and striving to improve the global competitiveness of Chinese Academy of Sciences

Data source: Compiled from International Evaluation of Funding and Management of the National Natural Science Foundation of China Synthesis Evidence Report (Center for Science & Technology Evaluation, NCSTE, 2010) and related materials.

Major projects and plans in China of the support for basic research is shown in Figure 8-1.

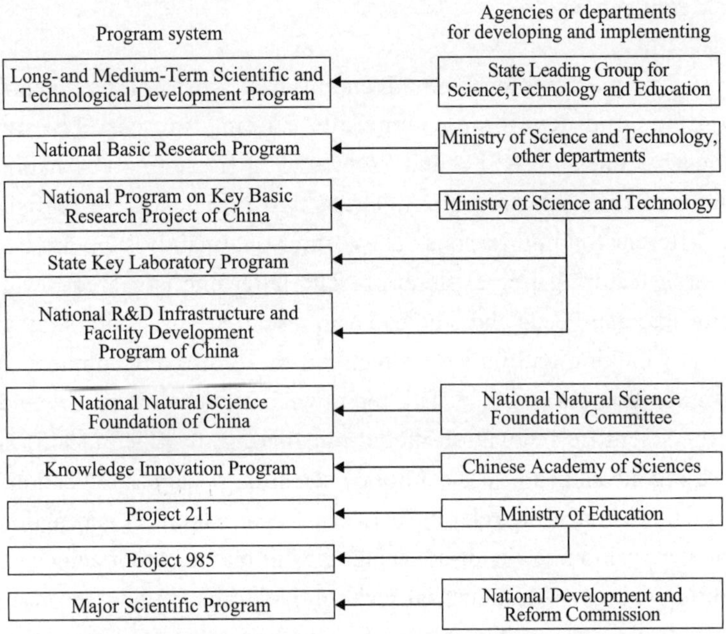

Figure 8-1 Major plans and projects of the support for basic research

Data source: Li et al., 2007.

In the Chinese basic research system, national investment is the main source. The Ministry of Science and Technology, Ministry of Education, Natural Science Foundation of China and Chinese Academy of Sciences fund basic research by different means. National investment in basic research is reflected in two parts: the scientific funds featuring free exploration and national scientific projects. Besides national investment within the fixed research expenditure, these two kinds of basic research are conducted by numerous competitive means such as the national and provincial scientific foundations, National Program on Key Basic Research Project of China (Program 973), and all kinds of talent support programs (such as "Thousand Talents Program", Changjiang Scholars Program of China and "Hundred Talents Program" of Chinese Academy of Sciences). Knowledge

Innovation Project of Chinese Academy of Sciences, Project 985 and Project 211 of the Ministry of Education also play an important role in basic research inputs.

China puts emphasis on advancing basic research by scientific programs and developing numerous funding sources. The most noticeable ones are NSFC and Program 973 (including key national scientific research programs). These two projects are established for different funding reasons. The former one mainly supports basic research featuring free exploration. The latter one is geared towards major national needs and studies basic research issues with strategic, forward-looking and holistic effects on the future development and scientific progress of China. This top-down basic research model abides by the scientific innovation model put forward by Bush (1945) and has a big advantage in the control of scientific resources allocation. It also demonstrates the relation between free exploration and national strategic positioning. Its disadvantage lies in that it cannot achieve the mobile synergy with industrial technological innovation. Because of the fact that the transformation of basic research achievements into productivity is influenced by many factors, its efficiency and effect have large uncertainty. So this may result in the low contribution of basic research to the industrial technological innovation (Figure 8-2).

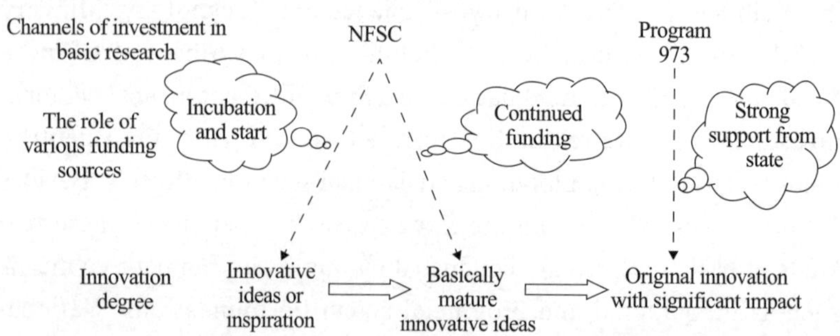

Figure 8-2 Coordination of major basic research projects in major original innovation

Data source: Zhou, 2008.

Since the policy of reform and opening-up, influenced by the US scientific and economic development model, China has attached more importance to basic research. A series of development strategies and policies have been launched.

Before the Decision on the Reform of the Science and Technology Management System released by China in 1985 began to officially implement the scientific institutional reform, the scientific innovation system was planned by the central authorities, not driven by the market. It was driven by objectives instead of the diffusion. R&D capacity was weak. A large number of scientific resources were concentrated in the Chinese Academy of Sciences, academic institutions under different ministries of the central government, academic institutions under the local governments, institutions of higher education and defense research institutes (OECD, 2008).

Since the State Leading Group for Science, Technology and Education was established to coordinate national scientific innovation policies in 1983, China has initiated a series of plans, policies, laws and regulations to promote the status of basic research. At the beginning of the 1990s, it began to implement the strategy of strengthening basic research. It also made adjustments in the scientific planning system including the Knowledge Innovation Project of the Chinese Academy of Sciences. The National Plan for Long- and Medium-Term Scientific and Technological Development (2006–2020) explicitly points out that as a rapidly developing country, China should attach importance to the role of basic research in serving national objectives and it also needs to make use of basic research to solve the key and bottleneck issues in future development. By the time China proposed a national strategy of independent innovation in 2006, the scientific innovation policies in China have formed into a comprehensive system.

Table 8-2 compares different scientific plans and policies in terms of the descriptions and focuses of basic research.

One feature of China's national innovation system is that projects grow at an increasing rate. Large numbers of comprehensive and goal-specific plans are one major measure for China to implement innovative policies. However, the government's emphasis on basic research do not bring about sufficient investment in basic research expenditure. At the beginning of the 21st century, China began to direct scientific goals to strategic adjustments whose economic structures are supported by independent innovation. It also put enterprises at the forefront of the national innovation system and stimulated more enterprises to fund R&D activities. However, this R&D expenditure is more applied to product development instead of basic research.

Table 8-2 Different strategic focuses of major projects and policies in advancing basic research in China

Plans or policies	Year	Goals	Focuses on basic research	Comments
National Key Technologies R&D Program of China①	1982	Concentrating resources on key and common technologies needed by industrial and social development and exploring new frontiers	Geared to the needs of national economy and social development and mainly addressing major scientific issues in economic and social development	Supporting industry-university-research institute cooperation and prioritizing the development of agriculture, biology, communications technology and sustainable development
Major Scientific Project	1983	Large or big investment modern key instruments and equipment needed in scientific research processes	—	Belonging to the National R&D Infrastructure and Facility Development Program of China, including disciplines (domains) of nuclear physics, astronomy, geography, biology and information science

① National Key Technologies R&D Program of China was renamed National Key Technology Research and Development Program of the Ministry of Science and Technology of China from 2006.

Continued

Plans or policies	Year	Goals	Focuses on basic research	Comments
State Key Laboratory Program	1984	Supporting universities, scientific research institutes and enterprises in qualified laboratory construction	Geared to the national strategic development goals and the international competition, laying the groundwork for strengthening scientific reserves and original innovation capacity, and conducting basic research, applied basic research (including high-tech research before competition) and fundamental work	Committed to high-level research, identifying and fostering outstanding scientists, and conducting high-level academic exchanges
Decision on the Reform of the Science and Technology Management System by the Central Committee of the Communist Party of China	1985	Promoting R&D results from laboratories to the industry; "science and technology work must be geared to the needs of economic development"	Gradually implementing a trial system of scientific fund in basic research and part of applied research	Focusing on the industrialization of scientific research achievements
National Natural Science Foundation of China (NSFC)	1986	Supporting basic research, adhering to free exploration and playing a leading role	Forming a funding pattern of research projects, talent projects and environmental condition projects	Targeted at national objectives and the global science frontiers and encouraging the combination of free exploration and macro regulation
National High-tech R&D Program of China (Program 863)	1986	Enhancing the international competitiveness and R&D capacity of China in the field of [illegible]	—	Designating 19 key high-tech fields as development focuses

Continued

Plans or policies	Year	Goals	Focuses on basic research	Comments
National "Climbing Program"	1991	A major national basic research program aiming to strengthen basic research	Geared to national goals of basic research, and prioritizing the fundamental theory and technological base for solving major issues in national economic and social development	Including Major Key Basic Research Programs, Major National Engineering Basic Research Programs and Major National Science and Technology Programs
Decision of the Central Committee of the Communist Party of China and the State Council Concerning the Acceleration of Advancement of Science and Technology	1995	Climbing the peak of science and technology and revitalizing the nation through science and education	Continuing to increase basic research investment and the proportion of basic research in R&D; basic research should not only aim at national objectives, but also keep up with the global science frontiers	Demonstrating China's determination to make its basic science research stand out on the global arena
National Science and Technology Major Project	Unknown	Revolving around national objectives, further highlighting the focuses, and selecting a number of major strategic products, key generic technology or major projects as major programs	National plan for Long- and Medium-Term Scientific and Technological Development (2006–2020)	16 major programs, involving information, biology and other strategic industries, energy, resources, environment and health

Continued

Plans or policies	Year	Goals	Focuses on basic research	Comments
National Program on Key Basic Research Project of China (Program 973)	1997	Solving key scientific issues in economic construction, social development and national security and making a batch of major original innovations in the mainstream global scientific development	Supporting basic research in designated fields	Designating areas such as agriculture, energy, information, resources and environment, population and health, and new materials, and encouraging comprehensive research with multidisciplinary and multi-sector participation
Technology Innovation Fund for Technology-based SMEs	1999	Encouraging, nurturing, supporting and promoting technology innovation special government fund for technology-based SMEs	—	Including free funding and discount loans, etc.
Special Social Welfare Research Program of Scientific Research Institutes	2000	Supporting scientific research institutes in social welfare research, improving the level of social welfare research and services, and promoting the reform and development of scientific research institutes	—	Belonging to the National R&D Infrastructure and Facility Development Program of China

Continued

Plans or policies	Year	Goals	Focuses on basic research	Comments
Key Program for International S&T Cooperation Projects of China	2001	Using global key scientific resources and R&D technology to build a platform for international S&T cooperation	—	Belonging to the National R&D Infrastructure and Facility Development Program of China
National R&D Infrastructure and Facility Development Program of China ①	2002 (trial implementation) 2005 (official implementation)	Improving national S&T innovation capacity and enhancing the international competitiveness	Researching the experimental base and large-scale scientific instruments and equipment sharing platform, and forming a high-level national basic research experimental base across the fields	Focusing on the infrastructure of scientific innovation
Opinions on Further Strengthening Original Innovation Capacity	2002	Further strengthening the original innovation capacity of basic research and high-tech cutting-edge research in China	Putting forward opinions on how to strengthen basic research in terms of S&T management system, evaluation system, talents, scientific research base, scientific research equipment and data sharing, cultural environment innovation, international cooperation and so on	Putting forward suggestions and deployment for basic research from the perspective of improving original scientific research capacity for the first time

①The "National R&D Infrastructure and Facility Development Program of China" is one of the National Basic Research Plan during the 11th FYP, named as the four main national programs with National Basic Research Program, National Science and Technology Support Program, High-Tech Research and Development Program.

Continued

Plans or policies	Year	Goals	Focuses on basic research	Comments
National Plan for Long- and Medium-Term Scientific and Technological Development (2006–2020)	2005	By 2020, R&D spending will account for 2.5% of GDP; the annual R&D spending will increase to 900 billion RMB (112 billion USD) and the contribution rate of science and technology to China's economic development will amount to 60%; the degree of dependence on foreign technology will be reduced from 50% to 30%	The development of basic research should adhere to the principle of serving national objectives and encouraging free exploration; the deployment of basic research is conducted in terms of disciplinary development, science frontier issues, major national strategic needs and major scientific research plans	Leading the future is the proactive deployment of cutting-edge technology and basic research; strengthening basic science and research in cutting-edge technologies, especially cross-disciplinary research
National Basic Research Plan during the 11th FYP	2006	Adhering to the freedom of scientists to conduct independent innovation explorations and national targeted and strategic basic research tasks	Basic research is an important approach to realizing national strategies for development	Encouraging eligible enterprises to invest in basic research and conduct basic research, and encouraging social forces to contribute to basic research

Continued

Plans or policies	Year	Goals	Focuses on basic research	Comments
Key Projects in the National Science & Technology Program during the 11th FYP	2006	Proactively deploying basic research and research in cutting-edge technologies, promoting the continuous S&T innovation capacity and making a batch of major innovations; focusing on strengthening research of application and development, promoting the transformation, spread and diffusion of S&T achievements, and accelerating S&T industrialization	Basic research programs include National Natural Science Foundation of China (NSFC) and National Program on Key Basic Research Project of China (Program 973); NSFC mainly supports free explorative basic research; Program 973 mainly supports basic research and major scientific research programs geared to major national strategic needs	Clearly putting forward "stable support and proactive deployment of basic research"
Knowledge Innovation Project of Chinese Academy of Sciences	1998	Improving the S&T innovation capacity of Chinese Academy of Sciences and the entire country and making major contributions in S&T innovation	Strengthening basic scientific research, strategic high-tech innovation and technological research serving the social development	Increasing the inputs in basic research and improving the effect of allocation in Chinese Academy of Sciences

Continued

Plans or policies	Year	Goals	Focuses on basic research	Comments
Project 211	1993	Establishing a number of institutions of higher education and key disciplines as bases for training top-notch professional talents and solving major scientific issues in economic construction and social development through efforts of 10 years or more	—	Making universities contribute more to national development of economy, society and S&T by strengthening scientific education and scientific research
Project 985	1998	Guided by key areas and major needs of the country and the industrial development, geared to national S&T development strategies and disciplinary frontiers, establishing a batch of innovation platforms for competitive disciplines in universities with clear-cut professional characteristics	—	Focusing on basic research and talents cultivation oriented towards the world

Continued

Plans or policies	Year	Goals	Focuses on basic research	Comments
Key Projects in the National Science & Technology Program during the 12th FYP	2011	Continuing to increase spending on basic research and research in cutting-edge technologies, continuing to strengthen basic research, further encouraging the exploration of science frontiers, adhering to the guidance of demand, and striving to address major scientific issues which restrict the social and economic development	Aiming at the global science frontier, proactively deploying basic research and research in cutting-edge technologies, encouraging free exploration, increasing S&T accumulation and further enhancing the original innovation capacity	Basic research and research in cutting-edge technologies lay the important foundation for improving the original innovation capacity and long-term development of S&T capacity, which are the sources of S&T progress and innovation

Data source: Compiled from the open information on the website of the Ministry of Science and Technology, OECD (2008) and related materials.

8.2 The role of enterprises in major research programs in China

Taking into account the accessibility of data, feasibility for historical comparison and whether enterprises can participate[①], we select four national scientific programs that are closely connected with basic research. They are the National Natural Science Foundation of China, Program 973, Major National Scientific Research Program and State Key Laboratory Program. We aim to analyze the input

① In basic research planning system in China, Project 211 and Project 985 of the Ministry of Education and Knowledge Innovation Project of the Chinese Academy of Sciences play a significant role in the national fiscal inputs. The basic research achievements funded by these projects have made tremendous contributions to technological innovation in China. However, because these programs are mainly limited to basic research in institutions of higher education and the Chinese Academy of Sciences, enterprises cannot get involved. So it influences the analysis of their comparisons with other national scientific programs in this part. These programs are not included in this analysis. Furthermore, National High-tech R&D Program of China (Program 863) also involves some basic research, but its major objectives are the industrialization of high-tech. In order to maintain the consistency of the analysis perspective, this program is also excluded.

model of basic research resources in China, specifically the participation of enterprises and its trend. Table 8-3 reflects the historical changes of national finance in the four major basic research programs.

Table 8-3 Changes of national finance in major basic research projects (billion RMB)

Year	National Natural Science Foundation of China	National Basic Research Program (Program 973)	State Key Laboratory Program	Major National Scientific Research Program
1996	0.615	—	0.060	—
1997	0.788	—	0.065	—
1998	0.984	0.300	0.125	—
1999	1.004	0.400	0.105	—
2000	1.284	0.500	0.125	—
2001	1.598	0.589	0.130	—
2002	1.968	0.686	0.130	—
2003	2.049	0.800	0.130	—
2004	2.250	0.897	0.130	—
2005	2.701	0.983	0.134	—
2006	3.620	1.354	0.216	0.375
2007	4.331	1.646	1.600	0.353
2008	4.937	1.900	2.168	0.396
2009	6.004	2.600	2.917	0.700
2010	10.381	4.000	2.759	1.282
2011	14.043	3.092	2.961	1.408
2012	17.000	2.678	3.378	1.322
2013	16.162	2.828	2.891	1.227

Data source: Compiled from *China Statistical Yearbook on Science and Technology* over the years, Annual Report of National Scientific Programs of the Ministry of Science and Technology, Statistical Yearbook of Chinese Academy of Sciences and related materials.

8.2.1 National Natural Science Foundation of China

The National Natural Science Foundation of China (NSFC) is a direct result of the 1985 Decision on the Reform of the Science and Technology Management System by the Central Committee of the

Communist Party of China. It is an important measure for strengthening basic research and the earliest, largest and most systemic basic research support plan of China. Its establishment and model are both deeply influenced by the NSF of the US. NSFC's predecessor, the Science Foundation Committee of the Chinese Academy of Sciences was established on November 4, 1981. It began to fund basic research projects from 1982. On February 14, 1986, the State Council issued the Notice Concerning the Establishment of the National Natural Science Foundation of China. It was directly administered by the State Council.

The strategic positioning of the NSFC was designated as "to direct, coordinate and fund basic research and some applied research". Then in the enlarged meeting of the Party Group of the Foundation in 2004, the strategic positioning was clearly phrased as "supporting basic research, adhering to free exploration and playing a guiding role", and was formally written into the Development Plan of National Natural Science Foundation of China during the 11th FYP. The State Council passed the Regulations of the National Natural Science Foundation of China on February 14, 2007 and strengthened the management of funding and evaluation. In 2010, the new strategic positioning was expressed as "more emphasis on basic research, the leading edge and talents" (Table 8-4).

In earlier times, the NSFC occupied most of the expenditure of the government-financed basic research. As China's scientific and economic development has increasing needs towards basic research, the government has explored new channels in the expenditure on scientific research activities. Scientific programs like Program 973 and Project 211 were launched successively. However, in terms of basic research that encourages free exploration, the NSFC remains the main channel.

Since the 9th FYP, the NSFC has increased its funding. Large numbers of experimental basic research programs are funded. Development programs all emphasize scientific research and the cultivation of talents (Table 8-5).

Table 8-4 Correlation between NSFC and national basic research programs in different periods

Span	Deployment of national basic research tasks	Major tasks of NSFC	Funding priorities of NSFC
During the 8th FYP (1991–1995)	(1) Supporting scientists to freely choose the subjects (2) Choosing the key subjects for basic research based on the achievements scientists have made in choosing the subjects (3) Making possible breakthroughs or conducting major basic research with significant application prospects	(1) (2) (3)	Emphasizing that basic research should serve economic development
During the 9th FYP (1996–2000)	(1) Further adjusting and optimizing disciplinary layout (2) Improving and developing the mechanism of the NSCF (3) Continuing to organize and implement a batch of key programs of basic research with significant scientific value and application prospects (4) Strengthening scientific infrastructure construction featuring scientific equipment (5) Conducting basic research geared to significant decision-making issues in the development of economy, society and S&T	(1) (2) (3) (5)	Targeting national objectives and the leading edge of global science, and encouraging the combination of free exploration and macro regulation

Continued

Span	Deployment of national basic research tasks	Major tasks of NSFC	Funding priorities of NSFC
During the 10th FYP (2001–2005)	(1) Major scientific issues (2) Discipline construction (3) Explorative research	(2)(3)	In terms of the sources of subjects, adhering to the mechanism of two driving forces, which are "issues derived from science" and "issued derived from social and economic development"
During the 11th FYP (2006–2010)	(1) Disciplinary development (2) Frontier scientific issues (3) Basic research geared to major national strategic needs (4) Major scientific research programs	(1)(2)	—

Data source: International Evaluation of Funding and Management of the National Natural Science Foundation of China Synthesis Evidence Report, National Center for Science & Technology Evaluation, 2010.

Table 8-5 Major objectives and funding plans of NSFC in different periods

Span	Description of the development plan	Direction and objectives	Funding plan
Development Plan of National Natural Science Foundation of China during the 9th FYP (1996–2000)	Consistently developing basic research and prioritizing national objectives	Properly arranging the proportion of basic research and applied basic research (at 3 : 7)	Funding 18,000 projects, 600 to 700 key projects, and 60 major projects within 5 years
Development Plan of National Natural Science Foundation of China during the 10th FYP (2001–2005)	Revolving around the scientific frontier and national strategic needs, combining major scientific issues in the national economic and social development and national security, and supporting scientists in conducting innovative research	Increasing funding and moderately expanding the scale of funding; promoting the coordinated development of disciplines and forming competitive research	During the 10th FYP, National Science Fund for Distinguished Young Scholars plans to support about 900 people, about 100 innovative research groups and each year it will support 1,200 general programs of Science Fund for Distinguished Young Scholars
Development Plan of National Natural Science Foundation of China during the 11th FYP (2006–2010)	Promoting the balanced, coordinated and sustainable disciplinary development, fostering a batch of outstanding scientists with global influence and innovation teams aiming at the global science frontier, promoting the overall performance and international competitiveness of basic research, and striving for major breakthroughs in some important areas	Focusing on original innovation, promoting independent innovation capacity, deploying priority areas, and enhancing the overall performance of the key areas	Science Fund for Distinguished Young Scholars plans to fund about 900 people, 100 newly-initiated innovation group projects, 100 basic science talents training bases and 1,800 key programs

Continued

Span	Description of the development plan	Direction and objectives	Funding plan
Development Plan of National Natural Science Foundation of China during the 12th FYP (2011–2015)	Promoting several mainstream disciplines to emerge in the global arena and creating a batch of excellent scientists and innovation teams with global influence; promoting the overall level of basic research in China, significantly enhancing the global influence of basic research and the independent innovation capacity of several key fields of science	Original innovation strategy, innovative talents strategy, openness and cooperation strategy, innovative atmosphere strategy, outstanding management strategy	Unknown

In terms of the scale of funding, the NSFC has remained one of the major funding channels in China's basic research since its establishment. Its funding has accounted for about 20% to 40% of the national basic research expenditure (reaching 47.86% in 2012). Figure 8-3 shows the trend of basic research expenditure increase and the scale of funding for the NSFC from 1998 to 2013. From 1987 to 2013, basic research spending increased from 450 million RMB to 55.68 billion RMB, up by 124 times. National financing for the NSFC increased from 100 million RMB to 16.162 billion RMB, up by 162 times.

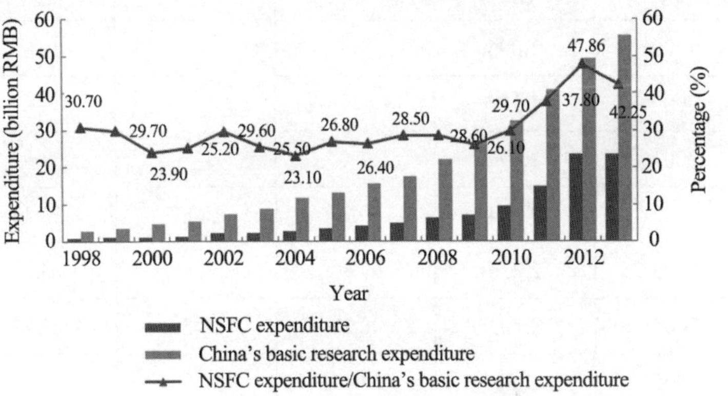

Figure 8-3 Total amounts of National Natural Science Foundation and its proportion in national basic research expenditure from 1998 to 2013

Data source: *China Statistical Yearbook on Science and Technology* over the years and the Annual Statistical Report of NSFC.

Different from Program 973, the NSFC is a major field for scientists to acquire the knowledge purely for their interest in research because of the fact that the NSFC regards free exploration as its prime consideration in funding (in these years it begins to take economic and social development into account in funding). So the subjects for funding are mainly restricted to institutions of higher education and public research institutes (mainly Chinese Academy of Sciences). The proportion in expenditure stays at over 98%, with a low participation of enterprises (Table 8-6 and Table 8-7).

Table 8-6 Changes of approval of projects funding in National Natural Science Foundation of China from 1998 to 2013

Year	Total funds in projects (billion RMB)	Percentage of funds received by institutions of higher education (%)	Percentage of funds received by research institutes (%)	Percentage of funds received by Chinese Academy of Sciences (%)
1998	0.889	—	—	—
1999	1.006	—	—	—
2000	1.119	—	—	—
2001	1.399	—	—	—
2002	2.184	69.27	28.87	23.13
2003	2.236	67.82	30.87	23.26
2004	2.702	70.28	28.29	23.44
2005	3.52	73.49	25.03	20.27
2006	4.11	72.92	25.35	21.07
2007	4.971	72.64	25.77	19.07
2008	6.309	73.41	25.18	20.54
2009	7.054	75.37	23.33	19.28
2010	9.653	75.85	22.70	18.70
2011	14.958	79.65	18.75	13.88
2012	23.656	77.56	20.90	15.28
2013	23.524	77.97	20.54	13.82

Data source: *China Statistical Yearbook on Science and Technology* over the years, the website of the National Natural Science Foundation of China (NSFC).

Table 8-7 Changes of investment and its structure of National Natural Science Foundation of China from 2001 to 2013

Item	2001	2002	2003	2004	2005	2006	2007	2008	2009	2010	2011	2012	2013
Number of General Programs funded	4,435	5,808	6,359	7,711	9,111	10,271	7,713	8,924	10,061	13,030	15,329	16,891	16,194
Funding for General Programs (billion RMB)	0.798	1.156	1.322	1.675	2.259	2.686	2.275	2.886	3.305	4.525	8.989	12.480	12.000
Percentage of institutions of higher education and scientific research institutes funded by General Programs (%)	98.99	98.64	98.63	98.74	98.6	98.47	98.38	98.66	99.98	98.43	98.28	98.25	98.31

Continued

Item	2001	2002	2003	2004	2005	2006	2007	2008	2009	2010	2011	2012	2013
Percentage of funds allocated to institutions of higher educations and scientific research institutes by General Programs (%)	99.05	98.79	98.73	98.81	98.63	98.56	98.50	98.79	99.99	98.57	98.37	98.44	98.52
Number of projects funded by Key Programs	124	208	252	224	303	277	373	425	391	436	497	538	564
Funds allocated by Key Programs (billion RMB)	0.180	0.312	0.360	0.333	0.488	0.443	0.635	0.780	0.724	0.946	1.425	1.567	1.663
Percentage of institutions of higher education and scientific Research institutes funded by Key Programs (%)	100	100	99.21	100	100	99.28	98.39	99.29	99.95	99.08	99.60	99.07	99.29

Continued

Item	2001	2002	2003	2004	2005	2006	2007	2008	2009	2010	2011	2012	2013
Percentage of funds allocated to institutions of higher educations and scientific research institutes by Key Programs (%)	100	100	99.33	100	100	99.44	98.44	99.35	99.95	99.11	99.65	99.08	99.30

Data source: The website of the National Natural Science Foundation of China (NSFC).

Although projects applied by some enterprises are funded, most of these enterprises are the enterprises that had been transformed from scientific research institutes, and the programs they acquired can only be categorized as general or youth projects, while few are key projects or outstanding projects. And the funding intensity is not high either (Table 8-8).

Table 8-8 Enterprise-responsible projects funded through the National Natural Science Foundation of China from 1999 to 2011

Year	Enterprise as the responsible unit of the program funded	Number of programs	Category
1999	Huajian Electronic Co., Ltd	1	Key Program
	Chengdu Organic Chemicals Co., Ltd, Chinese Academy of Sciences	5	General Program
	No.710 Institute of Aviation Industry Corporation of China	1	General Program
	Guangzhou Chemistry Co., Ltd, Chinese Academy of Sciences	4	General Program
	Guangzhou Electronic Technology Co., Ltd, Chinese Academy of Sciences	1	General Program
	Beijing Institute of Aeronautical Materials, Aviation Industry Corporation of China	1	General Program
	No.22 Research Institute of China Electronics Technology Group Corporation	1	General Program
	Shanghai CASB Biotechnology Co., Ltd (CASB)	1	General Program
	No.4 Research Institute of China Aerospace Science and Technology Corporation	1	General Program
	Xi'an Research Institute of China Coal Technology & Engineering Group Corp	2	General Program
	General Research Institute for Nonferrous Metals	4	General Program
	Central Iron & Steel Research Institute	6	General Program
2000	Shanghai CASB Biotechnology Co., Ltd (CASB)	1	General Program
	Changsha Research Institute of Mining and Metallurgy CO., Ltd	1	General Program
	No.22 Research Institute of China Electronics Technology Group Corporation	1	General Program
	Research Institute of Petroleum Exploration & Development, PetroChina	1	General Program
	Chengdu Organic Chemicals Co., Ltd, Chinese Academy of Sciences	3	General Program
	Research Institute of Petroleum Processing, SINOPEC	1	General Program

Continued

Year	Enterprise as the responsible unit of the program funded	Number of programs	Category
2000	General Research Institute for Nonferrous Metals	5	4 General Programs, 1 Youth Program
	Central Iron & Steel Research Institute	5	General Program
	Beijing General Research Institute of Mining and Metallurgy	2	1 General Programs,1 Youth Program
	Jinling Petrochemical Co., Ltd, SINOPEC	1	Youth Program
	Chengdu Organic Chemicals Co., Ltd, Chinese Academy of Sciences	1	Youth Program
	Chengdu Information Technology Co., Ltd, Chinese Academy of Sciences	1	Youth Program
	Guangzhou Chemistry Co., Ltd, Chinese Academy of Sciences	4	Youth Program
2001	Changsha Research Institute of Mining and Metallurgy Co., Ltd	1	Key Program
	Chengdu Organic Chemicals Co., Ltd, Chinese Academy of Sciences	4	2 General Programs, 2 Youth Programs
	Research Institute of Petroleum Exploration & Development, PetroChina	1	General Program
	Shenyang Aircraft Design & Research Institute, Aviation Industry Corporation of China	1	General Program
	No.12 Research Institute of China Electronics Technology Group Corporation	1	General Program
	No.53 Research Institute of China North Industries Group Corporation	1	General Program
	Central Iron & Steel Research Institute	6	4 General Programs, 2 Youth Programs
	General Research Institute for Nonferrous Metals	1	General Program
2002	Sinoma Science & Technology Co., Ltd	1	Key Program
	Research Institute of Tubular Goods, China National Petroleum Corporation	1	Key Program

Continued

Year	Enterprise as the responsible unit of the program funded	Number of programs	Category
2002	Beijing General Research Institute of Mining and Metallurgy	4	1 Key Program, 2 General Programs, 1 Youth Program
	Chengdu Organic Chemicals Co., Ltd, Chinese Academy of Sciences	8	General Program
	Guangzhou Chemistry Co., Ltd, Chinese Academy of Sciences	4	General Program
	Jinling Petrochemical Co., Ltd, SINOPEC	1	General Program
	Beijing Solar Energy Research Institute Group Co., Ltd	1	General Program
	Chengdu Information Technology Co., Ltd, Chinese Academy of Sciences	1	General Program
	Hangzhou Water Treatment Technology Development Center Co., Ltd	1	General Program
	No.13 Research Institute of China Electronics Technology Group Corporation	1	General Program
	Central Iron & Steel Research Institute	5	General Program
	Shanghai CASB Biotechnology Co., Ltd (CASB)	1	General Program
	Huajian Electronic Co., Ltd	1	General Program
	No. 4 Research Institute of China Aerospace Science and Technology Corporation	1	General Program
	China Science Publishing Group	1	Youth Program
	Xi'an Research Institute of China Coal Technology & Engineering Group Corp	2	Youth Program
	Guangzhou Chemistry Co., Ltd, Chinese Academy of Sciences	1	Youth Program
	General Research Institute for Nonferrous Metals	7	5 General Programs, 2 Youth Programs
2003	Central Iron & Steel Research Institute	9	2 Key Programs, 7 Youth Programs

Continued

Year	Enterprise as the responsible unit of the program funded	Number of programs	Category
2003	Chengdu Organic Chemicals Co., Ltd, Chinese Academy of Sciences	9	5 General Programs, 4 Youth Programs
	Xi'an Research Institute of China Coal Technology & Engineering Group Corp	2	General Program
	Beijing Institute of Aeronautical Materials, Aviation Industry Corporation of China	1	General Program
	China Merchants Chongqing Communications Technology Research & Design Institute Co., Ltd	1	General Program
	Chengdu Information Technology Co., Ltd, Chinese Academy of Sciences	2	General Program
	Guangzhou Electronic Technology Co., Ltd, Chinese Academy of Sciences	1	General Program
	Henan Transportation Research Institute Co., Ltd	1	General Program
	No. 22 Research Institute of China Electronics Technology Group Corporation	1	General Program
	Huajian Electronic Co., Ltd	1	General Program
	Research Institute of Petroleum Exploration & Development, PetroChina	2	General Program
	General Research Institute for Nonferrous Metals	4	2 General Programs, 2 Youth Programs
	Beijing General Research Institute of Mining and Metallurgy	1	General Program
	No. 41 Institute of China Electronics Technology Group Corporation	1	Youth Program
	Changsha Research Institute of Mining and Metallurgy CO., Ltd	1	Youth Program
	Guangzhou Chemistry Co., Ltd, Chinese Academy of Sciences	1	Youth Program
2004	Chengdu Organic Chemicals Co., Ltd, Chinese Academy of Sciences	4	General Program
	No.22 Research Institute of China Electronics Technology Group Corporation	1	General Program
	Beijing Institute of Aeronautical Materials, Aviation Industry Corporation of China	2	General Program

Continued

Year	Enterprise as the responsible unit of the program funded	Number of programs	Category
2004	Zhengzhou Tobacco Research Institute, China National Tobacco Corporation	1	General Program
	Oilfield Construction and Designing Institute, Daqing Oilfield	1	General Program
	Huajian Electronic Co., Ltd	1	General Program
	Guangzhou Chemistry Co., Ltd, Chinese Academy of Sciences	1	General Program
	No.12 Research Institute of China Electronics Technology Group Corporation	1	General Program
	Central Iron & Steel Research Institute	3	General Program
	Beijing General Research Institute of Mining and Metallurgy	3	General Program
	General Research Institute for Nonferrous Metals	3	General Program
2005	The 710th Institute of China Aerospace Science and Technology Group	1	General Program
	Research Institute of Petroleum Exploration & Development, PetroChina	1	General Program
	No.12 Research Institute of China Electronics Technology Group Corporation	1	General Program
	Beijing Aeronautical Manufacturing Technology Research Institute, Aviation Industry Corporation of China	1	General Program
	China National Petroleum Corporation	1	General Program
	China Nonferrous Metal Industry's Foreign Engineering and Construction Co., Ltd	2	General Program
	General Research Institute for Nonferrous Metals	2	General Program
	China National Nuclear Corporation	18	General Program
	Aviation Industry Corporation of China	1	General Program
	Beijing Aeronautical Manufacturing Technology Research Institute, Aviation Industry Corporation of China	1	General Program
	China North Industries Group Corporation	2	General Program
	Central Iron & Steel Research Institute	4	General Program

Continued

Year	Enterprise as the responsible unit of the program funded	Number of programs	Category
2006	Research Institute of Petroleum Exploration & Development, PetroChina	4	2 General Programs 2 Youth Programs
	Beijing Aeronautical Manufacturing Technology Research Institute, Aviation Industry Corporation of China	1	General Program
	Institute of Porous Flow & Fluid Mechanics, China National Petroleum Corporation & Chinese Academy of Sciences	1	General Program
	Wuxi Institute of Petroleum Geology, Petroleum Exploration and Production Research Institute, SINOPEC	1	General Program
	No. 36 Research Institute of China Electronics Technology Group Corporation	1	General Program
	No. 4 Research Institute of China Aerospace Science and Technology Corporation	1	General Program
	Beijing General Research Institute of Mining and Metallurgy	1	General Program
	China Electric Power Research Institute	1	General Program
	Tianjin Institute of Pharmaceutical Research	1	General Program
	Guangzhou Municipal Engineering Design & Research Institute	1	General Program
	Central Iron & Steel Research Institute	4	General Program
	General Research Institute for Nonferrous Metals	2	General Program
2007	Beijing General Research Institute of Mining and Metallurgy	3	1 Key Program, 2 Youth Programs
	No.715 Research Institute of China Shipbuilding Industry Corporation	1	General Program

Continued

Year	Enterprise as the responsible unit of the program funded	Number of programs	Category
2007	No.711 Research Institute of China Shipbuilding Industry Corporation	1	General Program
	No.22 Research Institute of China Electronics Technology Group Corporation	2	General Program
	Beijing Institute of Aeronautical Materials, Aviation Industry Corporation of China	1	General Program
	Beijing Aeronautical Manufacturing Technology Research Institute, Aviation Industry Corporation of China	1	General Program
	Chengdu Organic Chemicals Co., Ltd, Chinese Academy of Sciences	2	General Program
	Shanghai Research Institute of Petrochemical Technology, SINOPEC	1	General Program
	Petroleum Exploration and Production Research Institute, SINOPEC	3	General Program
	Research Institute of Exploration and Development, Changqing Oilfield Company, PetroChina	1	General Program
	Research Institute of Petroleum Exploration & Development, PetroChina	3	General Program
	Zhengzhou Tobacco Research Institute, China National Tobacco Corporation	2	General Program
	Chengdu Information Technology Co., Ltd, Chinese Academy of Sciences	1	General Program
	Technique R&D Center, Haier Group	1	General Program
	Xi'an Research Institute of China Coal Technology & Engineering Group Corp	1	General Program
	General Research Institute for Nonferrous Metals	7	5 General Programs, 2 Youth Programs
	Central Iron & Steel Research Institute	6	General Program
2008	Beijing Aeronautical Manufacturing Technology Research Institute, Aviation Industry Corporation of China	1	General Program
	Research Institute of Petroleum Exploration & Development, PetroChina	1	General Program

Continued

Year	Enterprise as the responsible unit of the program funded	Number of programs	Category
2008	No.13 Research Institute of China Electronics Technology Group Corporation	1	General Program
	No.22 Institute of China Electronics Technology Group Corporation	1	General Program
	Zhengzhou Tobacco Research Institute, China National Tobacco Corporation	1	General Program
	Jinling Petrochemical Co., Ltd, SINOPEC	1	General Program
	Chengdu Organic Chemicals Co., Ltd, Chinese Academy of Sciences	4	General Program
	Central Iron & Steel Research Institute	4	1 General Program, 3 Youth Programs
	General Research Institute for Nonferrous Metals	6	4 General Programs, 2 Youth Programs
	Beijing General Research Institute of Mining and Metallurgy	3	1 General Program, 2 Youth Programs
	Beijing Aeronautical Manufacturing Technology Research Institute, Aviation Industry Corporation of China	1	Youth Program
	Petroleum Exploration and Production Research Institute, SINOPEC	1	Youth Program
	Research Institute of Petroleum Exploration & Development, PetroChina	1	Youth Program
	Jiangsu Research Institute of Building Science Co., Ltd	1	Youth Program
	No.55 Research Institute of China Electronics Technology Group Corporation	1	Youth Program
	Shanghai Huaguan Biochip Co., Ltd	1	Youth Program
	Chengdu Organic Chemicals Co., Ltd, Chinese Academy of Sciences	2	Youth Program
2009	Beijing Aeronautical Manufacturing Technology Research Institute, Aviation Industry Corporation of China	2	General Program
	Chinese National Human Genome Center (CNHG)	1	General Program

Continued

Year	Enterprise as the responsible unit of the program funded	Number of programs	Category
2009	Petroleum Exploration and Production Research Institute, SINOPEO	1	General Program
	Guangzhou Chemistry Co., Ltd, Chinese Academy of Sciences	1	General Program
	Sinosteel Luoyang Institute of Refractory Research Co., Ltd	1	General Program
	Guangzhou Electric Apparatus Research Institute	1	General Program
	No.22 Institute of China Electronics Technology Group Corporation	1	General Program
	Zhengzhou Nonferrous Metals Research Institute Co., Ltd, Aluminium Corporation of China	1	General Program
	Jiangsu Research Institute of Building Science Co., Ltd	1	General Program
	Central Iron & Steel Research Institute	4	3 General Programs, 1 Youth Program
	Beijing General Research Institute of Mining and Metallurgy	3	1 General Program, 2 Youth Programs
	General Research Institute for Nonferrous Metals	5	1 General Program, 4 Youth Programs
	Chengdu Organic Chemicals Co., Ltd, Chinese Academy of Sciences	4	General Program
	Beijing Institute of Aeronautical Materials, Aviation Industry Corporation of China	1	Youth Program
	Research Institute of Petroleum Exploration & Development, PetroChina	1	Youth Program
	Zhengzhou Tobacco Research Institute, China National Tobacco Corporation	2	Youth Program
	No.22 Institute of China Electronics Technology Group Corporation	1	Youth Program
	Jiangsu Research Institute of Building Science Co., Ltd	1	Youth Program
2010	Beijing Institute of Aeronautical Materials, Aviation Industry Corporation of China	2	General Program

Continued

Year	Enterprise as the responsible unit of the program funded	Number of programs	Category
2010	Petroleum Exploration and Production Research Institute, SINOPEC	3	2 General Programs, 1 Youth Program
	Research Institute of Petroleum Exploration & Development, PetroChina	2	1 General Program, 1 Youth Program
	Guangzhou Chemistry Co., Ltd, Chinese Academy of Sciences	3	1 General Program, 2 Youth Programs
	Beijing Institute of Aeronautical Materials, Aviation Industry Corporation of China	2	General Program
	No.55 Research Institute of China Electronics Technology Group Corporation	1	General Program
	Shanghai Research Institute of Petrochemical Technology, SINOPEC	1	General Program
	Chengdu Organic Chemicals Co., Ltd, Chinese Academy of Sciences	3	2 General Programs, 1 Youth Program
	China Merchants Chongqing Communications Technology Research & Design Institute Co., Ltd	3	2 General Programs, 1 Youth Program
	Beijing Sinoma Synthetic Crystals Co., Ltd	1	General Program
	Zhengzhou Tobacco Research Institute, China National Tobacco Corporation	4	Youth Program
	No.22 Institute of China Electronics Technology Group Corporation	2	Youth Program
	Sinosteel Luoyang Institute of Refractory Research Co., Ltd	1	Youth Program
	Jiangsu Research Institute of Building Science Co., Ltd	1	Youth Program
	General Research Institute for Nonferrous Metals	7	1 General Program, 6 Youth Programs
	Central Iron & Steel Research Institute	7	4 General Programs, 3 Youth Programs

Continued

Year	Enterprise as the responsible unit of the program funded	Number of programs	Category
2010	Beijing General Research Institute of Mining and Metallurgy	3	1 General Program, 2 Youth Programs
2011	Petroleum Exploration and Production Research Institute, SINOPEC	3	1 General Program 2 Youth Programs
	Research Institute of Petroleum Exploration & Development, PetroChina	1	General Program
	China Southern Grid Electric Power Research Institute Co., Ltd	1	General Program
	Guangzhou Chemistry Co., Ltd, Chinese Academy of Sciences	2	General Program
	Central Iron & Steel Research Institute	4	3 General Programs 1 Youth Program
	Sinosteel Luoyang Institute of Refractory Research Co., Ltd	1	General Program
	Zhengzhou Tobacco Research Institute, China National Tobacco Corporation	5	1 General Program1, 4 Youth Programs
	Changsha Research Institute of Mining and Metallurgy CO., Ltd	1	General Program
	General Research Institute for Nonferrous Metals	8	1 General Program, 7 Youth Programs
	Shenyang Aircraft Design & Research Institute, Aviation Industry Corporation of China	1	General Program
	Guangzhou Municipal Engineering Design & Research Institute	1	General Program
	Shenzhen Huada Gene Research Institute	1	General Program
	Guangzhou Municipal Engineering Design & Research Institute	1	General Program
	Chengdu Organic Chemicals Co., Ltd, Chinese Academy of Sciences	4	3 General Programs, 1 Youth Program

Continued

Year	Enterprise as the responsible unit of the program funded	Number of programs	Category
2011	No.24 Research Institute of China Electronics Technology Group Corporation	1	General Program
	No.22 Research Institute of China Electronics Technology Group Corporation	4	2 General Programs, 2 Youth Programs
	No.41 Research Institute of China Electronics Technology Group Corporation	1	Youth Program
	No.3 Research Institute of China Electronics Technology Group Corporation	1	Youth Program
	Beijing Institute of Aeronautical Materials, Aviation Industry Corporation of China	1	Youth Program
	No.55 Research Institute of China Electronics Technology Group Corporation	1	Youth Program
	Jiangsu Research Institute of Building Science Co., Ltd	1	Youth Program
	Beijing General Research Institute of Mining & Metallurgy	4	Youth Program
	Aerodynamics Research Institute, Aviation Industry Corporation of China	1	Youth Program
	Shanghai Research Institute of Petrochemical Technology, SINOPEC	1	Youth Program
	China Academy of Urban Planning & Design in Beijing	1	Youth Program
	Beijing General Municipal Engineering Design & Research Institute	1	Youth Program
	Hydro China Huadong Engineering Corporation	2	Youth Program
	Xi'an Research Institute of China Coal Technology & Engineering Group Corp	1	Youth Program
	China Coal Research Institute (CCRI)	2	Youth Program

Data source: Compiled from the Annual Report released by the National Natural Science Foundation of China (NSFC) and its Funding Search System.

Based on accessible statistics, from 1999 to 2011, 432 enterprise-responsible projects were funded. On average, 33 projects were funded on a year-to-year basis. Most of them were concentrated in several state-owned enterprises such as the China National Nuclear Corporation, Central Iron & Steel Research Institute, General Research Institute for Nonferrous Metals, Chengdu Organic Chemicals Co., Ltd, and Chinese Academy of Sciences (Figure 8-4). It was connected with the funding approach of the National Natural Science Foundation of China (which spanned 3 to 5 years and the requirements were mostly scientific papers or patents).

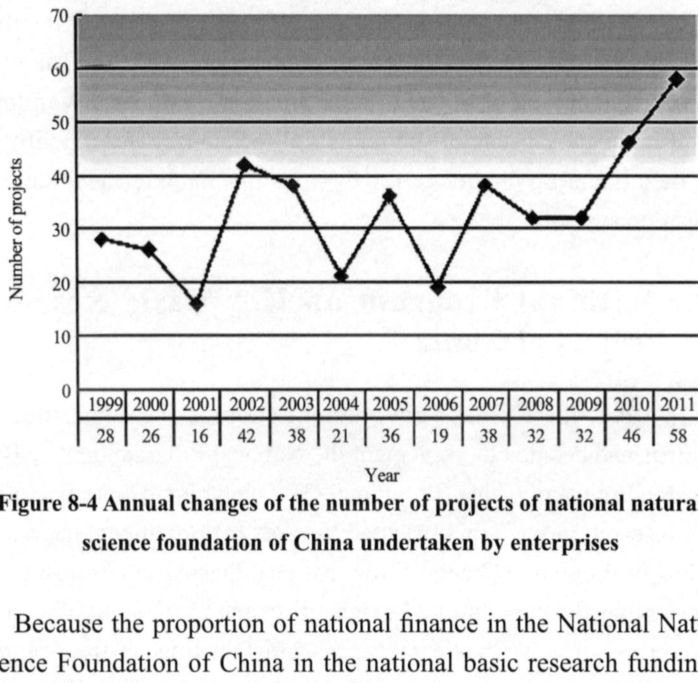

Figure 8-4 Annual changes of the number of projects of national natural science foundation of China undertaken by enterprises

Because the proportion of national finance in the National Natural Science Foundation of China in the national basic research funding is relatively higher (it even reached 38% in 2011), how to combine free exploration of the National Natural Science Foundation of China (later adding national needs) and core industrial technological innovation is a meaningful exploratory topic. Some people believe that more basic research should be funded as scientific funds, because when our country goes through the change from the "catch-up" stage to the independent innovation stage, more basic research will be needed. When China's science is close to the cutting-edge level, the importance of free exploration is correspondingly enhanced. In leading OECD countries (such as the United States, Germany, Canada, Finland, the Netherlands, Sweden and the United Kingdom), the proportion of funding in free explorative research in national R&D input is increased and corresponding plans are formulated. The "Overall Science and Innovation Plan" of the United States will double its funding for key basic research institutes in 10 years. The "Excellence Initiative Plan" of German plans to provide funding for research institutes and scientific research projects from 2005 to 2011 in order to establish flagship universities and increase funding for four major research institutes and German Science Foundation at an annual rate of 3%. Therefore, we believe that natural science funds should be integrated with topics facing innovative needs of the industrial technology. Admittedly, this will be a daunting challenge for the existing funding model and the evaluation system.

8.2.2 National Program on Key Basic Research Project of China

In 1997, the Chinese government adopted the suggestions of scientists and decided to implement the National Program on Key Basic Research Project of China (Program 973). The program aims at strategic basic research in key areas oriented towards major national needs. The funding of Program 973 comes from national finance and its positioning is "orienting towards national needs, directing at national objectives, conducting strategic basic research, and focusing on the effective integration and interaction with innovation chain". Program 973 and the Major Scientific Research Plan implemented a system that was headed by prime scientists. The five-year funding intensity totaled 20 million to 40 million RMB. It is reputed as a flagship plan of Chinese basic research. It is a major strategic approach to strengthening basic research and enhancing the national independent innovation capacity.

Since it was initiated in 1997, Program 973 has stuck to the positioning of "basic research facing national strategic needs and the combination of major national needs and cutting-edge basic science". The criteria to support the project are all based on the following questions: Are major scientific issues in the interest of national economic and social development? Can the scientific foundation be laid for future key new technologies? Can cutting-edge disciplinary development trends be demonstrated or does the project occupy a place in international science frontier as a fundamental indicator and an important standard? Under the auspices of Program 973, a large number of excellent scientific research achievements have been made in China, especially leading in areas such as nanotechnology, life sciences and information.

Program 973 has executed the deployment of national needs for basic research well, established two driving-forces for basic research funding systems with the free, explorative National Natural Science Foundation of China, and improved the layout of basic research. From its implementation in 1998 to 2012, Program 973 has revolved around agriculture, energy, information, resource and environment, population and health, material as well as comprehensive, cross-disciplinary and cutting-edge sciences (Table 8-9).

Table 8-9 Changes of expenditure on projects of Program 973 from 1998 to 2012

Year	Number of projects approved	Funding (billion RMB)
1998	15	0.3
1999	44	0.4
2000	27	0.5
2001	20	0.589
2002	26	0.7
2003	27	0.8

Continued

Year	Number of projects approved	Funding (billion RMB)
2004	33	0.9
2005	39	0.983
2006	67	1.35
2007	81	1.65
2008	79	1.9
2009	123	2.6
2010	94	4
2011	98	—
2012	99	—

Note: Major Scientific Research Plan was initiated in 2006. 40 projects were approved in 2006 and 42 projects were approved in 2007. Statistics of funds allocated to institutions include the Major Scientific Research Plan. The statistics of funds in 2011 and 2012 are inaccessible.

Program 973 is typical basic research directed by national plans and its topics must revolve around national strategic needs and solve major issues in social and economic development. Although Program 973 abides by "revolving around major national needs to conduct basic research" and changes the thinking pattern of "literature to literature" in the past, it has shortened the distance between basic research and the industrial sector, solved a series of major scientific issues in national needs and played a fundamental role in enhancing industrial technological innovation. Taking oil extraction as an example, scientists have put forward the principle of "the molecular designing of surfactants used for displacing an oil reservoir", and also designed and produced inexpensive, high-efficient, pollution-free surfactants. After the on-site experiment on Daqing Oilfield, oil recovery efficiency was raised by 15% to 20% compared with being powered by water, and the cost was noticeably decreased. ①

① The implementation of Program 973 has enhanced the international influence of basic research in China, *People's Daily*, June 27, 2006.

From the perspective of the industrial technological innovation, Program 973 resembles the National Natural Science Foundation of China and abides by the scientific information model of Bush. It adopts the top-down funding model and prioritizes national objectives, but it still has a long way to go in meeting the needs of the core industrial technology. Although Program 973 has boosted the scientific research output of China (papers and patents), it's of more significance for this program to cultivate larger numbers of youth talents. But not many of them work in enterprises. In terms of the implementation of Program 973, 44 projects are funded with enterprises serving as prime scientific units. But most of the enterprises that have been funded are state-owned enterprises or enterprises that have gone through institutional reforms (Table 8-10 and Figure 8-5). ①

Table 8-10 Funding for enterprise-responsible projects in Program 973 from 1998 to 2011

Responsible enterprise	Number of projects
General Research Institute for Nonferrous Metals	1
Beijing Zhongyan Tongrentang Medicine R&D Co., Ltd	1
Central Iron & Steel Research Institute	3
State Grid Corporation of China	1
Yiling Pharmaceutical Group	1
China Coal Research Institute (CCRI)	2
China National Offshore Oil Corporation	1
Aviation Industry Corporation of China	1
China National Nuclear Corporation	2
China Building Materials Academy	1
Commercial Aircraft Corporation of China Co., Ltd	1

① Commercial Aircraft Corporation of China Co., Ltd and Semiconductor Manufacturing International Corporation undertook Program 973. It was in 2009 that for the first time the responsible unit was not traditional large-scale, state-owned enterprises.

Continued

Responsible enterprise	Number of projects
China Petrochemical Corporation (SINOPEC)	10
China National Petroleum Corporation (PetroChina)	17
Semiconductor Manufacturing International Corporation	2
In total	44

Data source: The Ministry of Science and Technology, the website of Program 973.

The funding level for a single project in Program 973 is huge, which is about 20 to 30 million RMB. But comprehensively speaking, Chinese enterprises still lack knowledgeable talents who can meet the needs (Table 8-11). Taking 2009 as an example, of the 26 projects invested by China, scientific research institutes received 1.117 billion RMB (43%); institutions of higher education received 1.372 billion RMB (53%); institutions with enterprises as prime scientific units only received 3 projects, with a funding of 104 million RMB, which was only 4% of the total investment. More enterprises pin their hope on the transfer of achievements in Program 973, rather than receive national investment in conducting significant basic research.

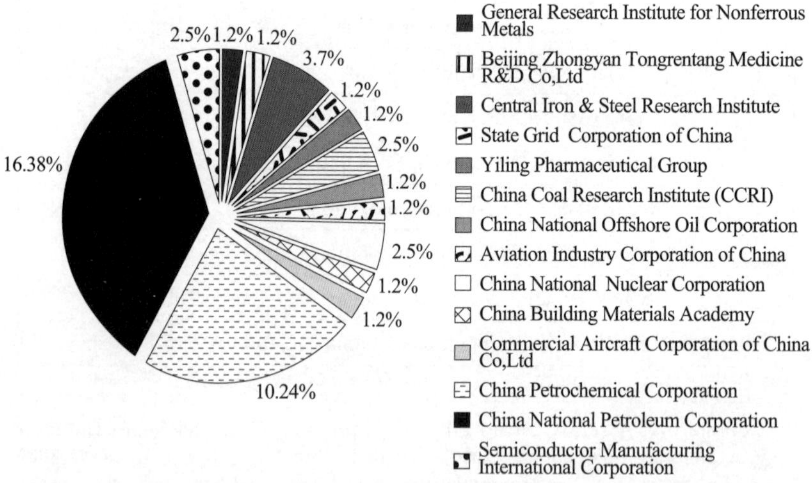

Figure 8-5 Distribution of enterprise-responsible projects in Program 973

Table 8-11 Number of Enterprise-responsible Projects Getting Funded in Program 973 from 1998 to 2012

Year	Number of Projects	Percentage (%)
1998	1	6.7
1999	3	6.8
2000	2	7.4
2001	1	5
2002	1	3.8
2003	1	3.7
2004	4	12.1
2005	3	7.7
2006	5	7.5
2007	3	3.7
2008	0	0
2009	6	4.9
2010	8	8.5
2011	2	2
2012	4	4

Data source: The Ministry of Science and Technology of China, the website of Program 973.

Admittedly, the situation is improving. As demonstrated from Program 973 in 2010 (Table 8-12), we find that some private enterprises began to engage in basic research by means of cooperating with public research institutes and universities. Among emerging industries, such as solar energy, high speed trains and Internet of Things, some high-tech enterprises began to undertake basic research projects.

Table 8-12 Sub-topics undertaken by enterprises in Program 973 in 2011

Program 973	Sub-topic	Unit
Basic research on low-carbon fuels and chemicals converted from low- and middle-rank coal	Reaction process, rationale and control of catalyzing and converting low- and middle-rank coal into methane and oil	ENN S&T Development Co., Ltd Institute of Coal Chemistry, Chinese Academy of Sciences
Basic research on broad-spectrum, high-efficient, thin-film solar cell	Base research on a new type of silicon-based, thin-film absorbent based on full solar spectrum	Graduate School of University of Chinese Academy of Sciences Nankai University Baoding Tianwei Group Co., Ltd
High-performance computer architecture principle, supporting technology and the application of cloud computing services application	Key technologies and application of high-performance extensible mobile service	Beijing University of Posts and Telecommunications China Mobile Research Institute
Basic research on high-performance and trustworthy virtual computing environment	Comprehensive testing and demonstration application of virtual computing environment	National Computer Network and Information Security Management Center Harbin Institute of Technology Hangzhou Ali Information & Services Co., Ltd
Fundamental theory and practical research of the Internet of Things	Verification and applied demonstration research of the Internet of Things	Wuxi Sensing Net Industrialization Research Institute State Grid Information & Telecommunications Co., Ltd Beijing University of Posts and Telecommunications

Continued

Program 973	Sub-topic	Unit
Light and high-temperature TiAl inter-metallic alloys and the science and technology foundation of their preparation process	High cleanliness melting of high-temperature TiAl alloys smelting and the science and technology foundation of their processing	North Western Polytechnic University University of Science & Technology Beijing Harbin Institute of Technology Baoshan Iron & Steel Co., Ltd
	Key science and technology foundation of investment casting of high-temperature TiAl alloys investment casting	Central Iron & Steel Research Institute (CISRI) Shanghai Jiao Tong University Danyang Precision Alloy Factory Co., Ltd Northwestern Polytechnic University
The key scientific issues in the processing of high-performance Kevlar fiber	Unknown	China BlueStar Chengrand Research Institute of Chemical Industry Suzhou Zhaoda Specially Fiber Technical Co., Ltd China ShenMa Group Co., Ltd Hebei Silicon Valley Chemical Co., Ltd China BlueStar Chengrand Research Institute of Chemical Industry

Continued

Program 973	Sub-topic	Unit
Key scientific issues in complex air separation units of ultra large and low-energy-consumption equipment sets	Associated designing theory of complex processing in multivariable units	Zhejiang University Hangyang Group
	Performance experiment of ultra large air separation equipment, simulation and integration	Zhejiang University Hangyang Group
High-performance roller bearing basic research	Dynamic principle of the interaction of complex interface system in bearing and bearing failure	Xi'an Jiaotong University Wafangdian Bearing Group Corp. Xinjiang University
	Research on the service performance of high-speed, heavy-duty, high-precision bearing	Xi'an Jiaotong University Wafangdian Bearing Group Corp.
Research on the theory and method of self-danger avoidance in complex low-altitude airspace	Simulated validation of self-danger avoidance in complex low-altitude airspace	Sichuan Jiuzhou Electric Group Co., Ltd Beihang University

Data source: The website of the Ministry of Science and Technology of China.

8.2.3 Major National Scientific Research Plan

Major scientific research plan is one kind of important organizational form of modern scientific research activities and usually refers to the large-scale, cross-disciplinary research team. It revolves around the overall objectives, is equipped with high-intensity resources and carries out systematic research in order to make major scientific breakthroughs. Since 2006, the Ministry of Science and Technology has launched the "Major National Scientific Research Plan" under Program 973 ①. Its funding all comes from national financing and it is a basic research development plan to lead the future development of 15 years and provides a driving force for science and technology according to the deployment of the Outline of the Program for Long- and Medium-Term National Scientific and Technological Development (2006–2020). The Major National Scientific Research Plan gears towards the global trends of scientific development and China's major strategic needs, and aims to lead and drive S&T development in China, rapidly enhance China's continuous innovation ability and make key headway. In 2006, it launched four research projects of protein, quantum control, nanotech science research, development and reproduction. In 2010, it launched two research projects for global change and stem cells.

The Major National Scientific Research Plan coordinated by the

① Both National Major Scientific Research Plan and Program 973 belong to the framework of the National Key Basic Research Development Plan. According to the interpretation of the Ministry of Science and Technology, Program 973 is a major basic research plan oriented towards major national needs and with strategic, forward-looking, comprehensive and leading effect on future development of our country and the progress of science and technology. It mainly revolves around agriculture, energy, information, resources and environment, population and health, materials, manufacturing and engineering science, comprehensive and major cutting-edge scientific areas. National Major Scientific Research Plan deployed by the National Plan for Long- and Medium-Term Scientific and Technological Development (2006–2020) is a basic research plan, which leads the future development and is the driving force for the development of science and technology. It mainly revolves around research on nanotechnology, quantum control, protein, development and reproduction, stem cells, global climate change and other fields.

Ministry of Science and Technology sticks to the proactive deployment and abides by the guidelines of Program 973. It is conducted by the Ministry of Education and Chinese Academy of Sciences. When the plan was launched, the Ministry of Science and Technology deployed and implemented a batch of major projects in accordance with the principle of "top-down design, overall arrangement, combination of advantages and dynamic adjustment". In the past 6 years, the plan has approved 156 major projects and arranged funding of more than 3 billion RMB. It has made major headway in research areas such as protein structure, quantum communication, nanotech materials and stem cells (Table 8-13).

Table 8-13 Investment of Major National Scientific Research Plan and its distribution

Year	Funding (billion RMB)	Number of projects
2006	0.375	40
2007	0.353	42
2008	0.396	35
2009	0.700	36
2010	1.282	53
2011	1.560	64

Data source: The website of the Ministry of Science and Technology of China.

In April, 2012, the Ministry of Science and Technology issued a special plan of six major scientific research plans including nanotech research during the 12th FYP. It has put forward more demanding requirements for basic research in areas such as nanotechnology, quantum control, protein, development and reproduction, stem cell and global climate change (Table 8-14).

Table 8-14 Overall planning of Major National Scientific Research Plan during the 12th FYP

Domain	Overall planning	Development objectives	Major areas for research
Nanotechnology research	Deepening basic research and promoting industrialization, combining nanotechnology with economy, promoting the integration of basic research, applied research and technological transfer	Making major original achievements in nano materials, nanotech devices and systems, biomedicine and measurement, etc.	Nano basic theory, nanotech materials with advanced functions, nano information materials and devices, nano biology and medicine, environmental nano materials, energy nano materials
Quantum control research	Maintaining the leading place in the practical quantum cryptography and quantum communications technology, iron-based superconductor and topological insulators	Making great breakthroughs in new materials, new principle and prototype equipment, achieving world-class results in quantum information, associated electronic system, small quantum system and artificial band gap system	Quantum information processing, associated electronic system, small quantum system and artificial band gap system
Protein research	Focusing on researching protein structure and functions, interaction and dynamic change, supported by protein research technique and method innovation	Revealing the biological mechanism of a number of spatial structure of membrane protein and important protein complexes, making a batch of far-reaching landmark achievements, maintaining China's world-leading place in proteome	Structural biology, proteomics, new technologies and new methods of studying protein, protein synthesis, degradation and regulation mechanism, protein biological functions, systemic biology, protein drug targets and molecular diagnosis

Continued

Domain	Overall planning	Development objectives	Major areas for research
Development and reproduction research	Solving fundamental and cutting-edge key scientific issues in development and reproduction, integrating new technologies, new methods and new models in multiple disciplines, conducting the layout according to three research directions and a systemic platform	Making significant breakthroughs in spermatogenesis sophisticated genetic and epigenetic regulation, embryo and organ development, assisted-reproductive technology, expounding the pathogenesis of several major development and reproduction diseases	Embryo and organ development, reproduction regulation, basic testing in major development and reproduction diseases and systemic platform of development and reproduction research
Stem cell research	Deepening the transformation and application research of stem cells, accelerating basic and clinical research on stem cells	Exploring ways to stem cell treatment of major diseases, making key breakthroughs in acquiring dry property of stem cells, maintaining and controlling regulatory mechanism, revealing the principle of the interaction between micro-environment and stem cells	Molecular mechanism of cell reprogramming, directed differentiation of pluripotent stem cells and the mechanism, interaction between stem cells and micro-environment and its mechanism, basis of stem cells treating visual system, diseases and clinical translational research
Global climate change research	Researching and developing global systemic model, dramatically enhancing China's research level on global climate change and providing scientific support for global sustainable development	Making major original achievements in key scientific issues such as the basic rule of global change, the influence of human activities on global change, the effects and adaptation of climate change and earth system model	The fact, process and mechanism of global climate change, the influence of human activities on global change, the impact and adaptation of climate change, earth system model

Data source: The website of the Ministry of Science and Technology of China.

The Major National Scientific Research Plan has exerted an important effect on the scientific field. For the past few years, it has made prominent research achievements, which is of vital significance to improving the independent innovation ability of the national science and technology. However, from the perspective of its objectives and the evaluation system, what it focuses on is cutting-edge technological innovation. It is driven by science instead of innovation. Therefore, enterprises have not been fully engaged in this process.

8.2.4 State Key Laboratory Program

In 1984, in order to accelerate the socialist modernization process of China, revolve around strategic objectives of national development, face the international competition and enhance scientific reserves and original innovation capacity, the former State Planning Commission joined hands with other departments in activating the State Key Laboratory Program. Through many years' efforts, by the end of 2013, China had established 265 State Key Laboratories on the strength of the systems of Chinese Academy of Sciences, institutions of higher education and enterprises (Table 8-15). Interim procedures for the construction and management of State Key Laboratories specifies that a State Key Laboratory is an indispensable component of the national innovation system and a significant base for our country to organize high-level basic research and applied research, gather and foster outstanding scientists and conduct academic exchanges.

Table 8-15 Operation of State Key Laboratory

Year	Number of laboratories	Research funding raised (billion RMB)	Expenditure on R&D (thousand RMB)	Number of R&D projects	Funding for R&D projects (thousand RMB)
2002	164	2.157	195.4	11,170	186.0
2003	162	2.648	219.2	12,509	243.3

Continued

Year	Number of laboratories	Research funding raised (billion RMB)	Expenditure on R&D (thousand RMB)	Number of R&D projects	Funding for R&D projects (thousand RMB)
2004	158	2.997	207.3	12,030	286.0
2005	179	3.728	331.9	12,965	327.4
2006	195	8.394	655.3	23,309	499.1
2007	192	5.830	413.8	17,440	603.0
2008	218	12.146	567.9	21,860	820.0
2009	218	9.751	716.9	16,165	656.6
2010	220	12.862	1,478.6	26,675	1,022.1
2011	266	14.174	972.2	28,264	1,292.6
2012	266	17.114	1,203.5	35,619	1,603.8
2013	265	16.488	1,163.0	39,529	2,150.8

Data source: *China Statistical Yearbook on Science and Technology* (2003–2014); Bian , 2011.
Note: The number of laboratories is the accumulated number (laboratories in operation during that year).

The State Key Laboratories have a clear mission: serving as important research institutes, which in turn serve national strategic objectives, engage in the core work of original innovation, undertake cutting-edge basic research and conduct the transfer of high-tech. In respect of the research positioning and the orientation of State Key Laboratories with 220 laboratories in operation in 2010, 20% of them focused on basic research, and work undertaken by this kind of laboratories belonged to explorative basic scientific research; 50% of them focused on applied basic research and this kind of laboratories attached vital importance to the trend of international scientific development and addressed significant scientific issues in national economic and social development. Besides, about 30% of the laboratories focused on engineering basic research, which oriented towards major needs of national economic and social development, directly undertook research tasks in major projects or programs or issues of related scientific principle (Bian, 2011). This distribution of resources was relatively rational and it also integrated original scientific innovation and applied science research which might influence industrial technological innovation.

Relying on enterprises and institutes with transformed ownership to build State Key Laboratories is an important measure made by the Ministry of Science and Technology during the 11th FYP and also a feasible way to enhance the independent innovation ability of enterprises. At present, the Ministry of Science and Technology has approved the construction of 96 enterprise-State Key Laboratories. In 2007, it approved the first 38 enterprise-State Key Laboratories. In 2009, it approved the second batch of 56 enterprise-State Key Laboratories. In 2010, 2 enterprise-State Key Laboratories were approved. These laboratories were mainly funded through Program 973 in the form of projects and distributed in 8 disciplines, namely, manufacturing, resources, materials, medicine, information, engineering, agriculture and energy. They mainly supported enterprises in conducting research on the applied foundation, cutting-edge technology, common technology, key technology and major technological integration and formulating industry standards. Among them, 45 were constructed relying on enterprises directly under the central government and their subsidiaries. 25 were set up based on local state-owned enterprises. And 26 were built based on private enterprises. State-owned enterprises have raised 3.71 billion RMB to strengthen their innovation ability, on average investing over 80 million RMB in each laboratory.

According to incomplete statistics, by the end of 2011, there had been 7,600 fixed employees in State Key Laboratories. Since 2006, enterprise-State Key Laboratories have hosted and undertaken over 1500 national level subjects including Program 973 and Program 863. They have published over 11,000 high-level research papers, lodged 6,800 applications for patents and inventions and more than 2,100 inventions and patents have been authorized ①.

The Chinese government's move to step up the construction of State Key Laboratories in enterprises is to a certain extent influenced by the innovative practices of the European countries, the United States and Japan, because many world-known enterprises have their labs or research institutes and rely on continuous innovation to lead their development. Establishing enterprise-State Key Laboratories can enhance the endogenous innovation capability of enterprises and industries, which will promote the transformation of scientific and technological achievements, and integrate S&T with economy. Practice shows that enterprise-State Key Laboratories have become an important part of the national innovation system in China and the strong driving force for improving innovation ability of enterprises and industries. But the question now is, the enterprise-State Key Laboratory construction mainly relies on funds raised only by enterprises. National-level support is conducted in the form of projects; therefore, it influences laboratories' focus on long- and medium-term basic research, which leads enterprises' labs to do more imitations than original ones.

① China has established 96 enterprise-State Key Laboratories, China News Service Website, November 25, 2011.

8.3 Summary

In order to improve China's S&T competitiveness, in the past two decades, the Chinese government has formulated many basic research development programs, implemented various financing plans, and promoted significant progress in scientific research. But in China, the big issue is that the total input is increasing but the intensity is low (4% to 6% for a long time), with a big gap with innovation-leading countries. A lower proportion of basic research is unable to support the demand for knowledge in terms of China's rapid S&T and economic development. Furthermore, from the perspective of the allocation of basic research funds, China adopts the "top-down" model and focuses on guiding basic research with free exploration and national planning; therefore, industry-driven basic research is neglected and the core technological innovation capacity of local industries is still weak.

So, strengthening the industry-driven basic research is a feasible solution to the "innovation dilemma". And it can improve the endogenous scientific ability of industry through the transfer and transformation of S&T achievements. Although China has provided more policy supports for enterprises to be the main body of technological innovation, more R&D investment goes to experimental development, rather than basic research and applied research.

Enterprises' involvement in basic research mainly takes three forms: The first is that enterprises strengthen the exploration of basic research based on their own funds and provide the knowledge foundation for the future technological innovation. They usually start by setting up internal laboratories. This is a common feature shared by enterprises in the United States, Germany, Japan and so on. In respect of R&D scale of basic research, large enterprises generally have a relatively complete research system. And the proportion of investment in basic research is higher in small- and medium-sized enterprises. Second, enterprises can outsource basic research tasks to public research institutes and universities or collaborate with external knowledge institutions. As R&D costs rise and scientific fields are further expanded, this model enjoys more popularity among enterprises. Third, enterprises participate in the national S&T plan and make use of external funds to improve the internal scientific capacity in the form of independent R&D and joint efforts. Among Chinese enterprises, most of them adopt the second model, while only a few of them adopt the first and the third ones.

References

Bian, S. Role of national laboratories' research in original innovation. Beijing: Graduate School of Chinese Academy of Sciences, 2011. [卞松保．北京：国家实验室在原始创新中的作用研究．中国科学院研究生院，2011.]

Bush, V. Science: The Endless Frontier. Washington, DC: US Government Printing Office, 1945.

Li, Z., et al. *Basic Research: Source of Comprehensive National Strength*. Beijing: Higher Education Press, 2007. [李主其，等．基础研究：综合国力之源．北京：高等教育出版社，2007．]

OECD. Reviews of innovation policy: China, 2008.

Zhou, J. *Foundation and Source of Innovation: Investment, Evaluation and Coordination of Basic Research*. Beijing: Science Press, 2008. [周寄中．创新的基础和源泉：基础研究的投入、评估和协调．北京：科学出版社，2008．]

Chapter 9

Industry-Driven Basic Research: Biomedicine Industry

Large-scale statistics are not easily obtained publicly, and therefore, many scholars adapt the method of case study in finding out how basic research impacts technological innovation and competitiveness within a certain industry. For example, in 1998 the US Committee for Economic Development (CED) compiled a report named America's Basic Research: Prosperity Through Discovery, which analyzed the great contributions made by basic research towards technological innovation in industries including laser, X-ray, semiconductor, GPS and the Internet. It also gave a detailed elaboration about the basic research carried out by universities (like Harvard University, Columbia University, etc.) and enterprises (like IBM, Xerox, Pfizer, Merck & Co Inc, Proctor & Gamble, etc.) and the contributions of such research in promoting industrial technology and competitiveness. In China, Liu (2007) studied the impact of organic chemistry on the industry of synthetic fuel and the relationship between basic research and technological innovation in General Electric (GE) by means of conducting case studies.

Therefore, the next three chapters (Chapters 9–11) employ the method of "TRACES" in analyzing major events along the course of technological innovation in three industries (biomedicine, TFT-LCD and carbon fiber) with the aim of providing evidence for the history of interaction between technological innovation and basic research in these industries and exploring the pattern of resource input in industry-driven basic research from the perspective of policies.

9.1 Research methods and data sources

A case study is not only the first choice to explain "how" and "why", but also well-suited for research areas, which are new or inadequately explored and which focus on finding explanations and discoveries (Yin, 2003). The approach of a case study can help to capture, trace and manage new phenomenon and problems that emerge from practice. In addition, in-depth dissection of a case enables researchers to better examine questions arising from the research framework. In the case study of each industry, we adopt three approaches, namely, data collection, follow-up study and in-depth interview within the industry of the following three chapters, so as to form a diversified source of information, as a result making the foundation of case study more solid and valid.

Analysis of historical events is an effective form of a case study. We have carried out interviews with experts to collect necessary information. In these in-depth interviews, scientists and technicians who are engaged in R&D in renowned universities and institutions in fields of biomedicine, TFT-LCD and carbon fiber answered questions where there are no definite solutions in public documentation and other questions we are concerned about, which helps to further improve this case study. Main sources of data about resources input into China's basic research in this chapter include: (1) the series of China Statistical Yearbook on Science and Technology from 1991 to 2011 and two national communiqués of R&D resource inventory; (2) database of the National Natural Science Foundation of China (NSFC); (3) list of projects under the National Basic Research Program of China (Program 973) from 1998 to 2012 on the official website of the Ministry of Science and Technology; (4) SCI papers database in Thomson Reuters' ISI Web of Knowledge and Derwent Innovations Index (DII). In the data analysis of projects, papers and patents, key words like "carbon fiber", "LCD", "FPD" and "biotechnology" are used to carry out retrieval within a set period.

9.2 Basic research and technological innovation of the biomedicine industry

Biomedicine mainly refers to biochemical drugs, bioengineering drugs and medical biological products. As innovations in new chemical drugs face mounting challenges, biotechnology drugs have become a crucial source of innovative drugs (DSTSD, 2009). The consultancy firm, Frost & Sullivan pointed out in a 2009 report that the sales of biomedicine will be more than one third of global drugs industry in 2020. For many years, the sales of biotechnology drugs have been keeping a growth rate of more than 15%, two times the sales of all drugs.[①]

The biomedicine industry is about the production of drugs, reagents, medical equipment or related products by employing technologies of recombinant DNA, monoclonal

① Development Plan of Medical Industry during the 12th FYP. Research Report by Chinese Economic Review, http://www.chinacir.com.cn/ztyj/04/index.html.

antibodies, cell culture, bioreactor, protein engineering, cloning, stem cells, bioinformatics, high-throughput screening and so on (Yao, 2010). It is a typical emerging high-tech industry, which relies heavily upon S&T-based innovation, and the key to its development is to achieve breakthroughs in the basic research of life sciences (biotechnology) and translate them into a product of value (Liu, 2008). With the in-depth research now emerging, scientific knowledge needed in this industry is becoming more complex and the cost of investment keeps increasing. It is a considerably long period from the budding of technology to the creation of product in the biomedicine industry, and during the whole process, it also needs enormous technological accumulation (including basic research and creation of patents), and a coalition of global biotechnology and pharmaceutical industries and a large amount of funds are all indispensable. In terms of the organization model, to attract new technologies from universities and laboratories is key to the development of this industry (Oliver, 2004), and the industry-university-research institute cooperation is one of the viable measures (Niosi, 2003).

As biomedicine is an emerging industry, the development of the biomedicine industry is influenced by many factors, including technology, market and governmental regulations. However, this section only looks into the role of basic research in this industry. Research capacities of China and other countries show little difference, but we find, from a comprehensive assessment of assorted information, that the biomedicine industry of China has not displayed the international competitiveness, which is expected from it. Liu (2008) believes one important reason is that despite its steady and expansive network of industry-university-research institute cooperation, it seems that the biomedicine industry of China lacks industrial basic research: The core place of the cooperation is mostly taken by universities and state-run institutions, and even a handful of enterprises that do occupy key positions are transformed from state-run institutions.

In 1928, A. Fleming, a British doctor, accidentally discovered penicillin. Although this was a significant scientific discovery, it took two other British scientists H. Florey and E. Chain 14 years to turn the newly discovered penicillin into an antibiotic ready to be used in clinical medicine. [①] The discovery of penicillin marks the opening of the history of biomedicine industry as well as the arrival of the times of antibiotics. In 1942, C. H. Waddington started epigenetics, which was then studied more systematically by R. Holiday, who pointed out that it was the study of heritable changes in gene expression that does not involve alterations in gene sequence. Epigenetics include the study of dosage compensation of X chromosome, DNA methylation, histone code, genetic memory and so on. In 1942, American scientist S. A. Waksman discovered streptomycin and soon applied it clinically, making streptomycin the second antibiotic used in clinical medicine. Unlike the accidental discovery of penicillin, the discovery of streptomycin was the result of elaborate and systematic long-term research, and it ushered in a golden age for finding antibiotics from microbial metabolism (Yao, 2010). In 1953, American scientist Watson and British scientist Crick raised the theory of the DNA double helix structure. This landmark achievement marked the beginning of the era of basic research of biomedicine, which focuses on molecular biology. In 1959, Japanese scientist Hamao Umezawa not only discovered several antibiotics of clinical value like kanamycin, but also came up with the concept of enzyme inhibitors, thus opening up a new era of finding substances of biological activity in secondary metabolite of microorganism. Till the 1950s, antibiotics, amino acids and enzyme inhibitors had become the key products in the biomedicine industry (Figure 9-1).

① The three of them won the Nobel Prize in Medicine in 1946.

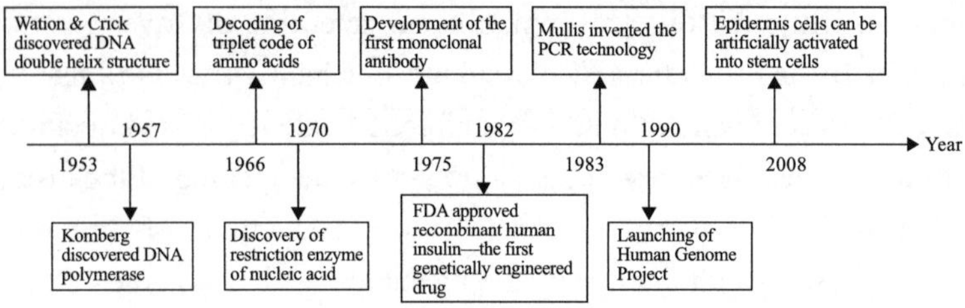

Figure 9-1 Major basic research and findings in the history of modern biotechnological pharmaceutics

The development of the biomedicine industry presented a perfect track for industrial innovation based on basic research. Biopharmaceutical studies based on biology had been conducted sporadically as early as the 18th century. However, basic research and application on a large scale did not take off until the application of insulin in the treatment of diabetes in the 1930s. Before the 20th century, basic research of bioscience was carried out in most cases by individuals, driven by curiosity. In the early and middle 20th century, countries like the United States started to carry out basic research of biology oriented by national demand in universities and research institutes and this gradually became a trend. Prompted by basic science, the industry of biotechnology began to take shape around 1975, and in the ensuing decade breakthroughs were witnessed in both laboratories and the business field. Genentech, the first biotechnology enterprise in the world, was established in April 1976, ① and then enterprises of this kind were set up one by one around the globe. From the perspective of industrial economics, Genentech is the first biotechnology enterprise successfully established and to become commercialized, and the first to be listed. Its biotechnology products took the lead in finding inroads into the market, and it is currently one of the most successful

① In 1976, Robert Swanson (partner of KPCB) and Professor Herbert Boyer (biochemist based in University of California, founder of the technology of DNA recombination and 1976 Nobel laureate) set up Genentech. The establishment of Genentech is a good example of scientists starting a business, which is also a common feature of this industry.

biotechnology enterprises in the world. The founding of Genentech also blazed the trail for the emergence of other partners and promised the glory of biotechnology industry of the United States. This clearly shows that the establishment and progress of one single enterprise can give impetus to a whole industry.

Because early biotechnology enterprises were all start-ups funded on science, they did a lot of industry-driven basic research. They not only contributed a lot to basic research in bioscience, but also gave a tremendous boost to the commercialization of research findings. ① Active involvement in basic research of biology not only helped these enterprises in their formation but also played a crucial part in their future development, because many venture investors in fact set great scores by the potential research capacity of these enterprises. In the fourth year after its establishment, Genentech went public and was traded as "DNA". The Wall Street took no time in recognizing the precursor of basic research in the industry of biotechnology. Since then a pattern was derived in the biomedicine industry, that is, enterprises would use basic research as the evidence of their innovative capability to win over external funding (including government funds and venture capitals) for their future development. The pattern was also present in the IT industry. In 1980, the United States passed the Bayh-Dole Act, which advanced technological transfer and commercialization of universities and research institutes, and promoted the transformation of biotechnology research from basic science to industrialization. During this period in the United States, investors and the government were very supportive of the biotechnology industry. In the meantime, biotechnology enterprises focused their efforts on R&D: Not only did they invest heavily in it, they also employed the best scientists to do the research (DSTSD, 2009). In the last decade of the 20th century, it had become an ironclad law in the biomedicine industry that the enterprises shall be involved in the basic research. Accordingly, many

① In 1971, Cetus (acquired by Chiron in 1991) was founded based on PCR technology. However, it was not the first enterprise to commercialize biotechnology.

significant research findings came from the enterprises. In 1997, the Roslyn Institute and the British biotechnology company PPL Therapeutics cloned Dolly the sheep. In February, 2001, the American firm Celera published the refined mapping of human genome and its preliminary analysis findings. ① On November 25, 2001, the Advanced Cell Technology based in Massachusetts, the United States cultivated early human embryos by cloning, which is a major breakthrough in therapeutic cloning research. In recent years, stem cell has become a new scientific area in biotechnology, and its huge business potential attracts a lot of private research institutes. Among them, American companies Aastrom Bioscience and Geron as well as British company Reneuron take a safe lead in applicable basic research. On January 23, 2009, Geron was given approval from the US Food and Drug Administration (FDA) to carry out trials using embryonic stem cells to treat patients with bone marrow damage.

Generally speaking, the biomedicine industry relies heavily on industry-driven bioscience R&D, so basic research and the technological innovation it brings play very important roles (see Figure 9-2).

① In the same year, a joint public association comprising laboratories of many countries also published the human genome map and the analysis in *Nature*.

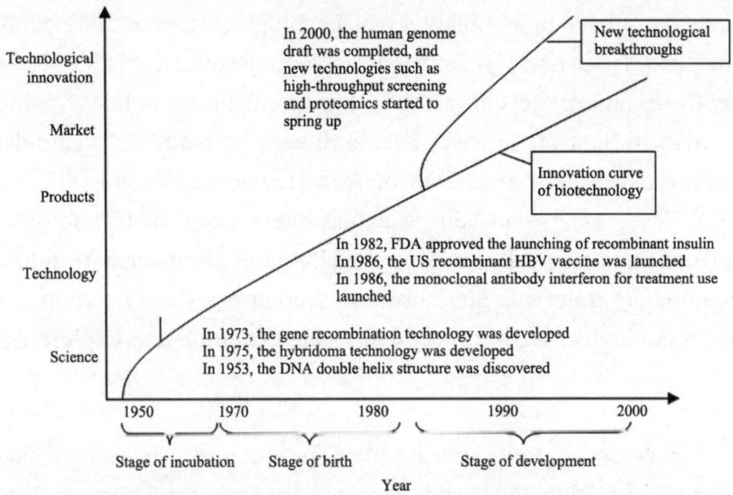

Figure 9-2 The roadmap of development trend of biotechnology

Data source: DSTSD, 2009.

In this industry, it is common for enterprises to actively take part in scientific research, which facilitates the process of turning numerous research achievements into industrial application. Although the value and economic benefits of many major recoveries in basic biomedicine are usually inconspicuous in the short term, huge returns will be reaped once breakthroughs occur in research. For example, in the United States, the industry of biotechnology has already developed for nearly 40 years, and is still not generally profitable. Despite the expectation of achieving returns from investment in 8 to 14 years, research about American biotechnology shows that many years of patience and efforts are entailed to realize the expected returns. The American case makes it clear that there are two crucial ingredients to be successful for any biotechnology enterprise: one being the powerful driving force of scientific technology, the other being the early success of commercialization of the technology. Scientists are the central resource for the biotechnology industry. In the burgeoning stage of this industry, which was the early 1970s to middle 1980s, it was in the laboratories that the fruit of biotechnology first budded and groundbreaking progress was made. What's more, many a biotechnology enterprise, for example Genentech, is set up by scientists from some institute (DSTSD, 2009). Therefore, enterprise operating in the biomedicine industry, which is heavily reliant on science, should always be ready to accumulate dynamic assets that are based on knowledge innovation (DSTSD, 2009). These assets can help to explore new areas for technological application, offer guidance to middle- and long-term trends of technological transition and ensure that companies can keep up with new technologies, lead the technological revolution and explore new markets.

Papers and patents can be used as indicators to assess basic research and product R&D and to analyze the layout and trend of R&D. They are effective indicators in mapping out the law of technological development, reflecting changes in the trends of technological development and assessing technological prowess in related areas of countries and regions. With the view to make clear the global distribution of basic research and technological innovation capacity in the field of biomedicine industry, we made an analysis of the papers and patents using SCI and Derwent Innovations Index, or DII, to find out whether basic research has an influence in the innovation capacity of industrial technology. Since a wide range of knowledge is involved in the industry of biotechnology, to simplify the analysis of papers and patents, we selected 9 key areas that have made major contributions to and hold pivot significance in this industry based on related literature. ①

(1) Table 9-1 shows the top 10 countries (or regions) whose published papers in SCI during 1981 to 2012 are the highest in number. Judged from the statistics, the United States takes the top spots in

① Time of retrieval: June 8, 2012. Range of retrieval: 1981–2012. Because there is an 18-month-long period of review for patents, the numbers shown for each year is the applications of 18 months ago. However, this has no consequence for the result of analysis in this chapter. The same is with the analysis below.

all areas, indicating its absolute predominance in basic research of bioscience. As for China, it makes it into the global top 10 in terms of capacity of scientific research in 8 areas, except monoclonal antibodies. However, when we compare China with the United States, a wide gap is observed (for example, the respective number of papers on vaccines from China is only one tenth of that of the United States). Admittedly, Chinese scientists have a disadvantage in using English language, which surely has an impact on their number of SCI papers.

Table 9-1 Top 10 countries or regions in terms of the number of SCI papers published in key areas of biotechnology during 1981–2012

Item	Ranking									
	1	2	3	4	5	6	7	8	9	10
Vaccine										
Item	US	UK	France	Germany	Australia	China	Canada	Japan	Nether-lands	Italy
Papers	57,547	12,287	7,714	7,006	5,749	5,516	5,449	5,270	4,444	4,436
Percentage	39.98%	8.54%	5.36%	4.87%	3.99%	3.83%	3.79%	3.66%	3.09%	3.08%
Monoclonal antibody										
Item	US	Japan	UK	Germany	France	Italy	Canada	Nether-lands	Australia	Italy
Papers	89,767	24,598	18,924	18,067	14,672	9,821	8,999	8,711	6,078	5,537
Percentage	41.50%	11.37%	8.75%	8.35%	6.78%	4.54%	4.16%	4.03%	2.81%	2.56%
Biochip										
Item	US	Germany	China	France	Chinese Taiwan	Korea	Japan	Italy	Russia	Canada
Papers	441	172	162	144	133	118	88	56	56	51
Percentage	25.32%	9.87%	9.30%	8.27%	7.64%	6.67%	5.05%	3.22%	3.22%	2.93%

Continued

Item	Ranking									
	1	2	3	4	5	6	7	8	9	10
Genomics										
Item	US	Germany	Japan	UK	France	China	Canada	Italy	Spain	Nether-lands
Papers	66,753	14,007	139,956	12,599	10,757	8,556	8,158	6,695	5,024	4,840
Percentage	42.88%	9.00%	8.99%	8.09%	6.91%	5.50%	5.59%	4.3%	4.3%	3.11%
Proteomics										
Item	US	Germany	China	UK	France	Japan	Canada	Italy	Korea	Australia
Papers	5,775	2,731	1,678	1,429	1,216	970	914	914	724	699
Percentage	34.71%	14.00%	8.60%	7.32%	6.23%	4.97%	4.68%	3.71%	3.58%	3.26%
Tissue engineering										
Item	US	China	Germany	UK	Japan	Italy	Canada	Korea	Nether-lands	Switzer-land
Papers	14,188	3,411	2,882	2,526	2,419	1,426	1,366	1,307	1,015	935
Percentage	41.30%	9.93%	8.39%	7.35%	7.04%	4.15%	3.98%	3.80%	3.00%	2.72%

Continued

Item	Ranking									
	1	2	3	4	5	6	7	8	9	10
Stem cell										
Item	US	Germany	Japan	UK	Italy	France	China	Canada	Nether-lands	Spain
Papers	84,245	20,406	18,900	14,675	11,883	11,632	11,318	9,116	6,398	5,578
Percentage	39.80%	9.64%	8.93%	6.93%	5.61%	5.50%	5.35%	4.31%	3.02%	2.64%
Gene therapy										
Item	US	Germany	Japan	UK	China	France	Italy	Canada	Nether-lands	Spain
Papers	55,156	11,144	11,011	9,011	7,598	7,020	6,677	5,338	3,754	3,218
Percentage	44.49%	8.99%	8.88%	7.27%	6.13%	5.66%	5.39%	4.31%	3.03%	2.60%
Recombinant protein										
Item	US	Japan	Germany	UK	China	France	Canada	Italy	Australia	Korea
Papers	49,962	12,336	11,175	8,339	7,858	7,118	5,330	4,162	3,408	3,228
Percentage	42.26%	10.44%	9.45%	7.05%	6.64%	6.02%	4.51%	3.52%	2.88%	2.73%

Note: In the field of monoclonal antibody, the number of SCI papers published by China is 4,820 (accounting for 2.23%), ranking 12th after Sweden. The percentage of the papers of a certain country/region accounting for the globe total is shown in the "Percentage" column.

(2) As for the funding agency for scientific papers, Table 9-2 shows the top 10 institutes or enterprises that funded the largest number of SCI papers related to key biotechnology areas during 1981 to 2012. According to the figure, the US NIH and the NSFC are typical institutes that funded many papers, while their corporate counterparts are mainly big pharmaceutical enterprises (like GSK) or professional biotechnology enterprises (like Roche). With basic research of bioscience funded by national technological plans, the United States still enjoys clear superiority. In the meantime, the statuses of China and the EU are also rising rapidly. Enterprises that are involved with basic research of biology are for the most part from the United States, which indicates that it is still the leading country in combining basic research with industrial innovation in the sphere of biomedicine. The reason behind this is that the United States not only boasts a nation-oriented basic research input system, but also has a dynamic enterprise-funded basic research system.

Table 9-2 Major funding institutes or enterprises that published most SCI papers in key biotechnology areas during 1981–2012

Area	Funding institute or enterprise	Country/ Region	Number of papers
Vaccine	National Institutes of Health	US	2,910
	National Natural Science Foundation of China	China	647
	Bill and Melinda Gates Foundation	US	435
	Wellcome Trust	UK	416
	National Institute of Allergy and Infectious Diseases	US	403
	GlaxoSmithKline	UK	373
	EU	—	502
	Sanofi Pasteur	France	249
	Centers for Disease Control and Prevention	US	235

Continued

Area	Funding institute or enterprise	Country/ Region	Number of papers
Monoclonal antibody	National Institutes of Health	US	2,296
	National Natural Science Foundation of China	China	558
	National Cancer Institute	US	281
	German Research Foundation	Germany	216
	Ministry of Education, Culture, Sports, Science and Technology	Japan	213
	EU	—	188
	Roche	Switzerland	166
	Novartis	Switzerland	164
	Pfizer	US	155
Biochip	National Science Foundation	US	16
	National Natural Science Foundation of China	China	15
	The French National Center for Scientific Research	France	10
	Ministry of Education, Culture, Sports, Science and Technology	Japan	10
	Science Council of Taiwan	Chinese Taiwan	8
	National Institutes of Health	US	7
	Russian Foundation for Basic Research	Russia	7
	Science Foundation Ireland	Ireland	7
	EU	—	6
Genomics	National Institutes of Health	US	4,408
	National Science Foundation	US	1,487
	National Natural Science Foundation of China	China	1,314
	EU	—	770
	Wellcome Trust	UK	563
	National Cancer Institute	US	502

Continued

Area	Funding institute or enterprise	Country/ Region	Number of papers
Genomics	German Research Institute	Germany	399
Proteomics	National Institutes of Health	US	888
	EU	—	362
	National Natural Science Foundation of China	China	307
	National Science Foundation	US	264
	German Research Institute	Germany	178
	Wellcome Trust	UK	92
Tissue engineering	National Institutes of Health	US	1,627
	National Science Foundation	US	715
	National Natural Science Foundation of China	China	645
	EU	—	324
	American Heart Association	US	128
Stem cell	National Institutes of Health	US	4,152
	American Heart Association	US	3,781
	National Natural Science Foundation of China	China	1,972
	EU	—	1,060
	National Cancer Institute	US	693
	German Research Institute	Germany	687
	National Science Foundation	US	636
	Ministry of Education, Culture, Sports, Science and Technology	Japan	598
	Wellcome Trust	UK	565
Gene therapy	National Institutes of Health	US	2,401
	American Heart Association	US	2,044
	National Natural Science Foundation of China	China	1,323
	EU	—	867
	National Cancer Institute	US	624

Continued

Area	Funding institute or enterprise	Country/ Region	Number of papers
Gene therapy	German Research Institute	Germany	371
	Ministry of Education, Culture, Sports, Science and Technology	Japan	331
	Wellcome Trust	UK	260

(3) In terms of industrial patent application, the statistics from the OECD (Table 9-3) show that in the period of 2002–2009, countries that filed most PCT patents in the field of biotechnology are still those that do well in industrial basic research and lead global innovations, like the United States and Japan; China's performance was not stable, and its patent applications were fewer.

Table 9-3 Distribution of PCT patents filed during 2000–2011

Country/ Region	2000	2004	2005	2006	2007	2008	2009	2010	2011
US	6,111	4,480	4,479	4,631	4,444	4,095	3,856	4,227	4,297
Japan	907	1,436	1,465	1,210	1,116	1,075	1,063	1,226	1,167
Germany	1,007	897	661	660	876	672	616	688	665
France	426	408	315	365	421	446	488	468	540
UK	604	442	492	429	443	408	359	384	386
Korea	148	208	200	274	309	303	356	458	511
EU	3,031	2,816	2,567	2,617	3,072	2,841	2,593	—	—
OECD members	11,042	9,767	9,602	9,569	10,089	9,220	8,605	10,358	10,544
China	932	138	127	184	195	206	290	459	552

Data source: OECD, 2013.

Judging from the global trends of patent applications related to biotechnology (Table 9-4), during 1991–2012, areas such as vaccines, monoclonal antibodies and gene therapy remained the focus of corporate technological innovation, and at the same time, tissue engineering and biochips experienced rapid development.

Table 9-4 Global trend of patent applications related to key technologies of bioindustry during 1991–2012

Year	Vaccine	Monoclonal antibody	Recombinant protein	Gene therapy	Stem cell	Tissue engineering	Proteomics	Genetics	Biochip
1991	394	684	418	52	50	5	0	83	0
1992	448	566	467	72	54	12	0	92	0
1993	430	485	429	113	82	4	0	114	0
1994	405	515	442	163	71	11	0	134	1
1995	477	401	459	263	124	10	0	116	0
1996	523	354	500	380	118	13	0	159	0
1997	597	369	657	572	190	19	0	164	2
1998	720	409	1,006	667	206	15	1	296	0
1999	937	763	1,730	1,061	315	67	2	755	11
2000	1,376	1,259	2,743	1,726	471	123	7	1,291	32
2001	1,681	1,383	3,292	2,753	595	166	34	1,552	74
2002	1,803	1,828	4,093	3,789	778	205	115	2,070	110
2003	2,235	3,240	4,198	5,596	1,258	410	118	2,127	259
2004	2,098	2,309	2,919	4,032	1,189	340	131	1,391	415
2005	1,708	1,455	1,930	2,640	1,229	279	102	921	520
2006	1,743	1,454	1,937	1,933	1,149	291	78	882	360
2007	1,829	1,374	1,766	1,672	1,401	283	96	855	428
2008	2,108	1,547	1,924	1,795	1,943	403	94	842	433
2009	2,751	2,019	2,599	2,160	2,316	490	98	934	516
2010	2,165	1,663	2,106	1,818	1,941	407	82	738	445

Continued

Year	Vaccine	Monoclonal antibody	Recombinant protein	Gene therapy	Stem cell	Tissue engineering	Proteomics	Genetics	Biochip
2011	2,342	1,660	2,581	1,780	2,255	398	75	959	417
2012	283	215	253	225	359	55	5	100	31
Total	29,053	25,952	38,449	35,262	18,094	4,006	1,038	16,575	4,054

Data source: Derwent Innovations Index. The data of 2012 ware collected before June 8, 2012.

We further searched the patents of subdivided key technologies of global bioindustry starting from 1963 ① with "patent applicant" as the key words, as shown by Table 9-5. As it can be seen, most patents are applied by American, Japanese and Swiss enterprises. As for China, apart from two Shanghai-based gene-developing enterprises with good performance in gene therapy, there are only Zhejiang University and Tsinghua University that hold dozens of patents in the respective fields of biochip and tissue engineering.

Table 9-5 Major enterprises or institutes with most patent applications in key areas of biotechnology accumulated during 1963–2012

Area	Fund-receiving institute or enterprise	Country/individual	Number of patent applications
Vaccine	Smithkline Beecham Corp	US	924
	Department of Health and Human Services	US	841
	Novartis AG	Switzerland	401
	GlaxoSmithKline ②	US	302
	University of California	US	267

① The reason for choosing 1963 as the starting point is that the Derwent Innovations Index only includes information of patents after 1963. The same is with the chart below.

② In December, 2000, SmithKline Beecham and Glaxo Wellcome merged to become GSK.

Continued

Area	Fund-receiving institute or enterprise	Country/ individual	Number of patent applications
Vaccine	Pasteur Institute	France	236
	Corixa Corp	US	219
	Merck & Co Inc	US	175
	Akzo Nobel	Netherlands	174
	Chiron Corp	US	162
Monoclonal antibody	Genentech Inc	US	2,141
	Department of Health and Human Services	US	407
	University of California	US	300
	Incyte	US	272
	Curagen	US	208
	Zymogenetics	US	196
	Amgen Inc	US	154
	Hoffmann-La Roche AG	Switzerland	149
Biochip	Seiko-Epson Corp	Japan	1,546
	Toppan Printing Co., Ltd	Japan	89
	Samsung Electronics Co., Ltd	Korea	49
	Commissariat Energie Atomique (CEA)	France	44
	Yokogawa Electric Corp	Japan	40
	Sumitomo Electric	Japan	38
	Zhejiang University	China	38
	Canon	Japan	36
	Philips	Netherlands	35
Genetics	Genentech Inc	US	461
	University of California	US	252
	DuPont	US	211
	Epigenomics AG	Germany	159
	Genaissance	US	148

Continued

Area	Fund-receiving institute or enterprise	Country/ individual	Number of patent applications
Genetics	Smithkline Beecham Corp	US	147
	Incyte	US	136
	Department of Health and Human Services	US	133
	Affymetrix Inc	US	123
	Pioneer Hi-Bred Int Inc	US	123
Proteomics	Incyte	US	59
	Yue H	Individual	41
	Baughn M R	Individual	36
	Yao M G	Individual	34
	Gandhi A R	Individual	32
	Tang Y T	Individual	30
	Chawla N K	Individual	28
	Lal P G	Individual	28
	Lu D A M	Individual	28
	Nguyen D B	Individual	27
Tissue engineering	Genentech Inc	US	82
	Massachusetts Institute of Technology	US	69
	Ethicon	US	54
	Boston Scientific	US	78
	University of California	US	39
	Sirna Therapeutics	US	33
	University of Michigan	US	33
	Tsinghua University	China	26
	The University of Texas	US	26
Stem cell	Monsanto Technology	US	858
	University of California	US	215
	Stine Seed Farm Inc	Germany	208

Continued

Area	Fund-receiving institute or enterprise	Country/ individual	Number of patent applications
Stem cell	Gen Hospital Corp	US	143
	Massachusetts Institute of Technology	US	115
	Olympus Corp	Japan	106
	Deltagen	US	101
	Kyoto University	Japan	99
	Stanford University	US	98
	Eby W H	Individual	97
Gene therapy	Genentech Inc	US	1,913
	Human Genomes Sciences, Inc	US	671
	Millennium Pharmaceuticals	US	565
	University of California	US	502
	Shanghai Bode	China	399
	Department of Health and Human Services	US	397
	Bayer Group	Germany	381
	Incyte	US	349
	ISIS	US	348
	Shanghai BIOWINDOW	China	320

After the analysis, we can see that the United States holds absolute advantages in basic research and industrial innovation of the biomedicine industry: Most of the top 10 professional biomedicine enterprises are located in the United States, which is also the major market of biomedicine industry. Statistic results of the data make it clear that the importance that enterprises attach to basic research is crucial. According to the annual report of the biotechnology industry issued by the consultancy Ernst & Young on June 14, 2011, as of the end of 2010, there were in total about 4,700 biotechnology enterprises around the world, among them 622 were listed (most of these enterprises are based in the United States, Canada, Europe and Australia). The total revenue of these listed enterprises in 2010 was 84.6 billion USD, while the input into R&D reached a staggering amount of 22.8 billion USD. ① Highly-intensive R&D investment is the basis of unceasing innovation in the biomedicine industry in countries like the United States. In China, however, its strong scientific ability and weak capacity in industrial innovation pose a stark contrast: Papers are mainly from universities and institutes, while enterprises have limited capacity in patent application. Besides, data revealed that the National Natural Science Foundation of China plays an important role in basic research of biotechnology of China. In fact, the Chinese government always attaches great importance to the research of bioscience. Apart from the Natural Science Foundation and the Program 973, three out of six of investment areas in the National Key Scientific Research Plan which took effect since 2006 are related to biology (namely, studies on protein, development and reproduction, and stem cell).

① The data is cited from the Program for the Development of Biotechnology during the 12th FYP.

9.3 Basic research of the biomedicine industry in different countries

Advanced technology is a feature of the biomedicine industry, where high risk and huge profit exist side by side. How to guide biotechnology enterprises and pharmaceutical enterprises to improve their ability of developing new products through basic research and how to apply the findings of basic research conducted by public-funded institutes and universities to the development of new drugs by pharmaceutical enterprises have always been a headache for this industry. Therefore, not only should enterprises invest more heavily in R&D, the government should also give funding to industry-oriented basic research. This resource input pattern, namely the collaboration among government, industry and venture capital, is more evident in the development of the biomedicine industry of developed countries.

9.3.1 Financial support from the government is one of the most important factors in the development of biotechnology industry

After examining the policies of the United States, Japan and some European countries, we can see that increased input of technology and capital remains the focus of policy toward this industry. Governments of many countries deem biotechnology as the hope of the national economy in the 21st century. In terms of the national policy toward R&D, the US Government takes biotechnology as the second most important industry, only after IT. In 2009, the US National Research Council issued a report named A New Biology for the 21st Century: Ensuring the United States Leads the Coming Biology Revolution, urging the country into action of quickening the development of "a new biology". The US National Institutes of Health (NIH) stated that the budget for basic research of biotechnology will double in the next five years to 23 billion USD, and governments of over 40 states issued new policies to encourage the realization of scientific studies' commercial value. After the promulgation of the *American Recovery and Reinvestment Act* (ARRA) in January 2009, investment into basic research of new drugs got a huge boost. Out of 10.4 billion USD of NIH investment, 310 million USD went into cutting-edge scientific research of biomedicine, which was a significant signal to stimulate the growth of the biotechnology industry sent by the US government during the financial crisis. In May 2012, the United States issued the National Bio-economy Blueprint, in which biology was viewed as the main engine of technological innovation and economic development of the United States, proposing to intensify investment toward biological research, push the translation of biological findings in the laboratory to commercialization, and promote the sound development of cooperative and competitive relationship between public and private sectors. ①

① www.bioon.com/trends/news/522974. shtml, 2012-05-14.

Some European countries, including France, Ireland, Finland and Denmark, earmarked funds to support the biotechnology industry. In 2010, the British Biotechnology and Biological Sciences Research Council issued the Age of Bioscience: Strategic Plan 2010–2015, taking advanced bioscience and biotechnology as prior areas for support. In May of 2012, the Research Council of Norway issued a national biotechnology development plan named "BIOTEK21" (2012–2021). The main goals were to stimulate the innovative capacity of the industry, promote practical application of biotechnology knowledge in trade, industry and society through intensive basic research on biotechnology. The plan was aimed at trade and industrial sectors, independent research institutes and universities in Norway, with funding for all investment projects amounting to 10 million krones.

As for Asian countries, on December 26, 2002, Japan set the strategy of "building the country through the biotechnology industry" as a new national target. The plan was to invest heavily in the biotechnology industry and try to make it a pillar of national economy, following the track of the automobile and IT industries. In December 2004, Japan held the Strategic Conference of Biotechnology. During the conference, the strategic plan of Japanese biotechnology development was introduced and decisions were made to significantly increase national investment in R&D of biotechnology, the allocated amount of which was 440 billion yen. In 2006, governmental expenditure on research of biotechnology reached 880 billion yen, which was mainly used to consolidate the foundation of biotechnology and educate relevant talents, with the aim to make Japan's biotechnology world-class. At the same time, Japan eased the regulation on the review of new drugs (the review period was cut to 3 years, while in 2002 the required period was 5 to 7 years), so as to quicken the pace of industrialization of biotechnology. With the strong support from the government, the respective numbers of Japan's published papers and patent applications related to the biotechnology industry are now in top-ranking positions in the world, and Japan's biotechnology market is rapidly expanding. The S&T departments of

Korea mapped out a national plan, Bio-Vision 2016 (2006–2016), to give guidance and impetus to the development of its biotechnology industry. The Singaporean government plans to invest 1.7 billion USD within the next 5 years into the construction of the industry. This move caused many western biotechnology and biomedicine enterprises to re-arrange their R&D and production plans. In 2007, India issued its strategy of biotechnological development and claimed to commit itself to quadruple the investment into this industry within 5 years. ①

The plan for biomedicine innovation of each country is shown in Table 9-6.

① Ministry of Science and Technology of China, the Program for the Development of Biotechnology during the 12th FYP.

Table 9-6 Part of technology and development plans regarding the biomedicine industry in some countries

Country	Plan	Department in charge (Time of initiation)	Major scientific goals related to the biomedicine industry	Highlights related to the biomedicine industry
US	Human microbiome project (HMP)	National Institutes of Health (December 19, 2007)	Determining whether common core human microbiome is present in different individuals; deepening the understanding of the relationship between changes in human microbiome and human health; developing new technologies and bioinformatic tools to realize said goals	Developing a reference set of microbial genome sequences and performing preliminary characterization of the human microbiome; exploring the relationship between disease and changes in the human microbiome; developing new technologies and tools for computational analysis and establishing a resource analysis and coordination center
	Epigenomics Project (EP)	National Institutes of Health (January, 2008)	Coordinating and developing a series of reference epigenomes; assessing and exploring epigenomic mechanisms underlying conditions of development or aging, or responses to exposures (behavioral, social and stressful factors); developing new tools to analyze in vivo imaging of epigenetic changes in cells and tissues	Developing epigenetic technology; developing affiliated egigenetic technology; setting up reference epigenome mapping centers; discovering novel epigenetic marks in mammalian cells; discovering new ways of activating, silencing and stabilizing the gene expression of eukaryote

Continued

Country	Plan	Department in charge (Time of initiation)	Major scientific goals related to the biomedicine industry	Highlights related to the biomedicine industry
US	Multidisciplinary Project (MDP)	National Institutes of Health (January, 2009)	Combining the advantages of different disciplines to solve new problems which are beyond the solution of current biotechnology and medicine	Interdisciplinary research of neuropsychiatric phenomics and geroscience; research on neurotherapy; invention of drugs based on genomics, etc.
	Tissue Science & Engineering plan (TSEP)	National Science and Technology Council (NSTC) (June, 2007)	Understanding and controlling the cellular response; formulating biomaterial scaffolds and the tissue matrix environment; developing enabling tools; promoting scale-up, translation and commercialization	Understanding the cellular machinery; identifying, validating biomarkers and assays; advancing imaging technologies; assembling and maintaining complex tissue; defining cell/environment interactions; improving tissue preservation and storage
	NIH Roadmap	National Institutes of Health (2004)	Carrying out initiatives that have profound significance for research on biomedicine or behavioral science and whose findings can realize human health missions of all NIH institutes and centers	5 main roadmap programs including microbiome, protein capture/proteome tools, phenotype service and tools, inflammation and epigenomics

Continued

Country	Plan	Department in charge (Time of initiation)	Major scientific goals related to the biomedicine industry	Highlights related to the biomedicine industry
US	Nanotechnology in cancer and oncology	National Institutes of Health (2007)	Combining nanotechnology, cancer and research on oncology with molecular biomedicine to improve treatment	Prevention and control of cancers and tumors; early diagnosis and proteomics; imaging diagnosis; multifunctional treatment devices, etc.
	2006–2011 Strategic Plan: New Frontiers in Environmental Sciences and Human Health	National Institute of Environmental Health Sciences (2005)	Expanding the role of clinical research in environmental health sciences; using environmental toxicants to understand basic mechanisms in human biology; developing sensitive markers of environmental exposure, early (preclinical) biological responses, genetic susceptibility, etc.	Not clear
	Emerging Frontiers in Research and Innovation (EFRI)	National Science Foundation (2007)	Funding interdisciplinary programs with high risks; promoting basic understanding and controllability of scientists toward biology and manufacturing; bringing about more research areas and new industries	The first 9 programs related to biomedicine, which get funded include: reconnection of signals between biochips and bacteria cells; physical microenvironment regulating the differentiation of stem cells; biological and chemical research on transfer, proliferation and differentiation of cells, etc.

Continued

Country	Plan	Department in charge (Time of initiation)	Major scientific goals related to the biomedicine industry	Highlights related to the biomedicine industry
UK	BBSRC Delivery Plan (2008–2011)	Biotechnology and Biological Sciences Research Council (2007)	Promoting UK's capacity in research on basic systems biology and predictive biology; carrying out cooperative and complementary research through the "strategic platform of new technologies"	Study on aging; research on global health threats; nanotechnology and its application in molecular biology; response mode
	UK Biobank	Medical Research Council and Wellcome Trust (February, 2006)	Integrating research resources; identifying the relationship between environment & genetic factor and diseases; understanding the interaction between human genome, life style and the environment	Recruiting 500,000 people to take part in a follow-up survey spanning 10 years to collect genetic and environmental data of these participants for further study
France	Research and Innovation of Biotechnology of France (2007)	National Agency for Research (ANR) (February, 2007)	Strengthening cooperation between biotechnology enterprises and institutes; facilitating the translation of biotechnology from basic research to industrialization	Development and optimization of new therapeutic products; development and optimization of new vaccines and informative immunotherapy; early-stage clinical research on new biomedicine; products and tools of health diagnosis; promoting the optimization of biological processing and industrialization of biotechnological tools during the biomolecule and biomedicine production

Continued

Country	Plan	Department in charge (Time of initiation)	Major scientific goals related to the biomedicine industry	Highlights related to the biomedicine industry
France	Vision 2020—CNRS 12 Goals: Life Sciences	French National Center for Scientific Research (CNRS) (June, 2007)	Enhancing interdisciplinary research and key research areas; upgrading/improving basic facilities of research	The origin of planet and life; cognition, sociality and brain science (research about the development, structure and functions of neural network; research about the emergence and differentiation of cognitive and social abilities; research on dysfunction; nanoscience and nanotechnology)
	ANR Medical and Biological Plan (2006–2008)	National Agency for Research (ANR) (2006)	Increasing the output of knowledge to meet demands of society; enhancing corporate cooperation on research; enhancing research cooperation between the EU and international communities	Pathological and biological research on human diseases; systems biology; research and innovation of biotechnology, etc.
	2007–2010 Plan of French Biomedicine Agency	French Biomedicine Agency	Continuing the efforts towards the improvement of transplantable entryway; developing and ensuring the therapeutic effect and security of activities and practices; following moral principles when conducting research on embryos	Facilitating research on embryo and embryonic cells; focusing on research of adult stem cells; promoting cell therapy, etc.

Continued

Country	Plan	Department in charge (Time of initiation)	Major scientific goals related to the biomedicine industry	Highlights related to the biomedicine industry
Germany	High Technology Development Strategy of Germany	German Federal Government (August, 2006)	Systematically building the bridge between life sciences and economy; further shortening the innovation chain featured with industry-university-research institute cooperation and the R&D cycle from "concepts-research-pilot production test" to the final marketable products	Innovating medical technologies; strengthening research on systems biology and existing research groups in fields of biology and medicine
	German Drugs Project (2007–2011)	German Federal Government (2007)	Enhancing the relationship between stakeholders in the biopharmaceutical industry and encouraging the creation of new therapies	Intensifying basic research on the pathogenesis of diseases; integrating various forces along the value chain of the biopharmaceutical industry; cultivating active substances; developing innovative biomarkers
Japan	3rd Phase Basic Plan for Scientific technologies	Ministry of Education, Culture, Sports, Science and Technology (2005)	Curing diseases that afflict Japanese people; promoting scientific technologies to gain superiority in the global competition of the biomedicine industry	Clinically translational basic research and clinical research; new target therapy to treat cancer; new technologies to cure new diseases and secondary conditions, etc.

Continued

Country	Plan	Department in charge (Time of initiation)	Major scientific goals related to the biomedicine industry	Highlights related to the biomedicine industry
Japan	MECSST Life Sciences Research Plan (2007)	Ministry of Education, Culture, Sports, Science and Technology (January, 2007)	Improving the application of life sciences research findings in practices; promoting the translation of basic research findings related to cancer; improving comprehensive data bases and genetic resources; stepping up R&D of biotechnology in universities and institutes	Life process reproducing technology; clinically oriented research and transitional clinical research; new therapies treating cancer, including target therapy; improving the basis of life sciences of the highest level in the world, etc.
Korea	1st Five-Year Comprehensive Plan for the Development of Medicine	Health and Welfare Department (December 21, 2005)	Raising the international competitiveness of Korea in the area of modern pharmaceutical industry through the scientification, industrialization and internationalization of its drugs	Organizing industrial clusters of Korean medicine with high added value; formulating standards for medicated preparations; improving the R&D system of therapeutic technologies of Korean medicine
	Conception of 2025	Science & Technology Department (June, 2000)	Striving to make Korea an important research center in the Asia-Pacific region for life sciences and biotechnology	Developing scientific technologies that are suitable for life sciences and medicine of Korea; continuing research on brain; developing core technologies in preparation for the arrival of an aging society in 2010, etc.

Continued

Country	Plan	Department in charge (Time of initiation)	Major scientific goals related to the biomedicine industry	Highlights related to the biomedicine industry
Korea	Basic Plan for the Incubation of Bioengineering	Science & Technology Department (1994)	Concentrating R&D on areas with international competitiveness; acquiring core basic technologies; enhancing industrial competitiveness	Most programs are related to areas such as genes, systems biology, bioinformatics, new genetic drugs and utilization of biodiversity, etc.
	21 Century Frontier Research and Development Plan	The Korean government	Enhancing basic research on areas of human health, agriculture and microorganism and capacity of technological development of Korea	Research on human genome; technologies to utilize proteomics; applied research on cells; research on cerebral functions and diseases; development of substances regulating the functions of the human body; applied technology of genetic materials of microorganisms, etc.
India	National Strategic Plan of Biotechnology Development	Department of Biotechnology (DBT) (November, 2007)	Transforming life sciences knowledge into ecologically harmonious and competitive products needed by society in future biological economy	Improving the framework of biotechnology management; promoting inter-departmental cooperation; pushing the development of bioindustry, etc.

Continued

Country	Plan	Department in charge (Time of initiation)	Major scientific goals related to the biomedicine industry	Highlights related to the biomedicine industry
India	Indian Strategic Plan of Innovative Medicine	DBT and Department of Science and Technology (March, 2005)	Streamlining policies and permission processes related to genetic engineering and medical products; promoting scientific education of biotechnology; establishing a group of biotechnology research centers	Strategic plan of preventive and therapeutic biotechnology; strategic plan of regeneration, genomics and medicine; strategic plan of diagnostic biotechnology

Data source: DSTSD, 2009.

9.3.2 Intensified basic research oriented by industrial innovation is one of the contributing elements toward the development of the biomedicine industry

The United States is a master in combining basic research on biology with innovation of industrial technology. The success of its biotechnological industry is the fruit of multiple factors, including proactive national policies toward technology, a sound layout of public-funded basic research, enterprise-initiated basic research and abundant venture capitals. American biomedicine enterprises have always been at the global front line of searching for novel therapies and treatments, and they have the most intensified basic research among all American enterprises. During the past decade, these enterprises shouldered 19% of research input in the pharmaceuticals industry of the United States, and their investment in basic research was much heavier than the global average and that of the NIH. ① According to a survey conducted by Burrill & Company, in 2010, American biomedicine enterprises invested 67.4 billion USD in basic research on the R&D of new drugs and vaccines, with an increase of 1.5 billion USD from 2009. In January, 2010, *JAMA* magazine published the findings of the largest quantitative study about R&D in the American biomedicine industry by far, which revealed that the governments and private enterprises invested 100 billion USD in basic research on biotechnology every year. Among them, the enterprises contributed the largest share (58% of national biomedicine R&D input in 2007), followed by the federal government (33%). Biomedicine had gradually become the industry where basic research is the most centralized in the United States. This industry holds the single largest share of R&D in all American enterprises, accounting for nearly 20% of domestic R&D funds of all American enterprises.

---

① www.phrma.org.

France has the largest market for biopharmaceuticals in Europe, and is also the global stronghold for innovation in biotechnology. Highly capable researchers and engineers and heavy investment in basic research are the features of the biomedicine industry of France. In 2008, 22 thousand experts participated in the basic research on biomedicine in France, and 4.9 billion euros were invested in this field, which is why France is called the hotbed for basic research on biomedicine.

In Poland, the biopharmaceutical industry enjoys the highest growth nationwide. Poland has a group of research institutes that specialize in the R&D of biotechnology. They not only provide a good R&D platform for Polish biomedicine enterprises but also supply graduates majoring in biotechnology for them, greatly boosting the capacity for scientific research of biomedicine enterprises. ①

Japan is a newcomer to the field of biomedicine, but it is in a position now only after the United States and Germany. The development pattern of Japan is not unlike that of China or India, which in the early stage followed the procedures of R&D outsourcing—production of generic drugs—R&D through joint ventures—independent R&D of new drugs. These procedures remained unchanged for several decades until the Japanese government started to pay attention to biomedicine. The innovation system of Japan's biomedicine industry is distinct for its feature of "the principle of independent development", which is quite different from the network system developed by the United States. The latter has a large proportion of involvement from enterprises and universities, while the former is too much centered on large enterprises and has less participation from universities. This leads to the fact that universities only devote themselves to basic research, while some small enterprises find it hard to acquire venture capitals, and therefore fall to cooperate with universities. Compared with that of the United States, the biomedicine industry of Japan lacks the mechanism for institute-based commercialized innovation.

① http://info.pharmacy.hc360.com/2012/05/280939399462.shtml.

In 1984, the United States revised the Bayh-Dole Act, which allowed universities and patent inventors to profit from the commercialization of university research programs funded by taxpayers. This gave rise to the establishment of many small biotechnology enterprises. On the contrary, Japanese universities and academic institutions always lack interest in commercializing their findings. During the years from 1980 to 2000, there were 240 newly established biotechnology enterprises in Japan, while in the United States the number was 2,624 (DSTSD, 2009).

Among developing countries, India has a commendable biotechnology industry, which is partly due to its sound basic research. The Indian government values and encourages the participation of enterprises in research on life sciences and biotechnology, and stresses on relevant industrial development. As for now, India has more than 800 enterprises, over 50 R&D laboratories and 20,000 strong professionals working in the biotechnology field, joined by 6,500 postgraduates and 1,500 doctors annually. However, one of the shortfalls of this industry in India is that most enterprises lack the ability to carry out R&D (DSTSD, 2009). There are some biotechnology centers in India located in cities like Bangalore and Hyderabad. In 2009, the Department of Biotechnology of India (DBT) set up the Translational Health Science Technology Institute (THSTI) in the "bottom-up" manner. 30% of DBT budget would be used on public-private cooperative projects; big enterprises would be the main players in the cooperative projects on advanced biotechnology, while support for innovation in small- and medium-sized enterprises would be continued. DBT plans to invest 8 billion to 10 billion rupees into THSTI in the next 5 years. ①

The research institutes related to biomedicine of some countries are shown in Table 9-7.

① http://www.forbeschina.com/review/201105/0009482.shtml.

Table 9-7 National scientific research institutes related to biomedicine of each country

Institute	Country	Time of establishment	Scale (Number of staff)	Achievements in biomedicine research	Major research areas of biomedicine	Positioning	Funds
National Institutes of Health (NIH)	US	1930	27 institutes and centers (18,000)	116 Nobel Prize laureates during 1993–2007	Cancer; neuroscience; brain disorders in children; infectious diseases	Innovation in basic research and broad application of research findings; enhancing economic competitiveness and guarantee returns	25 billion USD per year
Medical Research Council (MRC)	UK	1913	29 institutes and 3 academies (4,232)	5 Nobel Prize laureates during 1936–2002	Human and ethnic study; neuroscience and mental health; immunology and infectious diseases; genetics ; molecular structure and dynamics; cytobiology; development and growth; medical physiology and progression of diseases	Knowledge and research; disseminating knowledge and technologies; improving the quality of life and economic competitiveness of the United Kingdom	500 million GBP per year

Continued

Institute	Country	Time of establishment	Scale (Number of staff)	Achievements in biomedicine research	Major research areas of biomedicine	Positioning	Funds
French National Center for Scientific Research (CNRS)	France	1939	6 scientific ministries and 2 affiliated academies (32,000)	—	—	Conducting and evaluating basic research and basic technological research; facilitating the translation of research findings	3 billion EUR per year
Max Planck Institute (MPG)	Germany	1948	82 institutes and research centers (12,600)	—	Development and evolution biology; immunobiology and biology of infectious disease; cognitive science; microbiology and ecology, etc.	Complementary, emerging and outstanding basic research	1.5 billion EUR per year

Continued

Institute	Country	Time of establishment	Scale (Number of staff)	Achievements in biomedicine research	Major research areas of biomedicine	Positioning	Funds
Chinese Academy of Sciences (CAS)	China	November, 1949	12 sub-academies, 100 affiliated institutes and more than 100 State Key Laboratories and engineering centers (more than 50,000)	Original research achievements including the synthesizing of chloramphenicol, penicillin, artemisinin, salvianolate and antofloxacin hydrochloride, etc.	Genomes of serious human diseases; integrated research on transcriptome and genome; research on the regulation mechanism of gene expression and the origination and treatment of diseases; research on protein interaction networks of serious human diseases	Original scientific innovation; innovation and integration of key technologies; contributing to economic development, national security and sustainable development of society	22.3 billion RMB in 2010
Council of Scientific and Industrial Research (CSIR)	India	1942	40 institutes and 80 research stations (23,000)	—	Frontiers of bioscience	Driving and facilitating the development of scientific and industrial research of India	200 million USD

Data source: DSTSD, 2009.

Because basic research on bioscience involves many disciplines and the knowledge systems become ever more complex, cooperation among enterprises, universities and research institutes becomes a new trend (see Table 9-8). A collaborative research ecosystem is formed by American biotechnology enterprises through cooperation with universities and research institutes on R&D. The capacity of research innovation of these enterprises is rapidly boosted through collaborative conduction of basic research with scholars of public institutes, thus giving the United States a leading edge over other countries in respect to basic research of the biomedicine industry all the time. Usually, biomedicine institutes of each country will set up clinical centers to carry out clinical trials on their findings and thus engage in translational study. For example, NIH of the United States, MRC of the United Kingdom and MPG of Germany have all established clinical research facilities to close the gap between institutes and enterprises and push the basic research findings in the field of biomedicine towards clinical application and commercialization (DSTSD, 2009). Besides, a favorable venture capital system is another important force in driving the development of basic research and innovation of the biomedicine industry. In 1977, Genentech succeeded in synthesizing somatostatin. This huge scientific breakthrough made by an enterprise not only shocked the academia and industry but also attracted venture capital enterprises. As it happened, Amgen is an enterprise set up in 1980 by a group of scientists and venture capital investors. In 1989 and 1991 respectively, it developed technology drugs EPO (Epogen) and G-CSF (Neupogen), which were approved by the US FDA in succession and became two most successful commercialized drugs in the world at that time.

Table 9-8 Interface between national scientific research institutes and universities and enterprises (Biomedicine)

Institute	Affiliated hospitals or development foundation	Patents	Support from small- and medium-sized enterprises	Establishment of spin-off enterprises	Cooperative research with large enterprises	Direct personnel exchange	Universities	Regional cooperation
US National Institutes of Health (NIH)	√	√	√	□	√	×	√	□
UK Medical Research Council (MRC)	√	√	√	√	√	√	√	□
Max Planck Institute (MPG)	√	√	□	√	□	√	□	□
French National Center for Scientific Research (CNRS)	□	√	□	√	√	√	□	√
Chinese Academy of Sciences (CAS)	□	√	□	√	√	√	□	√

Note: " √ " stands for "Obvious operation"; "×" stands for "Strictly prohibited"; " □ " stands for "No apparent sign or further study is needed".
Data source: DSTSD, 2009.

Corporate intervention in basic research is a distinct feature of the biomedicine industry. Biotechnology enterprises fall into two categories: enterprises that invest heavily in R&D and base themselves on research innovation, and enterprises that do not invest much in R&D and base themselves on product innovation. As a result, there are two stages in the process of innovation in the biomedicine industry. The early stage is featured by copious R&D activities and research innovations related to the development of products, with the main aim to obtain patents. The later stage is when the products are commercialized and product innovations are made, which mostly constitutes improving products after they get launched into the market (DSTSD, 2009). In 2010, the global R&D expenditure of the biopharmaceutical industry totaled 67.41 billion USD. Enterprises that ranked among the top 10 who invested the most in R&D had a R&D intensity of above 10% (Table 9-9). These enterprises with prominent basic research capacity are also frontrunners of the biomedicine industry.

Table 9-9 Top 10 enterprises with the most R&D investment in biomedicine during 2009–2011

Enterprise	Location of headquarters	2009 (billion USD)	2010 (billion USD)	2011 (billion USD)
Pfizer	US	7.4 (15.5%)	9.4	9.1 (13.5%)
Roche	US	8.7 (19.4%)	9.2	8.8 (19%)
Merck	US	—	8.12	8.4 (17.5%)
Novartis	US	7.06 (16.7%)	8.08	9.58 (16.3%)
JNJ	US	6.66 (11.3%)	6.84	7.5 (11.5%)
GSK	UK	5.59 (12.8%)	6.09	6.29 (14.2%)
Sanofi-Aventis	France	6.25 (15.3%)	5.94	6.24 (13.5%)
AZN	US	4.23 (13.5%)	5.3	5.5 (16.3%)
Eli Lilly	US	4.13 (19.8%)	4.88	5 (20.6%)
Bristol-Myers Squibb	US	3.48 (16.9%)	3.56	3.8 (17.9%)
Takeda	Japan	4.64 (29.5%)	—	—

Note: Number inside the brackets is the ratio of R&D expenditure to that year's sales revenue.
Data source: www. bioon. com/industry/drug/.

9.4 Basic research in China's biomedicine industry

Biotechnology is an area of great concern for China, and it is regarded as a "window of opportunity" for China to catch up with developed countries (Conlé, 2010). The Chinese government has been paying attention to the innovation and development of this industry for a long time. ① However, the Chinese biomedicine industry still lags far behind that of the developed world in terms of scale, market share and competitiveness of products. It reveals that China is strong in its capacity for doing scientific research but weak in industrial competitiveness. In addition to the constraints posed by regulation from the central government, a weak IPR system and an incomplete system

① The Program 973 (National Key Scientific Research Program included), NSFC and Program 863 all make the point of giving support to life sciences and biotechnology. Development of major new drugs is also one of the 16 major special projects of the Ministry of Science and Technology. Besides, many policies and scientific plans also pay attention to biomedicine.

to commercialize research findings, China also lacks basic research that is oriented by industrial development. The Chinese government has but only a limited support for biotechnology enterprises, and most of the funds are used in the commercialization of research findings; corporate R&D investment is low, while venture capitals are for the most part willing to invest in those enterprises but not in innovations (Liu, 2008).

9.4.1 Innovations and development of China's biomedicine industry

The development of China's biomedicine industry can be roughly divided into three major stages.

(1) Stage of exploring (1960–1995). This stage focused on the R&D of biological and biochemical medicines. The exploration into technology of biomedicine production began, and a number of biopharmaceutical institutes and enterprises were established. Major technological breakthroughs were scored in areas including artemisinin and recombinant human interferon. During this stage, funds often came from national scientific projects such as Program 973, National Science and Technology Research Program and Program 863; enterprises invested little in R&D, and the drugs produced generally imitated foreign products.

(2) Stage of burgeoning (1996–2005). A group of biomedicine industrial bases were established (such as Zhangjiang Hi-Tech Park and Zhongguancun Science Park) following the development pattern of foreign countries, integrating R&D and manufacturing. National scientific plans and special funds for the industrialization of high technology gave sustained and strong support to the advancement of key biomedicine technologies. Management system on medicine was also extended and improved. Prominent features of this stage include steady growth in the R&D input of biomedicine, increased proportion of corporate R&D investment and significant increase of biotechnology patents.

(3) Stage of developing (2006 till now). The National Plan for Long- and Medium-Term Scientific and Technological Development initiated a major national S&T special project entitled "Invention of major new drugs" (with a budget of 6.6 billion RMB). A series of technological and industrial policies aimed at promoting biotechnology R&D and pushing the development of the biomedicine industry were launched in succession, such as the Plan for the Development of Bio-industry during the 11th FYP (2006) and Policies to Accelerate the Development of Biological Industry (2009), in which biomedicine was classified as a key development area and special support was given in areas such as funding, building of basic capacity for innovation and taxation (Liu & Cheng, 2011).

For China, one possible approach to breakthroughs in biomedicine, an industry based on science, is to catch up in basic research on life sciences and translate research findings into valuable products (Liu, 2008). Although the biotechnology industry keeps an annual growth rate of 30%, and Chinese enterprises have successfully developed new biotechnology drugs, such as Gendicine, the first commercialized gene therapy drug in the world, generally speaking, China's biomedicine industry still lacks international competitiveness. In terms of global distribution of varieties of biotechnology drugs, the United States takes up a share of 63%, followed by Europe, which accounts for 25%, and Japan (7%), while China's share is quite small. In recent years, domestic biopharmaceutical enterprises have gradually closed the gap with foreign counterparts not only with respect to basic equipment but especially, in R&D and pilot plan tests and a large number of enterprises with strong technological capacity have sprung up. However, the overall gap between Chinese and foreign biomedicine enterprises are only deepening, and China's biomedicine enterprises have not yet found inroads into the international market.

Compared with other hi-tech industries of China, the biomedicine industry has a weaker export capacity. In 2010, the proportion of value of export delivery under the category of biological and biochemical products manufacturing in the total value of output was raised to 12.38%. But it was still much less than those of electronic and telecommunications devices (54.52%), computers and office equipment (76.57%) and medical devices and instruments (19.30%). In terms of the proportion of export sales volume of new products in the total output value, the sector of biological and biochemical products manufacturing, accounting for 9.67%, was lagging further behind other hi-tech sectors (47.15% for electronic and telecommunications devices, 63.11% for computers and office equipment and 18.70% for medical devices and instruments). China also falls far behind European countries and the United States in respect of industry and market. For a long period, China had been a mere follower in many areas of technologies and products, while countries such as the United States, Germany and Japan formed predominance and patent barrier through excessive patent applications. In 2010, the sales volume of biotechnological products of China was 112.8 billion RMB, accounting for 0.6% of world's total, whereas the share of American products was as high as 79.8% (NBS & MOST, 2011).

One important reason for China's feeble competitiveness in the biomedicine industry is that Chinese enterprises lack a strong innovative capacity of R&D (especially in basic research). Compared with their foreign counterparts, China's biopharmaceutical enterprises are not too late in their start-up. In addition, it can be seen from the previous bibliometric analysis that enterprises do not lag far behind the world frontrunners in terms of biotechnology R&D. However, it remains a challenge to industrialize technological patents of the Chinese universities and institutes. What is more obvious is that China lacks independent intellectual property rights in pharmaceutical biotechnology. According to statistics, in 2010, there were only 77 biopharmaceutical enterprises engaged in R&D activities, accounting for less than 9% of all biotechnology enterprises and just 8.3% of all pharmaceutical enterprises that conducted R&D (whose number was

929). The expenditure on R&D of biopharmaceutical enterprises was 1.328 billion RMB, accounting merely for about 10% of that of the whole pharmaceutical industry (12.263 billion RMB). The number of R&D personnel was even smaller, with 5374 employees, accounting for 9.7%.

Generally speaking, there is a considerable time lag between basic research and industrial development of China's biomedicine industry. In other words, the research of biotechnology in China started quite early and in general keeps up with the latest international developments; on the other hand, China lags behind other countries in terms of technological innovation and industrial development, which is mainly due to the weak overall industrial research capacity. The non-continuity of development results in historical gap. Most of the major multinational enterprises have experienced a long period of development, so they have already accumulated huge amounts of capital and formed a solid foundation for scientific research (for example, Johnson & Johnson, Pfizer and Roche Group were respectively founded in 1886, 1849 and 1896). In comparison, it has barely been two decades since Chinese pharmaceutical enterprises started to pay attention to and invest in the R&D of biotechnology, and the start-ups in the biotechnology industry are also short of long-term accumulation in basic research. As for the patents related to biotechnology, among all applications filed by foreign entities in China, 78% are from enterprises (7% from universities, 6% from institutes); while for those filed by Chinese entities, most are filed by universities and research institutes, only 25% are from Chinese enterprises (31% from universities, 26% from research institutes) (Liu & Cheng, 2011). For a long time, the basic research of China's biomedicine has not been able to go beyond the boundaries of research institutes and universities. Many brilliant research findings are kept within laboratories or the pilot test stage, failing to form industrialized production. Meanwhile, enterprises have no choice but to conduct parallel R&D in order to make independent innovation. This, together with weak R&D capability of domestic enterprises and their vulnerability from the risk of heavy and long-cycle investment, lengthens the period of transforming research findings to marketable products. For example, in the area of monoclonal antibodies, many basic research results can hardly find their way to facilitate production.

The results of the data analysis of papers above show that China experiences rapid development in several key areas of technology, among which biochips, genomics and gene therapy enjoy the highest speed of development, in a large part due to multiple sources of support for the development of biotechnology from the Chinese government in the form of funds, policies, etc. However, the international pace in the basic research on areas such as monoclonal antibodies, recombinant proteins and genomics appears to be slowing down, while China is still witnessing growth in these fields, which shows that the overall development of China's biotechnology is in the initial fast-growing stage, but it also indicates that China has lagged behind European countries and the United States in the stage of technological development. If basic research failed to play its due role in the development of China's biomedicine industry, multinationals might prolong their overwhelming dominance over China through technological advantage and tools like patents and standards that derive from it, confining China to the low value-added link of the international industrial chain. Therefore, the 12th FYP for National Economic and Social Development proposed to consolidate the foundation of technological innovation of the biomedicine industry and give full play to enterprises in technological innovation.

9.4.2 Policies and technological plans promoting basic research on biomedicine

China regards biomedicine as an emerging industry of national strategic importance and is dedicated to becoming a new global innovation and R&D center of life sciences; therefore, related basic research gets strong support from the national finance (Table 9-10). The Chinese government has already been aware of its advantage and now is actively arranging fast track for several approval procedures in order to preserve R&D findings.

In the 11th FYP for National Economic and Social Development released in October 2005, it was stated clearly that the pace will be quickened to develop the bioindustry to achieve new breakthroughs in key technologies and R&D of important products.

On February 9, 2006, the State Council issued the Outline of the National Program for Long- and Medium-Term Scientific and Technological Development (2006–2020), stating that biotechnology was one of the five strategic focal points in frontier fields of technology and scientific development that demanded critical breakthroughs in the next 15 years, and China would give strong support for its development. It also set higher requirements for basic research in fields such as functional genome, proteomics, stem cells and therapeutic cloning, tissue engineering, bio-catalysis, and conversion technologies.

In October, 2006, the Ministry of Science and Technology released the National Program for the Development of Basic Research during the 11th FYP, proposing to improve the coordinated layout and development between various sub-disciplines in the field of biomedicine within the period, including medical genetics, biomedical engineering, immunology, microbiology, virology, etc. The program also urged intensive support for emerging disciplines such as stem cell biology and tissue engineering, as well as enhanced integration between basic research and clinical research of life sciences.

On November 3, 2006, the Ministry of Science and Technology issued the National Program for Scientific and Technological Development during the 11th FYP. In the program, biotechnology and medical technology were classified into the advanced technological trends, which required prior deployment. It proposed to take modern biological high technology as a point of breakthrough and to intensify basic research on areas such as genomics, proteomics, biological nanoscience, manufacturing of vaccines and antibodies and transgenosis.

On April 8, 2007, the State Council released the Program for the Development of Bioindustry during the 11th FYP. In the program, bioindustry was regarded to be a new leading industry in the world economy, after the information industry. The program stated that it was of strategic significance to speed up the development of China's bio-industry and demanded to accelerate the development of biomedicine industry.

In May, 2007, the National Development and Reform Commission (NDRC) released the Program for the Development of High-Technology Industry during the 11th FYP, claiming to prioritize the development of biomedicine industry and give full play to the advantages of biotechnology and focus on the areas such as gene therapeutic drugs, synthetic drugs and biomedicine.

On June 2, 2009, the State Council released Policies to Accelerate the Development of Biological Industry. It ranked biomedicine and biopharmaceuticals as two key fields and demanded to accelerate the large-scale, aggregated and internationalized development of biological industry, so as to quicken the process of transforming it into a pillar industry in the hi-tech field and an emerging industry of national strategic importance.

On October 10, 2010, the State Council released the Decision to Accelerate the Development of Strategic Emerging Industries, listing bioindustry as one of the seven strategic emerging industries and proposed to develop biotechnology drugs, new vaccines and diagnostic reagents with the aim to push the biomedicine industry to a higher level.

On July 13, 2011, the Ministry of Science and Technology released the National Program for Scientific and Technological Development during the 11th FYP. In the program, biological and medical technologies were again classified as frontier technologies, and it was proposed to strengthen basic research and focus on the R&D of genomics, the new generation of sequencing technology, proteomics technology, stem cells technology, biosynthetic technology, molecular diagnostic and molecular imaging technology, bioinformatics technology, technologies of drug targets spotting and molecular drug design.

In November 2011, the Ministry of Science and Technology released the National Program for the Development of Medical Technology during the 12th FYP together with other 9 ministries. The program proposed to develop strategic emerging industries in the area of biomedicine and lay stress on the basic research of life sciences to solve scientific problems. It also urged to make overall arrangement for investments into medical science from the National Science and Technology Major Project, the National High Technology Research and Development Program (Program 863), the National Key Basic Research Program (Program 973), the National Science and Technology Pillar Program, the National Natural Science Foundation and funds from the PLA, ministries and local governments.

In November, 2011, the Ministry of Science and Technology released the Program for the Development of Biotechnology during the 12th FYP, stating that the goal of biotechnology development of China during 2011 to 2015 was to enhance forward-looking basic research with the aim of significantly boosting China's innovation capacity in biotechnology, raising the overall level of biotechnology into world class, and making some areas rank top among all countries. It also planned to develop an array of crucial technologies and manufacturing processes and create new drugs in key technological areas such as "omics" technology (including genomics, proteomics or metabolomics), synthetic biological technology, bioinformatics technology, stem cells and regenerative medical technology, technologies related to gene therapy and cell therapy, etc.

On December 2, 2011, the Ministry of Science and Technology released the Special Program for the Development of Modern Bio-manufacturing Technology during the 12th FYP, stating that by the end of 2015, an innovation system of modern bio-manufacturing technology shall take shape, and breakthroughs shall be achieved in an array of crucial technologies, so as to improve the technological progress and international competitiveness of the bio-manufacturing industry and make China one of the advanced countries in the world in terms of technological level of the bio-manufacturing industry.

In January, 2012, the Ministry of Industry and Information Technology issued the Program for Pharmaceutical Industry during the 12th FYP. Biopharmaceuticals would be the most important development field, and major breakthroughs were expected in fields such as nucleic acid drugs, gene therapy drugs, products of cell therapy including stem cells, blood products, humanized/human monoclonal antibody drugs, vaccines, genetically engineered protein and polypeptide drugs and so on. The special fund aimed to encourage the creation of major new drugs and projected that it would rise above 40 billion RMB from 19 billion RMB during the 11th FYP.

On May 30, 2012, the State Council released the National Program for the Development of Strategic Emerging Industries during the 12th FYP, proposing that the bioindustry shall be oriented to meet major demands of people's well-being, agricultural development and protection of resources and the environment, and that the development of generic crucial technologies like the utilization of biological resources as well as the processing of equipment shall be enhanced to quicken the pace in constructing a modern bioindustry system. According to experts, the bioindustry is due to become one of the major aspects in the restructuring of China's economy, and it can be expected that fields including biomedicine, bio-agriculture, bio-energy, bio-manufacturing and biotechnology service will experience a robust growth.

Table 9-10 Major technological projects and development plans for biomedicine industry of China

Project	Department in charge (Year of initiation)	Major scientific goals related to the biomedicine industry	Highlights related to the biomedicine industry
Outline of the National Program for Long- and Medium-Term Scientific and Technological Development (2006–2020)	Ministry of Science and Technology (2006)	Realizing national targets, discovering key basic knowledge and developing key applied technologies in areas of bioscience and biomedicine	Target discovery technology; flora and fauna species and drug molecular design technology; genetic manipulation and protein engineering technology; human tissue engineering technology based on stem cells; in addition, there are three major special projects
National Program for the Development of Basic Research during the 11th FYP	Ministry of Science and Technology (2006)	Arranging distribution of technological resources among biology, medicine, brain science and cognitive science (biological and medical technology; quantitative study of life process and systematic integration; brain science and cognitive science)	Basic research on human health and diseases; protein research; research on development and reproduction
National Program for the Development of Science and Technology	Ministry of Science and Technology (2006)	Predicting future development and making forward-looking deployment of frontier technologies and basic research (biological and medical technology; quantitative study of life process and systematic integration; brain science and cognitive science)	Invention of major new drugs; prevention and treatment of serious diseases such as AIDS and virus hepatitis; developing prevention and treatment technology of major diseases to promote the well-being of Chinese citizens

Continued

Project	Department in charge (Year of initiation)	Major scientific goals related to the biomedicine industry	Highlights related to the biomedicine industry
Program for the Development of Bioindustry during the 11th FYP	National Development and Reform Commission (2007)	Striving to make major economic indexes of China's bioindustry rank among the top countries of the world after 10 years' efforts; getting hold of independent intellectual property rights in key biotechnology areas that are crucial to the overall economic and social development and national security; significantly boosting the international competitiveness of China's bioindustry	Prioritizing the invention of new vaccines, diagnostic reagents, new drugs and new medical devices
Outline for the Development of National Science & Technology Pillar Program during the 11th FYP	Ministry of Science and Technology (2008)	Setting up a preliminary system of treatment and prevention techniques of serious diseases; identifying a group of common diagnostic techniques and comprehensive treatment and prevention techniques	Prevention and treatment of heart, cerebral and pulmonary vascular diseases; research on the prevention, early diagnosis and comprehensive treatment of common cancers; development of key technologies and products of major digital medical devices

Continued

Project	Department in charge (Year of initiation)	Major scientific goals related to the biomedicine industry	Highlights related to the biomedicine industry
Policies to Accelerate the Development of Biological Industry	General Office of the State Council (2009)	Guiding resources such as technology, talents and capital to concentrate on the bioindustry to promote innovation and industrialization of the biological industry and accelerate the large-scale, aggregated and internationalized development of the biological industry	Prioritizing the invention of new vaccines and diagnostic reagents that prevent and diagnose serious infectious diseases; rigorously developing biotechnology drugs, small molecular drugs and modern Chinese medicines that are highly effective on common diseases and serious diseases
Program for the Development of Biotechnology during the 12th FYP	Ministry of Science and Technology (2011)	Significantly boosting China's innovation capacity in biotechnology, raising the overall level of biotechnology into world class, and making some areas rank top among all countries	Research on "-omics" study, synthetic biology, bioinformatics, stem cell and regenerative medicine, gene therapy and cell therapy, molecular typing and individualized diagnosis and treatment, biochip and bio-imaging, drug target discovery and drug molecular design

Major national technological plans that support biomedicine include the National Natural Science Foundation of China, Program 973 and National Science and Technology Major Project. ① Due to the

① Besides, Program 863 classified medical biotechnology as a point of breakthrough for the development of biological high-tech industries at the very start, and there are special projects that support the development of biological and medical technologies of the Ministry of Science and Technology of China. However, due to the inaccessibility of all data and the focus on basic research in this chapter, these two programs are not analyzed here.

inaccessibility of some data, we focus on the analysis of support for basic research on biomedicine from NSFC and Program 973 (National Key Scientific Research Program included).

The National Natural Science Foundation of China attaches great importance to research on life sciences and medicine. It can be seen from Table 9-11 that starting from 2001, both the number of projects and funds of general and key projects of bioscience and biomedicine accounted for a large proportion of the National Natural Science Fund, which shows that China values basic research on bioscience.

With respect to the funding of key technologies of bioindustry (Table 9-12), funds mainly concentrate on areas such as vaccine, genomics, tissue engineering and stem cell.

with Program 973 (Table 9-13), the number of projects are rather stable, but currently there is only one enterprise that stands as the affiliated enterprise of a chief scientist (Shijiazhuang Yiling Pharmaceutical Co., Ltd).

Table 9-11 Overview of NSFC projects and funds of life sciences and medicine in previous years

Year	Number of general projects		Investment in general projects		Number of key projects		Investment in key projects	
	Number of projects	Percentage (%)	Funds (million RMB)	Percentage (%)	Number of projects	Percentage (%)	Funds (million RMB)	Percentage (%)
2001	1,557	44.17	276	43.19	26	20.97	38	21.11
2002	2,181	37.55	392	33.88	42	20.19	67	21.31
2003	2,438	38.34	447	33.81	71	28.18	91	25.33
2004	3,064	52.40	605	46.92	80	35.71	108	32.31
2005	3,344	48.85	802	46.09	87	28.71	126	25.86
2006	3,863	53.29	963	47.30	76	27.44	110	24.83
2007	2,890	37.47	805	35.37	113	30.29	179	28.11
2008	3,471	38.90	1,031	35.73	121	28.47	210	26.98
2009	3,981	39.57	1,181	35.74	114	29.16	200	27.62
2010	5,413	41.54	1,726	38.14	132	30.28	283	29.27
2011	6,112	39.87	3,458	38.47	160	32.19	444	31.16

Data source: The website of the National Natural Science Foundation of China.

Table 9-12 Overview of NSFC projects and funds of basic research on key technologies of bioindustry

Item		Year											
		2000	2001	2002	2003	2004	2005	2006	2007	2008	2009	2010	2011
Vaccine	Number of projects	16	8	34	36	47	44	50	49	55	63	73	79
	Funds (thousand RMB)	2,930	1,420	6,570	6,790	14,070	11,700	16,640	12,080	17,580	21,230	23,010	38,680
Monoclonal antibody	Number of projects	0	4	3	6	12	8	6	10	14	16	14	28
	Funds (thousand RMB)	0	630	640	880	2,410	1,710	1,520	2,200	5,000	4,270	5,670	10,670
Biochip	Number of projects	3	1	3	7	3	8	7	4	11	9	9	5
	Funds (thousand RMB)	610	190	1,360	1,700	740	3,710	3,750	1,130	3,290	2,440	2,990	2,730

Continued

Item		Year											
		2000	2001	2002	2003	2004	2005	2006	2007	2008	2009	2010	2011
Genomics	Number of projects	4	4	12	7	11	9	16	16	32	32	30	32
	Funds (thousand RMB)	4,050	4,050	3,010	1,280	3,340	6,790	3,510	1,1060	2,8540	1,7730	2,0370	28,520
Proteomics	Number of projects	0	0	11	16	41	46	50	66	66	65	71	110
	Funds (thousand RMB)	0	0	2,980	3,050	8,220	14,920	12,860	21,050	22,780	17,860	20,290	54,480
Tissue engineering	Number of projects	9	5	25	24	25	41	39	54	53	78	74	99
	Funds (thousand RMB)	1,510	630	4,510	4,530	5,240	11,110	9,200	19,270	19,740	17,530	23,770	44,790

Continued

Item		Year											
		2000	**2001**	**2002**	**2003**	**2004**	**2005**	**2006**	**2007**	**2008**	**2009**	**2010**	**2011**
Stem cell	Number of projects	17	16	67	88	97	125	147	203	283	349	491	632
	Funds (thousand RMB)	3,650	7,200	12,830	17,580	19,990	35,190	37,770	68,440	97,450	121,780	204,370	329,500
Gene therapy	Number of projects	0	6	34	46	53	52	65	62	65	42	52	87
	Funds (thousand RMB)	0	1,690	7,010	9,890	10,780	12,530	17,760	19,180	22,540	13,600	21,510	37,720
Recombinant protein	Number of projects	2	0	1	3	5	6	0	2	2	2	4	3
	Funds (thousand RMB)	310	0	160	600	770	1200	0	450	550	600	990	580

Data source: The website of the National Natural Science Foundation of China.

Table 9-13 Overviews of biotechnology projects in Program 973

Year	Number of projects	Number of corporate projects	Projects and their institutes
1998	3	0	Establishment of the theory and technological system of disease genomics (Shanghai Second Medical University, Institute of Basic Medical Sciences of Chinese Academy of Medical Sciences); Discovery and optimization of lead compounds of new drugs treating major diseases (Shanghai Institute of Materia Medica of Chinese Academy of Sciences); Basic research on the origination and development of cancer (Institute of Cancer Research of Chinese Academy of Medical Sciences)
1999	8	0	Basic and applied research on major cellular activities (Peking University); Basic research on brain function and serious brain diseases (Fudan University); Basic research on the prevention of major infectious diseases (Wuhan Institute of Virology of Chinese Academy of Medical Sciences); Basic research on early systemic damage and tissue repair of serious injuries (Third Military Medical University of Chinese PLA); Basic scientific problems of tissue engineering (Shanghai Second Medical University); Basic research on major scientific problems about prescriptions (Beijing University of Chinese Medicine, China Academy of Chinese Medical Sciences); Basic research on reproductive health (Kunming Institute of Botany of Chinese Academy of Sciences); Research on the functions, three-dimensional structure and folding principle of proteins (Tsinghua University, Institute of Biophysics of Chinese Academy of Sciences)

Continued

Year	Number of projects	Number of corporate projects	Projects and their institutes
2000	2	0	Basic research on the occurrence and prevention of cardiovascular and cerebrovascular diseases (Peking University); Basic research on the aging mechanism and prevention of age-related diseases (Chinese PLA General Hospital)
2001	5	0	Basic research and clinical application of stem cell (Shanghai Institutes for Biological Sciences of Chinese Academy of Sciences); Basic research on the molecular mechanism and application of specific immune response of antigen and immune tolerance (Second Military Medical University of Chinese PLA); Research on the differentiation of reproductive neural crest stem cell and plasticity of tissue stem cell in human embryo (Peking University); Research on proteomics of major human diseases (Academy of Military Medical Sciences of Chinese PLA, Fudan University); Research on the heredity and environmental controllability of birth defects in the Chinese population (Peking University, Shanghai Jiao Tong University)
2002	6	0	Research on the discovery and functions of new drug targets based on bioinformatics (School of Medicine of Shanghai Jiao Tong University); Basic research on body injury caused by environmental chemical pollutants and its prevention (Peking University); Cellular signal transduction system of inflammation and its regulatory mechanism (Institute of Biochemistry and Cell Biology under SIBB of CAS); Mechanism of the occurrence and metastasis of cancer and its molecular repression (Huazhong University of Science and Technology, Institute of Cancer Research of Chinese Academy of Medical Sciences); Research on some new and interdisciplinary questions of life sciences (Academy of Military Medical Sciences of Chinese PLA); Research on the interaction network of proteins that are important in regulating cell proliferation (University of Science and Technology of China)

Continued

Year	Number of projects	Number of corporate projects	Projects and their institutes
2003	8	0	Basic research on the application of neuro-damage repairing and functional reconstruction (Ministry of Education); Biological foundation of psychoactive drug substance dependence and its prevention and treatment (General Logistics Department of Chinese PLA); Basic research on the immune application of chronic failure of transplanted organs (Ministry of Health); Bioinformatics theory and application of gene function prediction (Ministry of Education); Basic research on key problems of bio-catalysis and biotransformation (Ministry of Education); Research on key scientific problems about treatment of important clinical medical information (Guangdong Provincial Department of Science and Technology, Health Office of General Logistics Department of Chinese PLA); Glyco-biology and glycol-chemistry: Typical carbohydrate structure, function and its regulating mechanism (Ministry of Education); Creation of GM higher animals and analysis of advanced brain function (Science and Technology Commission of Shanghai Municipality)
2004	8	0	Systematic biological research on polygenic diseases with complex traits (Science and Technology Commission of Shanghai Municipality); Basic research on the mechanism of prevention and treatment of cancer (Ministry of Health); Basic research on the application of gene therapy (Ministry of Education); Research on new drugs that are based on gene functions (Chinese Academy of Sciences); Basic research on extremophiles and their utilization (Chinese Academy of Sciences); Basic research on key chemical questions during the process of identification and regulation of organic system (Ministry of Education); Research on the structures and functions of bio-membranes and membrane proteins (Chinese Academy of Sciences); Research on the molecular structure and functional development of antigens (Science and Technology Commission of Shanghai Municipality)

Continued

Year	Number of projects	Number of corporate projects	Projects and their institutes
2005	13	0	Epigenetic mechanism of tumor and nervous system diseases (Chinese Academy of Sciences); Research on the pathogenesis of hereditary birth defects using model animals (Ministry of Education); Basic research on treatment of serious injuries and repairing of damaged tissue; Studies on major basic scientific questions of tissue engineering (Health Office of General Logistics Department of Chinese PLA); Functional brain imaging of feature binding and invariant perception (Chinese Academy of Sciences); Research on the pathogenic mechanism of major human pathogens (Ministry of Health); Basic research on the resistance mechanism of pathogens of major human infectious diseases (Ministry of Education); Basic research on the theories of collateral diseases and acupuncture therapy (State Administration of Traditional Chinese Medicine); Research on the laws of compatibility of medicines in prescriptions (State Administration of Traditional Chinese Medicine); Systematization and innovation of basic theories of TCM (State Administration of Traditional Chinese Medicine); Basic theory and key scientific problems in restoring visual performance; Physical and technological foundation of atomic frequency standards (Science and Technology Commission of Shanghai Municipality, Ministry of Education); Gene expression regulation of human non-coding RNA and its medium (Guangdong Provincial Department of Science and Technology, Ministry of Education)

Continued

Year	Number of projects	Number of corporate projects	Projects and their institutes
2006	16	0	Basic research on the pathogenesis, prevention and treatment of major vascular diseases (Institute of Basic Medical Sciences of Chinese Academy of Medical Sciences); Research on the pathogenesis of type-2 diabetes (Nanjing University, Chinese Academy of Sciences); Basic research on human fertility regulation and related major diseases (Institute of Zoology of Chinese Academy of Sciences); Adaptation mechanism and intervention measures against damages caused by environmental factors such as high-altitude hypoxia (Academy of Military Medical Science of Chinese PLA); Research on the biology and immune response mechanism of HIV (Chinese Academy of Medical Sciences); Basic research on infection and immunity (Wuhan University); Research on the molecular biology and pathogenesis of major animal pathogens (Huazhong Agricultural University); Basic research on the specificity of acupoints based on clinical practices (Chengdu University of TCM); Research on the basic theory of therapeutic evaluation of treatment based on syndrome differentiation in TCM practice (China Academy of Chinese Medical Sciences); Hereditary and innovative study on the theory of drug property of TCM (China Academy of Chinese Medical Sciences); Hereditary and innovative study on the theory of etiology and pathogenesis of TCM (Shanghai University of Traditional Chinese Medicine); Research on key technological problems of molecular imaging (Institute of Automation of Chinese Academy of Sciences); Research on some advanced scientific problems related to the processes of material creation and chemical conversion (Jilin University); Protein machine and the cellular mechanism (University of Science and Technology of China); Research on the plasticity of brain structure and function (Shanghai Institutes for Biological Sciences of Chinese Academy of Sciences); Research on the transposition mechanism and of PB transposon in mammals and its analysis method (Fudan University)

Continued

Year	Number of projects	Number of corporate projects	Projects and their institutes
2007	17	0	Basic research on the mechanism of aging and intervention (Chinese PLA General Hospital); Research on the mechanism of gene-environment interaction of serious birth defects in the Chinese population (Peking University); Basic research on the treatment and prevention of heart failure and malignant arrhythmia (Harbin Medical University); Basic research on the molecular mechanism and intervention strategy of serious heart diseases (Peking University); Research on the pathogenesis, treatment and prevention of major neurological blindness conditions in China (Sun Yat-sen University); Research on gene-environment interaction of depression and schizophrenia (Southeast University); Basic research on immunological recognition, immuno-regulation, and origination and intervention of immunity-related diseases (Second Military Medical University of Chinese PLA); Basic research on acupuncture anaesthesia based on clinical practices (Peking University); Research on basic problems related to theory of drug properties of TCM (Shandong University of Traditional Chinese Medicine); Basic research on the theory of special therapies in TCM (Shanghai University of Traditional Chinese Medicine); Basic research on the pathogenesis, clinical early warning and prognosis of hepatitis B aggravation (302 Military Hospital of China); Basic research on the clinical supervision, treatment and prevention of hepatitis B aggravation (Huazhong University of Science and Technology); Basic research on intestinal micro-ecology and infection (Zhejiang University, Second Military Medical University of Chinese PLA); Basic research on the treatment and prevention of malaria and schistosomiasis (Second Military Medical University of Chinese PLA); Scientific foundation of cell factory as a bio-refinery technology (Institute of Microbiology of Chinese Academy of Sciences); Basic study on process science of industrial biotechnology (Beijing University of Chemical Technology); Artificial selection and genome evolution (Kunming Institute of Zoology of Chinese Academy of Sciences); Tracking the origination of immune system back to hundreds of millions of years ago (Sun Yat-sen University)

Continued

Year	Number of projects	Number of corporate projects	Projects and their institutes
2008	16	0	Research on depression from the perspective of proteomics and peptidomics (Chongqing Science & Technology Commission); Origination and regulating mechanism of post translational modification of proteins and its biological and pathological effects (Science and Technology Commission of Shanghai Municipality, Ministry of Education); Research on computational biology based on the structures of and interactions between proteins (Chinese Academy of Sciences); High-resolution structure determination and efficient preparative technology of proteins (Chinese Academy of Sciences); Research on new technologies and approaches of in vivo genetic operations to study overall functions of proteins (Chinese Academy of Sciences); Structures and functions of membrane proteins which are related to major diseases (Ministry of Education); Functions and regulation of proteins in the microenvironment of cancer cells (Tianjin Municipal Science and Technology Commission, Ministry of Education); Functions and regulating mechanism of key proteins in lipid metabolism (Chinese Academy of Sciences); Research on porcine iPS cells and their differentiation and development (Tianjin Municipal Science and Technology Commission, Ministry of Education); Reprogramming mechanism of iPS cells (Chinese Academy of Sciences); Research on the molecular mechanism of development regulation of entodermal tissues and organs (Ministry of Education, Chongqing Science & Technology Commission); Basic research on neurodevelopment of depression and Alzheimer's disease (Chinese Academy of Sciences); Molecular cytogenetic mechanism of neuron growth and degenerative changes (Science & Technology Department of Sichuan Province, Ministry of Education); Research on the heredity and epigenetic regulation mechanism of formation of plant's vital organs (Chinese Academy of Sciences); Research on the molecular regulation mechanism of muscle and fat development of livestock (Ministry of Education); Molecular mechanism of male infertility and birth defects caused by environmental or genetic factors (Chinese Academy of Sciences)

Continued

Year	Number of projects	Number of corporate projects	Projects and their institutes
2009	13	0	Heterogeneous regulation mechanism of tumor-associated cell in the progression of cancer (Fourth Military Medical University of Chinese PLA); Malignant biological behavior of tumor invasion and transfer and related molecular intervention (Huazhong University of Science and Technology); Basic research on the origination, prevention and treatment of cerebrovascular diseases (Second Military Medical University of Chinese PLA); Neurobiological mechanism of psychological dependence of opiates (Fudan University); Basic research on diseases and injuries of respiratory system (Guangzhou Medical University); Relationship between signal transduction network of cells in inflammatory reaction and tumors (Xiamen University); Basic research on the pharmaceutically active substance in endemic plants and microorganisms in China (Kunming Institute of Botany of Chinese Academy of Sciences); Basic research on the application of organ transplant in immunology (Zhejiang University); Basic research on the infection, treatment and prevention of hepatitis C (Institute of Biophysics of Chinese Academy of Sciences of Chinese PLA); Research on the microevolution of major pathological bacteria (Academy of Military Medical Sciences of Chinese PLA); Systematic discovery of new non-coding RNA and its genes and the construction of "bicolor network" (Institute of Biophysics of Chinese Academy of Sciences); Research on the structural mechanism of epigenetics (Institute of Biophysics of Chinese Academy of Sciences); Basic and applied medical research on the decoding of chromatin (Fudan University)

Continued

Year	Number of projects	Number of corporate projects	Projects and their institutes
2010	29	0	Research on the immunological mechanism and intervention strategy of rheumatoid arthritis' onset (Peking University); Research on the clinical translation of leukemia based on systems biomedicine (Ruijin Hospital Affiliated with School of Medicine under Shanghai Jiao Tong University); Study on the selection of gastric cancer bio-marks and the effects of early warning and diagnosis based on a large population (Fourth Military Medical University of Chinese PLA); Research on functions of cancer stem cells during the origination and development of cancer and the mechanism (Third Military Medical University of Chinese PLA); Basic research on the origination, development and intervention of congenital heart disease (Fuwai Hospital of Chinese Academy of Medical Sciences); Clinical basic research on the development and heredity of schizophrenia (Shanghai Jiao Tong University); Research on the mechanism of signal transduction between interacting cells in the process of inflammation and its application (Shanghai Institutes for Biological Sciences of Chinese Academy of Sciences); Basic and applied research on polypeptide toxins excreted by venomous animals exclusively present in China (Hunan Normal University); Basic research on the application of gene target therapy (Sichuan University); Research on pathology, treatment and prevention of major food-borne parasitic diseases in China (Sun Yat-sen University); Research on the invading mechanism of major viruses (Wuhan Institute of Virology of Chinese Academy of Sciences); Research on the pandemic features and pathology of major zoonotic intracellular bacteria (Shihezi University); Research on several key questions about medical imaging of serious diseases based on Shanghai Synchrotron Radiation Facility (Shanghai Jiao Tong University); Biological research on proteins related to the congenital immunity of cells against viruses (Wuhan University); Biological research on major proteins related to immunity (Academy of Military Medical Science of Chinese PLA); Research on the structural foundation, regulation and functions of membrane transport by key proteins in information handling of nerve cells (Zhejiang University); Functions and mechanism of important protein groups related to cell growth regulation (Shanghai Institutes for Biological Sciences of

Continued

Year	Number of projects	Number of corporate projects	Projects and their institutes
2010	29	0	Chinese Academy of Sciences); Research on the functions and mechanism of protein groups related to the stability of genome and regulation of cell cycle (Peking University); New method to label protein based on the expansion of codons (Peking University); Regulation of pathological and physiological process of serious diseases by major lipid metabolites of cyto-membrane (Peking University); Analysis of vast mass spectrum data of proteome and its application in human genome annotation (Beijing Institute of Genomics of Chinese Academy of Sciences); Research on the functional network and regulation mechanism of proteins related to non-coding RNA of malignant tumors (Sun Yat-sen University); Pathology of premature ovarian failure and related serious diseases (Nankai University); Research on the molecular mechanism and biomarkers of premature ovarian failure (Nanjing University); Research on models and pathogenesis of several serious diseases based on iPS cells technology (Shanghai Jiao Tong University); Molecular regulation of development of thymus gland and immune cells (Institute of Zoology of Chinese Academy of Sciences); Research on the regulation mechanism and application of induced differentiation from stem cells to germ cells (Sun Yat-sen University); Research on the identification, lineage reprogramming and tracing of stem cells (Fudan University); Research on the maintenance of potency, differentiation control and immune regulation of progenitor cells (Shanghai Institutes for Biological Sciences of Chinese Academy of Sciences)

Continued

Year	Number of projects	Number of corporate projects	Projects and their institutes
2011	14	0	Research on Mechanism of epigenetics in the malignancys generation and development (Peking University); Basic research on the pathogenesis, diagnosis, treatment and intervention of atherosclerosis (Chinese Academy of Medical Sciences); Research on the molecular mechanism of pathophysiological changes of type-II diabetes (Shanghai Jiao Tang University); Basic research on pathology and intervention strategy of Parkinson's disease; Cellular epigenetic mechanism of the origination and development of malignant tumors (Capital Medical University); Basic research on the carcinogenic mechanism of viruses and its intervention (Sun Yat-sun University); Functions of glial cells during the damage repairing and functional reconstruction of the central nervous system and their significance (Zhejiang University); Basic research on the pathology and measures of intervention of sensor-neural deafness (Wenzhou Medical University); Basic research on pathology and intervention of myopia (Fudan University); Research on the molecular mechanism of inter-species infection and disease dissemination by major viruses (Institute of Microbiology, CAS); Molecular mechanism of latent infection of viruses (Wuhan University); Research on the laws of mutation and pathology of major pathogens (Chinese Academy of Medical Sciences); Basic research on the third-generation gene sequencing system which is based on micro- and nano-fabrication (Southeast University); Study of key scientific problems of multi-model molecular imaging (Institute of Automation, CAS); Imaging-based research on cerebral network and its clinical application

Continued

Year	Number of projects	Number of corporate projects	Projects and their institutes
2012	16	1	Pathogenesis of fatty liver syndrome and hyperlipidemia caused by dyslipidemia (Peking University); Pathogenesis and early diagnosis of common glomerular diseases (Nanjing General Hospital of Nanjing Military Command of Chinese PLA); Study on the progressive mechanism of chronic kidney disease (CKD) (Southern Medical University); Research on the mechanism of hypertension caused by environmental factors and metabolism and related intervening measures (Third Military Medical University of Chinese PLA); Research on the genetic basis and pathology of childhood autism (Central South University); Basic research on GPCR new drug targets related to serious cardiovascular diseases (Peking University); Research on the cellular and molecular mechanism of repair and regeneration of seriously damaged vital tissues and organs (General Hospital of Chinese PLA); Research on the mechanism of rapid acclimatization and long-term adaptation to high-altitude hypoxia (Academy of Military Medical Science of Chinese PLA); Research on the molecular mechanism and prevention of prostatic cancer (Second Military Medical University of Chinese PLA); Research on the law of concerted application of prescription components for the treatment of cardiovascular disease (Tianjin University of Traditional Chinese Medicine); Basic research on the laws of special effect of acupoints and key influential factors (Chengdu University of TCM); Research on the mechanism of diseases related to micro-angiopathy (Shijiazhuang Yiling Pharmaceutical Co., Ltd); Research on the molecular mechanism of interaction between major pathogens and host (Tongji University); Research on the functional genomes and molecular pathology of major animal pathogens (Huazhong Agricultural University); Basic research on inflammation caused by interaction between viruses and cells (Wuhan University); Research on the origination and persistence of persistent infection caused by major viruses (Fudan University)

Data source: The website of the Ministry of Science and Technology of China.

9.5 Summary

In this chapter, we have shown that biomedicine is an industry that relies heavily on industrial basic research. Taking genetically engineered drugs as an example, the upstream technologies (construction of the genetically modified bacteria) involve synthesis, purification and sequencing of the desired gene, cloning and introduction of the gene as well as culture and selection of the genetically modified bacteria. The downstream technologies involve purification and scale-up of the desired protein, and examination and assurance of product quality. Therefore, R&D and IPR are the most crucial factors (DSTSD, 2009). To find out how to orient basic research to meet the demands of industrial development, especially how to coordinate investment from the government and enterprises, is crucial to pushing biomedicine from basic research in laboratories to industrialization (Figure 9-3).

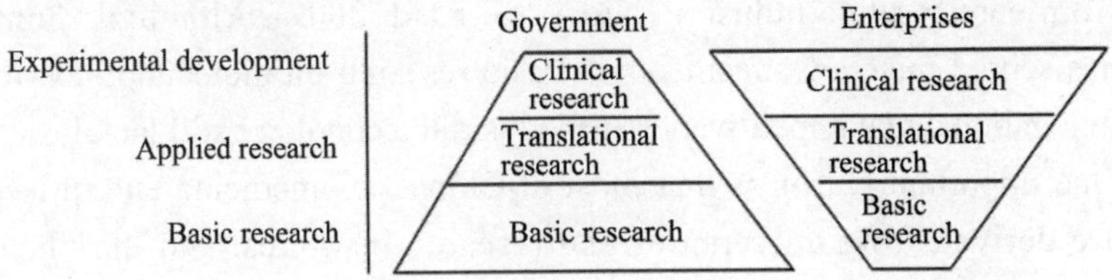

Figure 9-3 Coordination of the government and enterprises in biotechnology R&D activities

Because it is believed that biotechnology will bring about drastic changes in the organizational structure and value-creation process of many industries, both the rich world and emerging economies cast their eyes on the basic research and industrial development of biotechnology. During the two decades of development, the industry of biomedicine has always had the features of heavy investment, high risk, long investment cycle and huge returns, and it is highly dependent on basic research, so not all countries can do well in this field. Meanwhile, this industry shows typical American traits (Liu, 2008). The American government, together with enterprises, has constructed a sophisticated investment system and a rigorous management system for the basic research and industrial development of biotechnology. Despite that the United Kingdom has excellent biological research ability and Japan is good at the industrialization of biotechnology, the United States still leads this industry in the long run. Its superiority is the result of multiple factors: long-term support from the government for basic research on biology; close connection between the enterprises, research institutes and universities, the system of IPR protection and plentiful venture capitals (Pisano, 2002).

The biomedicine industry of China started in the 1970s. Thanks to the persistent research of scientists, China now holds a technological opportunity to catch up with the developed countries (Liu, 2008). The biotechnology industry is still in its early stage of development, which lowers the technological barrier preventing developing countries from entering this industry (Niosi & Reid, 2007). Although China has scored huge advancements in basic research on biotechnology, its innovation and competitiveness of industrial technology still lack luster. One important reason is that most of China's biomedicine enterprises are derived from universities and research institutes, and they lack the expertise to translate basic research findings into marketable products, while traditional pharmaceutical enterprises deviate from basic research on biotechnology. This explains why basic research on biomedicine has not helped to boost the international competitiveness of this industry in China yet. In fact, the major achievements of China in the biotechnology and biomedicine industries lay in the theoretical research of life sciences, while in the areas of biotechnology and bio-industry, there is still a long way to go, because in a mature stage of the biotechnology industry, it is the engineers and industrialists that should be playing the leading role, not the biological scientists. However, for the time being, the main force of China's research projects in biotechnology are biological scientists, while the enterprises often find it hard to take part in them. Admittedly, this has to do with the fact that biotechnology as a whole has just started to develop.

Biotechnology has broad applications in many industries, especially the pharmaceutical industry, which arouses great attention from Chinese policy-makers. Chinese research on biomedicine is for the most part in step with the rest of the world. Modern biotechnological pharmaceutical research in China began in the 1980s; in 1989, the first genetically engineered drug made in China was approved. By this day, China has already had more than 30 biotechnological drugs and vaccines put into production, over 300 biotechnological pharmaceutical R&D institutes and over 150 enterprises with production capacity of biotechnological drugs, with an annual revenue of more than 40 billion RMB. However, the development of the biomedicine industry in China is still restricted by the poor R&D ability of the enterprises; also, findings of basic research conducted in universities and research institutes are not translated to innovative products rapidly and effectively. Since 2000,

R&D in the fields of life sciences and biotechnology has been taking the main share of China's technological input. But judging from the present condition, the output efficiency is not satisfying (Conlé, 2010).

Therefore, the government should take the lead in establishing a platform for technological innovation featuring industry-university-research institute cooperation to make full use of advantageous resources of universities and enterprises and form interdisciplinary technological innovation teams. Long-term and stable cooperation between thriving enterprises and universities/research institutes with sound research capacity and R&D basis of biomedicine should be encouraged, so as to construct a system for technological innovation that is centered on enterprises, oriented by the market and combining industry, universities and research institutes. To solve the crucial cruxes of common technologies restricting the development of the biomedicine industry, the government should take the initiative to form a working mechanism featured with complementary technological advantages, division of work and cooperation in R&D, shared responsibility and shared achievements.

References

Conlé, M. Health biotechnology in China: National, regional, and sectoral dimensions. Duisburg Working Papers on East Asian Studies, 2010 (87).

Department of Science and Technology for Social Development (DSTSD) of the Ministry of Science and Technology of China. *Report on Development Strategy of Biomedicine*. Beijing: Science Press, 2009. [科学技术部社会发展科技司．生物医药发展战略报告．北京：科学出版社，2009．]

Liu, L. *Theory and Practice of Basic Research Policy*. Beijing: Tsinghua University Press, 2007. [刘立．基础研究政策的理论与实践．北京：清华大学出版社，2007．]

Liu, X. Industrial innovation in China: Achievements and Challenges. *China Soft Science*, 2002 (12): 109-113, 34. [柳卸林．我国产业创新的成就与挑战．中国软科学,2002(12):109-113,34．]

Liu, X. *Globalization, Pursuing and Innovation*. Beijing: Science Press, 2008. [柳卸林．全球化、追赶与创新．北京：科学出版社，2008

Liu, X. & Cheng, P. *Annula Report of Science and Technology Development of China 2010*. Beijing: Science Press, 2011. [柳卸林，程鹏．中国科技发展研究报告 2010．北京：科学出版社，2011．]

National Bureau of Statistics (NBS) & Ministry of Science and Technology (MoST). *China Statistical Yearbook on Science and Technology 2011*. Beijing: China Statistics Press, 2011. [国家统计局，科学技术部．中国科技统计年鉴 2011．北京：中国统计出版社，2011．]

Niosi, J. Alliances are not enough explaining rapid growth in biotechnology firms. *Research Policy*, 2003, 32 (5): 737-750.

Niose, J. & Reid, S. E. Biotechnology and nanotechnology: Science-based enabling technologies as windows of opportunities for LDCs?. *World Development*, 2007, 35 (3): 426-438.

OECD. Science, technology and R & D statistics, 2013.

Oliver, A. Biotechnology entrepreneurial scientists and their collaborations. *Research Policy*, 2004, 33 (4): 583-597.

Pisano, G. P. Pharmaceutical biotechnology//Steil, B., et al. *Technological Innovation and Economic Performance*. Princeton: Princeton University Press, 2002.

Yao, W. *Introduction to Biotechnology Pharmacy*. Beijing: China Medical Science Press, 2010. [姚文兵．生物技术制药概论．北京：中国医药科技出版社，2010．]

Yin, R. K. *Case Study Research: Design and Methods*. Thousand Oaks: Sage Publications, INC, 2003.

Chapter 10

Industry-Driven Basic Research: TFT-LCD Industry

Flat panel displays, represented by TFT-LCD (thin film transistor-liquid crystal display), have become a basic electronic device that is as important as semiconductor chips, and find wide application in various products from cell phones to TVs (Lu & Cai, 2010). Along with the rapid development of global consumer electronics industry, information processing comes to show features of digitization and mobilization, thus accentuating the value of flat panel displays. As a winner in the second generation of display technologies, LCD embodies the distinct features of its industrial basic research. Many early-stage breakthroughs came from the laboratories of American enterprises, who were the pioneers in the basic research and were first to produce LCD products. However, nowadays, the front runners in this industry are from Korea and Japan.

We think this industry is different from the biopharmaceutical industry in the process from basic research to innovation. Developed countries are forging ahead in both basic research and industries, but we like to see what happens in China and what kind of lessons we can learn from this industry of developed countries or regions.

10.1 What is TFT-LCD

Liquid crystal is the meso-phase between isotropic liquid and crystal. Early researchers of liquid crystal were mainly scientists from universities and research institutes. Austrian botanical physiologist F. Reinitzer was the first to officially discover the phenomenon. In 1888, when he was producing cholesterol benzoate, he found that this compound has two melting-points: At 145.5°C (293.9°F), it melts into a cloudy liquid, and at 178.5°C (353.3°F), it melts again and the cloudy liquid becomes clear; when the liquid cools, its color goes through multiple changes, from purple to orange, and then to green and so on. German crystal physicist O. Lehmann soon learned about this strange phenomenon. He believed there are indeed soft crystals in cholesterol, and published a paper named About Liquid Crystal. In 1890, German physicist L. Gatterman found that some new azoxybenzene compounds also have two melting-points. For the first time, a liquid crystal of known structure was discovered and found to show the "bi-refrigerant effect". The first person that figured out the nature of liquid crystal was G. Friedel, a French physicist. In 1922, he observed the molecular spatial array of liquid crystals and came up with rules of classification and naming of liquid crystals and their phases. He classified them in three types (nematics, smectics and cholesterics), but did not go into future studies. For nearly 40 years after that, research on liquid crystal stalled, because at that time people had not found any application of liquid crystal.

LCD holds advantages in terms of resolution, display area, energy consumption, size and fusion with digital technology that are incomparable by CRT. Key components of TFT-LCD include glass substrate, polarizer, color filter, liquid crystal, alignment film, backlight module, ITO film and driver IC. There are roughly three steps of manufacturing: manufacturing of TFT-array substrate, LCD cell assembly and LCD module assembly (Figure 10-1).

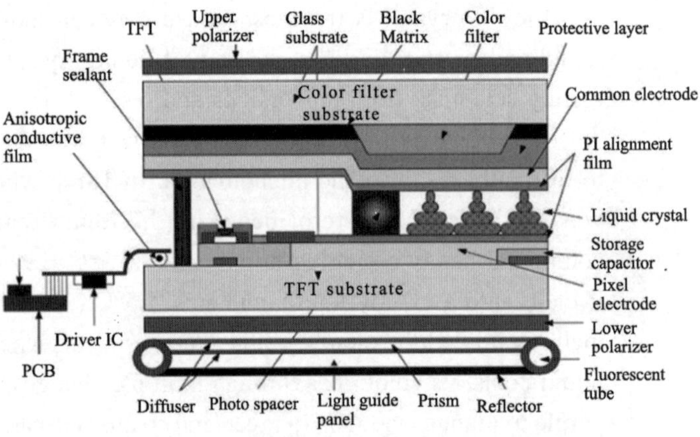

Figure 10-1 Schematic diagram of TFT-LCD

Data source: Beijing Tsinghua LC Technology Engineering Research Center.

TFT-LCD is an industry based on basic research and supported by various technologies. The questions encountered in R&D include expansion of viewing angle, increase of response speed, high-quality animation display technology, color-rendering technology and improvement of backlight. Research on these problems impacts the landscape of competition in this industry (Tian & Ye, 2010). Disciplinary systems involved including liquid crystal physics, crystal chemistry, organic chemistry, physical chemistry, electrochemistry, chemical bonding theory, optics, material science, colorization technology, circuit control and electronics. In-depth research and comprehensive application of said theories are required to improve the performance of LCD, including brightness, contrast, viewing angle, response speed, color rendering and so on (Table 10-1, Figure 10-2).

Table 10-1 Disciplinary systems involved in the development and innovation of LCD

Problems encountered in the understanding and development of LCD	Disciplinary systems involved
Understanding the definition of liquid crystal	Physical chemistry
Understanding the shape, size, structure and arrangement pattern of liquid crystal molecules	Atomic structure and molecular structure, chemical bonding theory, organic chemistry, crystal chemistry
What would happen if light enters liquid crystals, questions concerning polarization and color	Liquid crystal optics, electromagnetic theory, colormetry
What would happen if applying a voltage on liquid crystals	Eletro-chemistry, physical chemistry, electric field theory
Transparency and conductivity of ITO	Material structure of ceramics and solid state electronics theory
Driving modes of LCD (active or passive)	Electronics and circuit control
Principles and manufacturing of TFT	Microelectronics, manufacturing process of large-scale integration
Improving the display performance of TFT-LCD	Comprehensive application of various specialized knowledge and skills
Evaluation of display performance of LCD	Test method and visual theory

Continued

Problems encountered in the understanding and development of LCD	Disciplinary systems involved
Key materials and components of LCD	Organic material science, inorganic material science
Manufacturing of TFT-LCD, assembly of display modules	LCD manufacturing process, electronic packaging engineering

Data source: Tian & Ye, 2010.

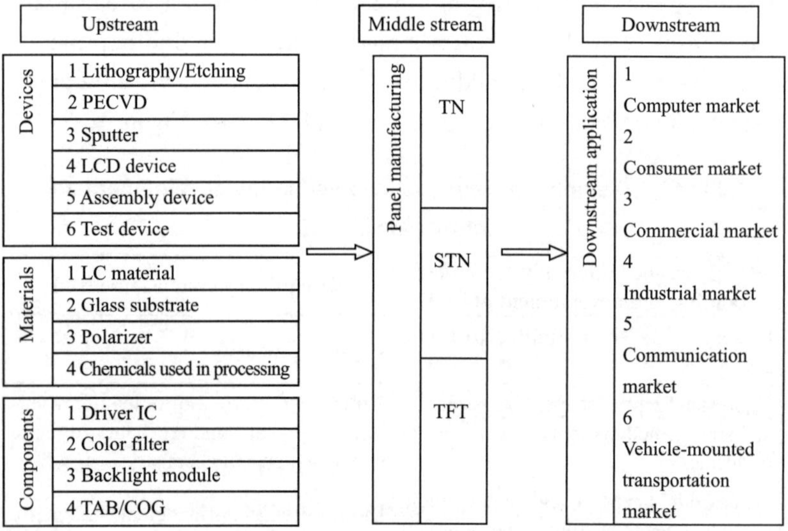

Figure 10-2 Structure diagram of the LCD industry

Data source: Hong, 2009.

10.2 History of research and innovations of TFT-LCD

During the 1930s to 1960s, the semiconductor electronics experienced substantial development. Inventions such as diode and triode appeared one after another, and electronic circuits including diodes and triodes were made into thin films and ultra-miniature sizes through photo-fabrication, or photolithography, which laid the foundation of fabrication processing for the application of LCD (Tian & Ye, 2010). It was during this period that enterprises started to assume the leading role in basic research and technological innovation of this field.

In 1963, Fergusson and others of the American Westinghouse Corporation created the "liquid crystal thermal imaging panel" to measure the thermal distribution in the surface of objects, making use of the phenomenon that the color of light reflected by cholesteric liquid crystals changes along with temperature. In 1968, when G. H. Heilmeire (based in the American company Radio Corporation of America, or RCA) studied how external electric field affects the internal electric field of liquid crystals, he came up with the concept of liquid crystal display technology, and then invented the LCD device with DS (dynamic scattering) mode. Soon after, American enterprises started to design digital LCD wristwatches. In May 1971 and 1972 respectively, Optel and Microma launched their digital LCD wristwatches with DS mode one after another. However, these products were found to be problematic in terms of the quality of liquid crystal and service life, which made them incapable of long-time service. Besides, they were also troubled by high driving voltage and slow response. In 1968, the RCA revealed the positive result of Heilmeire and his fellows in a radio report on a latest scientific achievement. This amazing discovery immediately raised the interest of the French Atomic Energy Commission (CEA), a multidisciplinary research team led by scientist P. G. de Gennes and comprising experts in crystallography, chemistry, material default, optics and nuclear resonance started to conduct basic research on physics of liquid crystals in Orsay, which is located in the suburb of Paris (de Gennes was awarded a Nobel Prize in physics in 1991). Stating from the 1980s, LCD technology evolved from STN-LCD to DSTN-LCD and then to TFT-LCD. Although the first 14-inch colored TFT-LCD was invented as early as 1990 and ushered in the era of home-use LCD, there was no sign in sight that LCD would be broadly applied to TVs until 1999. At that time, it was mainly used in small-size panels such as computer monitors, mobile TVs, PDAs and electron cameras. As the TFT-LCD technology matured and found large-scale commercial applications, this field experienced rapid development. An all-out competition began in frontiers including raw materials, manufacturing, product design and market price setting.

It was the Japanese enterprises that managed to fully integrate basic research on LCD with industrial innovation. In 1968, the 44-year old Japanese physicist Leo Esaki (he was awarded the 1937 Nobel Prize in physics for the invention of Esaki diode) was working in the laboratory of IBM. After getting wind of the LCD technology invented by the RCA, he immediately went back to Japan with the news. After that, enterprises including Panasonic, Hitachi, Toshiba, Mitsubishi Electric, Sanyo, Fujitsu and universities such as Tohoku University, Kyoto University and Tokyo Institute of Technology all started to invest heavily in basic research on LCD.[①] For example, the quality of early digital watch with DS mode LCD made by American enterprises was not satisfying. To solve this problem, the Japanese enterprise Seiko Group carried out in-depth research and found that the TN (twisted nematic) mode LCD was more applicable to this purpose, and soon made its digital LCD wrist watches stand out from other commercial competitors. Because at that time American enterprises failed to develop LCD wrist watches with better quality, Japanese enterprises courageously took up the challenge. Toshio Yamasaki, then the R&D director of Seiko Group, set out to crack the problem of how to realize the function of LCD under freezing point (back then the scope of working temperatures of LCD was quite narrow, which was 0–70℃).

After the successful establishment of the LCD industry, Japanese enterprises carried out active and arduous industrial basic research. According to market demands and the development orientation of products, they successively solved a series of technological problems concerning production equipment, process engineering, raw materials and so on. During the decade from the mid-1980s to the mid-1990s, Sharp dominated the patents of the LED industry in the world, became the world leader in setting up industrial standards of TFT-LCD, and realized the industrialization of LCD technology, thanks to its sound industrial foundation (Lu & Cai, 2010). It was exactly because they paid great at-

① The news spread by Leo Esaki about RCA's research on liquid crystals first raised the interest of Shunsuke Kobayashi, a scientist with the RIKEN Institute of Japan. He went to study in the United States in 1969 and later returned to Japan to conduct basic research on LCD. He then became the academic champion of Japan's liquid crystal research and was known as the "father of liquid crystal of Japan". He once served as a professor in Tokyo University of Science and Director of Liquid Crystal Research Institute.

tention to basic research that the Japanese enterprises became extremely sensitive to new changes in the external markets. They successfully opened the LCD market and for a long period dominated it. From 1960 to 1988, patents on LCD technology applied by Japanese enterprises added up to 208, much more than that of universities and research institutes (the number was 19). Although Korea makes most of LCD panels in the world, most of the advanced LCD technologies are held by Japanese enterprises (like Sharp and Seiko), which keep enhancing their leading position in the course of industrial technological development by transferring technologies to Korean and Chinese enterprises. The Japanese have almost applied all the patents on LCD technology and then transferred the technologies to Korean enterprises (whose LCD product ranks first in the world).

Sharp Corporation was the forerunner in Japanese LCD industry. It occupied the top spot in the early calculator market for a long time. It only lost the place to Casio in 1972 when the latter released a 6-digit mini calculator named "CASIO MINI", which was good in quality and reasonable in price. Sharp had to produce better products to rise to the challenge. When Kazuo Wada, Director of Sharp LCD Research Center, found that the LCD launched by the RCA was a passive display device (which does not emit light), which held obvious advantages over traditional florescent display devices, he organized the R&D on electronic calculators with LCD. However, Kazuo Wada encountered quite a few challenges due to the long manufacturing period of LCDs. Despite these challenges, thanks to the unceasing efforts of the corporate scientists, Sharp realized the commercialization of the first electronic calculator with an LCD in the world in June 1973. This device integrated components including electronic circuits, LCD and switches onto one substrate, and utilized an 8-digit DS mode LCD. With this innovative product, Sharp gained huge market advantages in next to no time (Tian & Ye, 2010). In 1983, Sharp succeeded in developing the EL-805 electronic enterprises using TN-LCD technology, thus becoming the first enterprise in the world to mass-produce LCD electronic computers. In 1986, Sharp made LCD its core business. Then, in 1990, Sharp developed the first 14-inch color TFT-LCD and set up the first specialized plant to produce LCDs in the world, ushering in the era of large-sized LCD industry. Supported by the Japanese government through the "government-industry-university-research institute" technological system, Sharp kept pushing forward basic research and technological innovation, introducing larger-sized TFT-LCP panels to the market step by step (Lu & Cai, 2010).

During the early stage of this emerging industry, Korean enterprises did not have much of an advantage, but they were fast in learning from foreign technologies and good at innovating on the basis of imitation. In Korea, the enterprises play an important role in basic research, and corporate R&D is usually closely associated with industrial innovation. This kind of basic research conducted to meet the industrial demands in a way helped the Korean enterprises to catch up with and surpass the early starters in the LCD industry—a repetition of what happened with the semiconductor industry. As new players, Korean enterprises not only rapidly made inroads into the LCD industry, which was dominated by a couple of Japanese technology tycoons, but went further to overtook them as the world champion in a matter of 10 years with amazing learning and innovating capacity (Lu & Cai, 2010). Samsung serves as a typical case: It followed the road of technological importation to independent innovation in the industry of LCD. Samsung first learned from Sharp, then assimilated its technology, and finally surpassed it. During the process, a knowledge association which was led by the government, robustly funded by consortia and composed of universities and colleges offered great technological support to Samsung. After absorbing and assimilating these Japanese technologies, the Korean enterprises realized low-cost production of large-size panels and applied LCD into TVs.

The enterprises based in Chinese Taiwan are the rising stars in the LCD industry. They value technological importation (the Japanese LCD enterprises transferred their technologies to Chinese Taiwan during the financial crisis in order to contend against Korean enterprises), and at the same time attach even more importance to technological R&D. For example, AU Optronics Corp set up the biggest photo-electricity R&D center in November 2002, focusing on the basic research on display technologies including TFT-LCD and LTPS. Although AU Optronics is a newcomer of the industry, and has been supported by Japanese enterprises in its gradual development, it has already developed a complete and independent technological system and acquired innovation ability in the field of LCD, thanks to its sustained attention on basic research and technological exploration.

After going through the development history of LCD's scientific research and technological innovation, we can see that it is a long way from doing LCD basic research to finding its application in display. American enterprises, and more importantly, Japanese enterprises, successfully closed the gap between scientific principles and industrialization (Table 10-2). At first, American enterprises (represented by the RCA) discovered new technological principles through basic research; then French scientists brought the basic research to a much higher level. However, it was the Japanese enterprises and universities that first acquired solid capacity for industrial basic research. Afterwards, Korean enterprises began to show their prowess: By virtue of their industrial basic research, they managed to surpass their Japanese counterparts. As for the enterprises based in Chinese Taiwan, although they overtook the early starters as the new market lacked terms of production, they lacked genuine basic research. As a result, although its industry of LCD is enormous in scale, the enterprises that still remain with real international competitiveness are only two: Chimei Innolux Corp and AU Optronics Corp. In fact, the Chinese mainland showed interest in LCD technology quite early. Soon after the emergence of the theory of liquid crystals, both Tsinghua University and Peking University boasted great authority for physical research theory of liquid crystals. But with enterprises rapidly translating scientific principles into technologies and due to the unlikelihood of getting research papers about liquid crystals published in journals of high impact factor like *Physical Review Letters*, both universities faced lack of funding in this field. Therefore, after the professors of related majors had retired, Tsinghua University and Peking University removed their liquid crystal physics and chemistry majors (Ouyang, 2011).

Table 10-2 Development history of LCD technology and the role of enterprises in basic research

Year	Major event	Enterprise	Note
1888	Austrian botanist Reinetzer discovered liquid crystal.		
1949	Wimer developed the first CdSe TFT in the world.	RCA	
About 1960	Williams built a flat panel display operated on optical switch and reflective light.	RCA	
About 1960	David Sarnoff Research Center under the RCA invented the TFT.	RCA	Some of the enterprises that conducted TFT research in the early stage was not aimed at the development of LCD, but to find out how to manufacture integrated circuits.
1963	Westinghouse launched a small-scale TFT research project.	Westinghouse	
1963	Williams found that azoxyanisole (liquid crystal powder with light yellow color) has a photo-electronic effect.	RCA	
1964	During the process of re-selecting liquid crystal materials, Heilmeire discovered the combination mode, which means that liquid crystals show dichroism, and then invented an LCD that turns milk white from transparent when applied a voltage–"dynamic scattering".	RCA	

Continued

Year	Major event	Enterprise	Note
1968	Heilmeire invented LCD (DS mode), made the first DSM LCD prototype (digital clock) and published related papers. A full-scale LCD TV model, some other prototypes with LCD, including a digital reader and digital electronic clock, were displayed.	RCA	It sparked great interest from Japanese industrial community on liquid crystals.
1970	Ferguson invented TN-LCD.	Westinghouse	In 1966, Ferguson left Westinghouse and founded his own enterprises ILIXCO.
Early 1970s	A competition between milk white DS LCD of the RCA and dark TN-LCD was unveiled, which ended with the crushing defeat suffered by the RCA.	RCA and ILIXCO	
1971	Lechner et al. proposed active matrix addressing.	RCA	
	Unsure of future development, the RCA called off its TFT project.	RCA	
	Schadt et al. developed TN liquid crystals.	F. Hoffmann-La Roche Ltd	
	Schiekel et al. invented ECB liquid crystals.	Siemense	
1972	Sharp bought an LCD patent from the RCA at a cost of 3 million USD.	Sharp	
1973	Seiko produced the first digital LCD watch.	Seiko	Seiko got the licensed patent from ILIXCO.
	Emstoff developed MOSFET LCD; TN-LCD started to become mainstream.	Hughes	
1979	Spear et al. based in Dundee University developed a-Si:H thin film.		
	Westinghouse terminated its TFT project.	Westinghouse	
1981	Snell et al. based in Dundee University developed TFT LCD with 5 × 7 dot matrix.		
	Scheffer et al. developed STN liquid crystal.	Brown, Boveri & Cie	

Continued

Year	Major event	Enterprise	Note
Early 1980s	The RCA, AT&T, GE, HP and Westinghouse all abandoned the development of panel display technology; the technology of panel display arrived at a crossroads in 1980s.		It was because the risks were high and projects were overstretched.
1982	Adopting the technology of active matrix display and replacing TFT by silicon chip, Seiko developed the first wrist TV with monochromatic display in the world, which, however, was not put into production.	Seiko	
Early 1980s	After working on it for a while, Casio, Citizen and Seiko realized that silicon chips would never be adopted in the mass production of LCD TVs, therefore they must face the challenge of adopting TFT.		
1983	Seiko developed micro LCD TVs using TFT.	Seiko	It turned the table for LCD.
1983	Corning started to produce TFT glass substrates.	Corning	
1983–1984	Large enterprises such as Panasonic, Hitachi and Toshiba started to develop and produce LCD TVs.		
1984	Seiko launched the first flat panel color TV (2 inches).	Seiko	However, in order to lower the cost, Seiko had to sell LCD to other manufacturers as components.
1986	Panasonic was the first to commercialize a-Si TFT LCD TV.	Panasonic	
1987	Sharp launched 3-inch LCD TVs.	Sharp	
1987	Samsung and LG started to research on TFT technology one after another.	Samsung and LG	
1988	Sharp developed the prototype of 14-inch a-Si TFT LCD TV; Toshiba and IBM announced to have developed bigger (14.26-inch) color LCD with higher definition.	Sharp, Toshiba and IBM	The realization of mass production is a challenge in manufacturing; the manufacturing process was costly but the output was low.

Continued

Year	Major event	Enterprise	Note
1989	Due to over-expansion in early years, Seiko voted down proposals of investment in new LCD production equipment.	Seiko	
1990–1991	NEC, DTI (a firm co-funded by IBM and Toshiba) and Sharp launched their first mass production line of large-size color TFT-LCD.	NEC, DTI and Sharp	From the announcement of prototype developing to the beginning of volume production, it only took Japan 3 years to establish the TFT-LCD industry.
1991	Two TFT-LCD enterprises were set up in Chinese Taiwan.		
1992	IBM rolled out Thinkpad 700C, the first laptop adopting color TFT-LCD in the world.	IBM	The introduction of Thinkpad 700C caused shortage of TFT-LCD across the industry.
1993	Applied Materials started to develop chemical vapor deposition (CVD) devices and worked closely with Sharp and Toshiba in the development of new technologies to reduce particle defect.	Applied Materials	CVD device is a crucial part in the manufacturing process.
1993	Applied Materials launched AKT-1600, the first CVD, greatly boosting the rate of qualified panels to 90%, from the previous 10%.	Applied Materials	The introduction of AKT-1600 broke the limit of product defect on the development perspective of TFT-LCD, and brought challenge to the dominant role of CRT for the first time.
1994	The 2nd generation of production lines started to be built up, each needing 4 AKT-1600 devices.		
1995	Samsung and LG built the first 2nd generation of volume production lines of TFT-LCD in succession.	Samsung and LG	
1996	Samsung built up the 3rd generation of production lines.	Samsung	

Continued

Year	Major event	Enter-prise	Note
1997–1998	LG and Samsung set up the 3.5th generation of production lines in succession.	Samsung and LG	After the onset of Asian financial crisis in July 1997, the global LCD market fell into a recession, while most Japanese manufacturers were in the red.
1998	Under the help of Japanese technologies, 6 enterprises based in Chinese Taiwan began the manufacturing of large-size TFT-LCD.		
1998	Samsung's shipment became the largest in the world.	Samsung	
2001–2002	LG and Samsung set up the 5th generation of production lines in succession.	Samsung and LG	Globally, TFT enterprises were still losing money.
2004	LG set up the 6th generation of production lines.	LG	
	Sharp launched the 65-model TFT-LCD TV.	Sharp	
2005–2006	Samsung set up two 7th generation of production lines in succession.	Samsung	
2006	LG and AUO set up the 7.5th generation of production lines in succession.	LG and AUO	
	LG Philips launched the 100-model TFT-LCD TV.	LG	
2006	Samsung set up the 8th generation of production lines.	Samsung	
2007	Sharp displayed the 108-model TFT-LCD TV.	Sharp	
2008	The first 8th generation of production lines in the world was put into production.	Sharp	
2009	Sharp set up the 10th generation of production lines.	Sharp	

Data source: Based on "*S* Curve, Technological Revolution and Innovation" (lectures given by Liu Xielin in the training classes in the Lenovo Academy); Tian & Ye, 2010.

The generations of LCD production lines are divided by the size of the glass substrate. Different generations have different sizes of products and slightly different production processes, but the core technologies used in production are relatively stable. During the 10 years of industrialization of TFT-LCD, there have been 7 major breakthroughs (Table 10-3), which means every 1.5 years for each breakthrough. However, after 2007, it became ever more difficult to lower the cost simply by enlarging the glass substrate. The reason is that after the 8th generation of LCD, it is possible that prices of manufacturing devices and auxiliary modules like photolithographic mask might take a hike, and it will be hard to lower the production cost while sticking to the previous manufacturing technology (Tian & Ye, 2010). Simply enlarging the glass substrate could no longer give enterprises the competitive edge. Therefore, it is imperative to conduct further industrial R&D.

Table 10-3 Evolution of LCD generations

Generation	Size of glass substrate (mm)	Roughly corresponding sizes of products	Time of introduction	Enterprises first to launch the line	Amount of investment
1G	300×400	Mobile or special products smaller than 9 inches	1991	Sharp	
2G	400×500	Mobile or special products smaller than 9 inches	1994		
3G	550×650	Mobile or special products smaller than 9 inches	1996		
4G	730×920	Mobile or special products smaller than 9 inches	2000		
4.5G		Mobile or special products smaller than 15 inches			
5G	1100×1250	8–32-inch monitors and TVs	2002	Samsung and LG launched the lines in succession	1.2 billion USD

Continued

Generation	Size of glass substrate (mm)	Roughly corresponding sizes of products	Time of introduction	Enterprises first to launch the line	Amount of investment
6G	1370×1670	18–37-inch monitors and TVs	February 2004 July 2004	Sharp, LG	2 billion USD
7G	1500×1800	32–42 inches TVs	November 2004	Jointed set up by Samsung and Sony	2.5 billion–3 billion USD
7.5G	1950×2250		2006	LG Philips and AUO launched the lines in succession	3.1 billion–3.3 billion USD
8G	2200×2600	32–60-inch TVs	2006	Samsung	3.5 billion–4 billion USD
9G	2400×2800				
10G	2600×3100	TVs bigger than 40 inches	2009	Sharp	4.5 billion–5 billion USD
11G	3000×3320	TVs bigger than 50 inches			

Although currently LCD dominates the market, many new technologies are still under development for creating speedier displays with higher resolution. Many R&D activities are still underway to develop new materials needed in LCD, with further research yet to be conducted and breakthroughs to be made. These materials include inorganic or organic film materials with special photo-electronic properties, high polymeric materials used in liquid crystal molecular orientation, semiconductor materials used for driving and backlight LED and so on. As a matter of fact, the much-hyped LED TVs in the Chinese mainland market are no more than LCD TVs with LED as backlight (in other words, still LCD TVs). Even the innovative OLED, which is universally recognized as having the most potential in the industry, still employs 70% (calculated by the amount of investment) of technologies same as LCD. Therefore, the development of the LCD industry still has a long way to go. There are numerous technological problems to be cracked before the industry turns mature and can entirely focus on production and marketing. As for now, there is still a considerable space for industrial development (Ouyang, 2010).

10.3 Distribution of basic research and innovation capacity in the LCD industry

To determine the global distribution of basic research and technological innovation capacity in the area of LCD industry, we carried out an analysis on the SCI and DII of papers and patents related to LCD technology, with the aim of finding out whether basic research has an impact on the capacity of industrial technological innovation.

(1) The top 10 countries or regions that published the most papers in SCI during 1981–2012 are listed in Table 10-4. We can see that the respective numbers of papers of the United States and Japan are quite close, and the papers of Korea and Chinese Taiwan also account for considerable proportions. As for the Chinese mainland, its number of SCI papers ranks 4th, with a proportion also nearing 10%. Considering Chinese scientists are not proficient in using English, it shows that they do have strong capacity in scientific research on liquid crystals.

Table 10-4 Top 10 countries or regions with the most SCI papers published about LCD technology during 1981–2012

Ranking	Country/Region	Number of papers	Global proportion
1	US	3,115	19.17%
2	Japan	2,702	16.30%
3	Korea	2,053	12.64%
4	Chinese Taiwan	1,840	11.32%
5	Chinese mainland	1,505	9.26%
6	Germany	766	4.71%
7	UK	765	4.71%
8	France	558	3.43%
9	Spain	388	2.39%
10	Netherlands	362	2.29%

(2) Table 10-5 shows the top 10 institutes or enterprises that offered most funds to publish SCI papers about LCD technology during 1981–2012. We can see that the Natural Science Foundation of Chinese Taiwan and the National Natural Science Foundation of China hold considerable advantages, and American and Korean institutes also occupy important positions, while the number of papers funded by other institutes is quite small.

Table 10-5 Major funding institutes or enterprises with the most SCI papers published in the area of LCD technology during 1981–2012

Funding institute or enterprise	Country/Region	Number of papers
Natural Science Foundation of Chinese Taiwan	Chinese Taiwan	161
National Natural Science Foundation of China (NSFC)	China	137
US National Science Foundation (NSF)	US	92
Ministry of Education, Culture, Sports, Science and Technology of Japan	Japan	68
Ministry of Science and Technology of China	China	50
Knowledge Economy Ministry of Korea	Korea	44
National Research Foundation of Korea	Korea	41
US National Science Board	US	28
Ministry of Education and Science of Spain	Spain	24

Continued

Funding institute or enterprise	Country/Region	Number of papers
Ministry of Education, Science and Technology of Korea	Korea	20
EU	—	19
Engineering and Physical Sciences Research Council (EPSRC)	UK	11
Samsung	Korea	19
Air Force Research Office	US	9
Council of Scientific and Industrial Research of New Delhi	India	9
German Chemical Industry Association	Germany	9

(3)We run a search on the patents of key technologies of global LCD industry with "patent applicant" as key word, and the results are shown in Table 10-6. We can see that patents of this industry are mainly held by enterprises based in Korea, Japan and Chinese Taiwan. This reflects the landscape of industrial competition: The LCD industry is still in the hands of enterprises that have advanced technologies.

Table 10-6 Major enterprises with most patent applications in the area of LCD technology accumulated during 1963–2012

Funded Institute	Country/Region	Number of patents
Samsung	Korea	9,418
Seiko Epson	Japan	7,567
LG Philips	Korea	5,430
Fuji	Japan	5,370
Sharp	Japan	4,562
Sony	Japan	1,994
AUO	Chinese Taiwan	1,797
Nitto Denko	Chinese Taiwan	1,201
Canon	Japan	1,187
Dai Nippon Printing	Japan	1,160

Continued

Funded Institute	Country/Region	Number of patents
Toppan Printing	Japan	1,152
Panasonic	Japan	1,114
Toshiba	Japan	1,022
Semiconductor Energy Laboratory	Japan	961
Sumitomo Chemical	Japan	891
Hitachi	Japan	878
ARUZE	Japan	854
Nikon	Japan	854
JSR	Japan	853
Innolux	Chinese Taiwan	841
Panasonic	Japan	769

The LCD panel industry is alternatively driven by scientific research and technological innovation, with industry-driven basic research as one of its distinct features. American scientists were the first to raise the concept of LCD technology; French and British scientists also made significant research findings in the early stage. The RCA could have become the leader in the LCD industry, but it overemphasized the shortcomings of the newly fledged LCD components and limited itself with traditional semiconductor products, which still remained successful at that time. Japanese enterprises overtook the development orientation of the LCD industry after a short time since they entered this industry, with much major basic research conducted by pioneering enterprises such as Sharp. Thanks to their excellent ability of learning and industrial basic research led by the LCD enterprises, the Korean enterprises occupied a huge market share soon after they entered the market. Owing to the rapid evolution of industrial technologies, it is hard for the academia to detect certain basic research induced by demands, and that is why the role of enterprises is so important. Japanese and Korean LCD enterprises all boast of excellent capacity of basic research, and have always been leading the direction of basic research in this industry. Although Chinese enterprises managed to quickly expand their production capacity in all manners after entering the industry, they have not acquired core technologies yet. China has published copious academic papers on LCD research, but its patent applications are clearly in shortage. One possible explanation is that there exists a huge gap between basic research on liquid crystal physics and the development of the LCD industry since the middle and later years of 1980s.

10.4 Basic research on LCD industry in developed countries and regions

Enterprises play a leading role in the LCD panel industry. Related to multiple disciplines, the industry's technological competition is dynamic. As a result, it is impossible for researchers at universities or research institutes alone to carry out the basic research on industrial innovation and development. Enterprises, therefore, constitute a key force in developing and transforming technologies. While Chinese enterprises are enthusiastic about all sorts of old and new LCD panel assemblies, Japanese and Korean enterprises focus on R&D and patents, which enables them to master leading technologies. Basic research, which is advocated by the government, is also an important driving force of the industry, and plays a key role in industrial development in Japan, Korea and Chinese Taiwan. Both Japan and Korea have made photoelectric technology and semiconductor technology their national strategic projects, which receive a funding worth 500 million to 1 billion USD. Meanwhile, private capital enters the industry supported by financial policies.

Japan's success derives from the industry-driven fundamental research in the form of cooperation among the government, enterprises, universities, and research institutes. In the early stage of development, the industry did not win much support from the Japanese government. Instead, it was advanced by means of some large projects. In 1979, the Japanese government offered a grant worth 770 million USD and started a "photoelectric fundamental research project" with the help of 16 enterprises led by the Optoelectronic Industry and Technology Development Association (OITDA) of the Ministry of International Trade and Industry. Then the 16 enterprises received a grant worth 820 million USD and started a 10-year fundamental research project on LCD industrialization (enterprises which received the grant must provide 70% research grant). Japanese Science and Technology Agency, Ministry of Posts and Telecommunications, Ministry of Culture also provided necessary assistance to their subordinate institutes and state universities. In recent years, with the rapid development of Korean and Chinese LCD industry and technologies, Japan started to integrate enterprises and academia to carry out applicable fundamental research and product development so as to maintain its leading position. Japan Society for the Promotion of Science (JSPS) has set No. 142 Special Committee to assist the research in the long run. Tokyo University of Agriculture and Technology, Tokyo Institute of Technology, Tohoku University and almost all electronics businesses (Sharp, Fujitsu, Hitachi, Panasonic, etc) have their own LCD institutes. This mechanism has boosted innovation by integrating enterprises, universities and research institutes. Six enterprises including Sharp, Toshiba, NEC, Hitachi, Panasonic and Mitsubishi established Advanced LCD Technologies Development Center, with a capital of 50.40 million JPY shared by the enterprises. The headquarters is based in Toshiba's Production and Technology Center. It specializes in energy-saving and next-generation technologies with the focus on LCD products in the next 4 to 5 years. Thanks to the technologies of the six enterprises, the development cycle is shortened to an extent that no other enterprise can rival. Investment risks are reduced too (Gao et al, 2005). Enterprises are major participants of Japanese government's LCD projects (Table 10-7).

Table 10-7 Japan's plans promoting the development of LCD panel

Project	Content	Participating organization(s)
Regional joint research undertakings (Aomori)	Development of TV with 30-inch or bigger sized display, power consumption less than 20W and price no more than 400 thousand yen	Tohoku University
30 IT undertakings jointly carried out by private enterprises	Cutting the price of 30-inch LCD TV to 200 thousand yen in 2005; launching 30–40-inch CNT-FED display and carrying out small-scale production of 60-inch display in 2004; launching OLED TV in 2006	Jointly developed by private enterprises
National projects	CNT-FED	Led by Mitsubishi
	LTPS	Led by Hitachi
	The cost objective of 30-inch LCD TV was 100 thousand yen	Sharp and Tokyo Equipment established the Future Vision company
Others	Improvement in the manufacturing technology of large-scale LCD panel and cutting of cost	Tohoku University set up the Institute of Industrial Technology
	OLED	Yamagata University
	R&D Plan for the next generation of PDP with lower power consumption	Fujitsu, Hitachi, Pioneer Corporation, NEC, PDP Panasonic, Fujitsu Hitachi PDP

Data source: Gao et al., 2005.

The Korean government has played a positive role in the LCD industry and technological development, and has set up a government-led resource-sharing mechanism to catch up with Japan. In 1990, the government established an "LCD Research Alliance" made up of universities, colleges and over 100 enterprises. The government invests 3 billion KRW each year into the alliance. From 1995 to 1998, the government and the industry invested 50 million USD to improve relevant functions of TFT-LCD, its mass production, volume production and so on. The research on TFT-LCD for TVs and PDP was one of the nation's seven major R&D projects of the Ministry of Industry and Resources, and it was granted a fund worth 3 billion KRW each year for 5 years. In 2002, Korea formulated a 10-year (2002–2012) "LCD Industrial Development Plan", aiming at the whole-chain development of the LCD industry worth several hundreds of thousands of dollars. Between 2001 and 2006, the government invested 5.1 billion KRW (supplemented by 2.1 KRW of private capitals) to build an "Engineering Technology Development Center" focused on LCD and OLED engineering technology research, which was included in the "Next-generation LCD Research Foundation Building Plan". In 2003, the government invested 4.25 million USD into LCD-related education and industrial consultancy. Besides, LCD enterprises in Korea are exempted from business tax for 7 years, which is also halved in the next 3 years. Imported equipment and raw materials are exempted from tariff. In terms of direct investment, the government invested over 3 billion USD in the form of direct investment and long-term loans into TFT-LCD of Samsung and LG via investment institutions like Korea Development Bank (KDB). ①

Chinese Taiwan is an important player of today's industry. It considered photo-electronics a key science in 1982 in the process of semiconductor and photo-electronic technological development. It protected the industry through the "Two Trillion & Twin Star" plan, provided key financing environment, and engaged businesses in R&D. Since the late 1980s, departments cooperated to boost the industrial development. Since 2000, it offered 10 billion TWD each year in the form of long-term interest-free loans to private investment institutions to invest in TFT-LCD businesses or directly to the businesses. The "6-

① Supporting strategies for LCD industry in Japan and Korea, *Science Times*, 2007-08-29.

year Development Key Plan (2002–2007)" granted subsidies worth 60 million USD to help private businesses set up R&D centers. A venture capital worth 3 billion USD was founded to develop core technologies in support of the new LCD industry.

10.5 Innovation, development and basic research investment of the LCD industry in the Chinese mainland

China's ability to carry out LCD basic research is strong. However, Chinese enterprises master few key technologies. To strengthen innovation-oriented basic research is a key to future technological policies.

10.5.1 Basic research and technological innovation of China's LCD industry

LCD was a new technology in the early 1960s. China's LCD study started in 1969, almost at the same time with the world. But China owns few key technologies. A reason is that its development is separated from production and that enterprises are not capable of basic research. In Japan, on the contrary, Panasonic, Hitachi, Toshiba, Mitsubishi

Electric, Sanyo, Fujitsu and other enterprises made efforts to invest in LCD research, so did Japanese key universities like Tohoku University, Kyoto University and Tokyo Institute of Technology. From 1960 to 1988, Japanese enterprises invented 208 LCD patents, among which universities and institutes claimed 19, demonstrating the leading role of enterprises (Wang, 2008). In general, Chinese LCD industry relies too much on foreign countries for core components and patented technologies.

It is both a strategic choice and the last resort for Chinese enterprises to enter the industry because when China finally did well in the cathode ray tube (CRT) technology, it is no longer the mainstream technology. CRT was replaced by TFT-LCD, meaning Chinese TV industry has lost its production capacity and cost advantage. TFT-LCD demands higher levels of automation, rendering the original staff no longer suitable for the new assembly. As a result, enterprises must train new staff. A technological breakthrough must be made to transform from CRT to TFT-LCD (Liu, 2008). With LCD dominating the industry, the LCD panel industry has long been shared equally by Korea, Japan and Chinese Taiwan. Giants like Samsung, LG, Chimei Innolux, AUO and Sharp almost monopolized all of the key technologies and the majority of output; therefore, they have a great say in the industry. The TFT-LCD industry of the Chinese mainland rose to break Japan's and Korea's monopoly of TFT panel supply. Chinese enterprises once again showed their ability to catch up fast. Now the Chinese mainland is the fourth largest TFT-LCD production center, next only to Korea, Chinese Taiwan and Japan. However, TFT-LCD is an industry with fast-developing technologies. Enterprises must be able to catch up continuously. Otherwise, they may fail despite past successes (Liu, 2008).

China's LCD industry has a strange phenomenon: Although scientists start basic research early and do not lack major accomplishments, the domestic display industry shows little interest in them. Traditional

TV enterprises have acquired enough market shares; therefore, they neither conduct research on new technologies nor follow university and institute studies. In fact, Chinese officials and some scientists realized early that the display is the core of tackling technological problems and therefore a strategically important industry. They have paid attention to TFT-LCD development. As early as the late 1970s, China entered the panel industry. No. 774 Factory under the original Ministry of Electronics Industry (Beijing Electronic Valve Factory and also the predecessor of BOE), No. 770 Factory (Changsha Shuguang Electronic Valve Factory), No. 713 Factory under Chinese Academy of Sciences (in Xingxiang, Henan Province) and Shanghai Electronic Valve Factory set up a trial production line for 4-inch mother glass TN-LCD to produce electronic watches, calculators and some other LCD products with panels. Tsinghua University, Changchun Institute of Physics and other institutes also began LCD R&D (Lu & Cai, 2010). ① As the Chinese TV industry became more competitive in CRT technology, few people were willing to explore a new technology whose future was at best vague. Therefore, China's investment in TFT-LCD development and its commercialization fell short, evidenced by the fact that there was no TFT-LCD project in the 10th and 11th FYPs. Before 2000, only scientific institutes conducted relevant basic research (Lu & Cai, 2010). As a result, China's panel technology again was dominated by foreign countries during the transitional period. And borrowing technologies rendered China at the end of the industrial chain. Short of basic research ability, China was less competitive in large LCD panel production. Its strength in CRT TV and display industries risked being replaced as new breakthroughs were made. Chinese enterprises are not equipped with the capacity of producing key components. They borrow equipment for cleansing, photo-etching and film formation and ultrahigh vacuum equipment. China's customs data show that in 2010 the Chinese mainland imported LCDs worth over 47 billion USD, next only to integrated circuits, oil and iron ore (Zhao, 2011).

① "640×200 super-twist LCD project" in the 8th FYP was completed by No. 770 Factory, Tsinghua University, and Nanjing No. 55 Institute.

In recent years, to acquire foreign advanced technologies through merger has become a major innovative strategy. In January 2003, BOE, which has been followed and studied LCD technologies for 10 years, purchased HYDIS under Korea's Hyundai Group with 380 million USD to enter the LCD industry. BOE acquired necessary technologies, patents, starter market and other strategic elements to enter the TFT-LCD industry through international acquisitions (Lu & Cai, 2010). Despite the industry's confidence in BOE's acquisitions and the enterprises's substantial growth in the early stage, its performance began to degrade in the second half of 2004. The enterprise had obvious financial crisis in 2005. BOE's technology source, BOE-HYDIS, a Korea-funded subsidiary was insolvent and went bankrupt, placing BOE in a predicament. In fact, many Chinese LCD enterprises have picked a similar path. For example, Changhong purchased Korean Orion Plasma Company and therefore had a complete set of plasma screen technologies and other core patents. LCD panel production lines of the 6th generation and higher were built, enabling mass production. As a result, Chinese large-screen TV production no longer depended on imports of LCD panels. BOE, TCL, IVO and other Chinese enterprises, which entered the LCD industry early, to a certain extent, accelerated high-generation factory building of Korean and Japanese enterprises. However, different from integrated circuits, LCD technologies of the 8th generation varied little from the 6th and the 7th generation. The difference lay in the size of glass substrate, transferring ways and equipment. High-generation production lines did not equal leadership. This has broken the monopoly on production, instead of technology. Apart from factories, Japan had monopoly on the supply of key components for LCD panels and drive circuits for the world.

The LCD industry's rapid development in the past decade benefited from the interaction between basic research and industrial innovation. In China, however, these two are separated. In advanced countries, not only universities and research institutes but also enterprises do well in LCD research. Chinese enterprises, in comparison, are weak in research. Past experience shows that without basic research ability and key technologies, China is likely to fall behind once there is another technology transformation. The government and enterprises highlight industrial chain building but overlook the need of original innovate. In the CRT time, Chinese TV had 50% of the world market. In the TFT-LCD time, its market share dropped to less than 20%, and LCD TV less than 8% (Tian & Ye, 2010). If Chinese TFT-LCD industry does not strengthen its basic research ability, its upgrading may be affected. China's microelectronics, photoelectron, core materials, equipment and special display fields may also lag behind advanced countries. Different from the past, corporate basic research must focus on a key issue—research orientation. To learn technological application process helps researchers to locate key technological issues and the follow-up research orientation. In the new display technology field, users and the industry have a higher demand for both application and basic research. To combine basic research and application is the way to advance the common development of the discipline and industry.

10.5.2 Policies and technological plans promoting basic research in the LCD industry

Having made related national policies, the Chinese government has been paying great attention to the industry of panel display (especially TFT-LCD) in recent years. It stated in many key plans and policies over the years to strengthen the guidance role of polices, enhance the strategic status of the panel display industries, accelerate the development of LCD and other new-type display industries and raise the capacity of independent innovation of core technologies of this industry. [①]

On December 30, 2005, the Ministry of Science and Technology issued the Outline for the National Program for Long- and Medium-Term Scientific and Technological Development (2006–2020), in which "high-definition large-size panel display" was recognized as one of the prior subjects in the major area "information industry and modern service industry". It also claimed to prioritize on the development of high-definition large-size display products, develop various panel and projection display technologies such as organic light emitting display (OLED), field emission display and laser display, and establish an industrial chain of panel display materials and components.

On October 25, 2006, the Ministry of Science and Technology issued the Outline for the National Key Basic Research Program (Program 973) during the 11th FYP, identifying basic research on photo-electronic devices as a major area of information technology to give support. It also pointed out that it was exactly because of lacking substantial basic research as a foundation that China still depended heavily on modeling from other countries in the production of high-tech products and short of independent innovation. What's more, the broad application of these products and immense market made basic research even more likely to be overlooked. In the National Program for the Development of Basic Research during the 11th FYP, which was released the same year, basic research on condensed matter and new effects was identified as a major research direction in the field of information technology during 2006–2010. It also claimed to focus the efforts on new materials, phenomenon, effects and theories so as to make major breakthroughs in these areas.

① These policies ignited an upsurge in the construction of domestic LCD production lines. BOE, CEC-Panda, Infovision Optelectronics and TCL all took quick actions. In a matter of only a few months, China suddenly entered the era of "higher generation" of LCD from the stage of import-dependency. However, the concentration of higher-generation of LCD programs immediately resulted in over capacity and waste of resources, which raised concern of departments in charge.

In November 2006, the Ministry of Science and Technology issued the National Program for Scientific and Technological Development during the 11th FYP, classifying "a new generation of panel display technology", as a major funded area under the category of "new material technology", which was an advanced technological field requiring beforehand deployment.

In March 2007, the Ministry of Information Industry issued the Program for the Development of Information Industry during the 11th FYP. In the section of goals of innovation capacity, it proposed to raise the research capacity of key technologies such as new-type display to internationally advanced level and develop some world-class product technologies, and acquire an array of core technologies and independent intellectual property rights. In the section of the development of core basic industries of China, it pledged to independently develop new-type panel display industries including TFT-LCD and PDP components, accelerate R&D and industrialization of new-type components, strengthen the ability to provide home-made key materials of panel display components and map out special policies to promote the development of the new-type display component industry.

In May 2007, the National Development and Reform Commission (NDRC) released the Program for the Development of High-Technology Industry during the 11th FYP, proposing to actively develop new-type display components such as TFT-LCD, acquire certain core technologies and accelerate the development of panel display materials.

On October 30, 2008, the Ministry of Industry and Information Technology released the Special Program for Basic Electronic Materials and Key Components during the 11th FYP, one of five special programs for information industries. It proposed to focus efforts on major breakthroughs of new-type display components during 2006–2011. It also pledged to prioritize the development of TFT-LCD and PDP to meet the demands of digitization, high definition and flat panel display, support leading enterprises with independent intellectual property rights and industrial foundation, facilitate the combination of R&D and industrialization, and strive to improve the capacity of independent innovation. It pledged efforts on the construction of R&D centers of LCD production process, technological engineering centers of LCD modules and R&D centers of PDP process. It also pledged to support the construction of 6th (and more advanced) generation of TFT-LCD panel production lines, accelerate the development and industrialization of home-made key accessories, and strive to make breakthroughs in color filters, glass substrate, polarizers, new-type backlight, certain production equipment and materials. The program also proposed to gradually complete the industrial chain and boost independent development capacity of panel displays.

In the Report on the Work of the Government 2006, the Chinese government proposed to list new-type displays as one of the special projects of industrialization of major national high technologies. The 2008 Report further proposed to continue the implementation of the said special projects.

In April 2009, the National Development and Reform Commission and the Ministry of Industry and Information Technology jointly issued the Restructuring and Invigorating Program for the Electronic Information Industry. It proposed to step up the support of taxation and financial policies, further improve and extend the validity period of preferential policies regarding LCD and other new-type display devices, make breakthroughs in key technologies of new-type display devices, enhance the industrial capacity for independent development and gradually take the initiative in the industrial development. It also made a point to support the construction of the 6th (and more advanced) generation of LCD panel production lines. In addition, the program declared to give interest subsidy to 6th to 8th generations of LCD panels, modules, accessory devices, manufacturing and test equipment since 2009. In the 2010–2012 Program for the Development of Panel Display Industry issued in the same year, it was also explicitly stated that as of 2012, China's panel display production technology should reach internationally advanced level and should have set up a technological innovation system with independent intellectual property rights in areas including process, materials and devices, with global share rising to above 20% from the current 5% regarding industrial scale, so as to support the transition and upgrading of China's color TV industry.

On May 8, 2010, the National Development and Reform Commission issued the 2010 Notice about the Decision to Continue Implementing the Special Project of Strategic Transition of the Color TV Industry (NDRC Hi-tech [2010] No. 1065). It set the R&D and industrialization of key accessory materials related to LCD as a priority, and proposed to give major support to the development of color filters (color resistors included), polarizers (including PVA films and TAC films), liquid crystals, driver IC, wet chemicals (mainly consisting of etchant, detergent, developing liquid, photo-resist and agent to lift off photo-resist), target materials, backlight (including brightness enhancing films and reflectors) in the TFT-LCD area.

On October 10, 2010, the State Council released the Decision to Accelerate the Development of Strategic Emerging Industries, listing new-type display as one important area in the development of the new-generation information technology industry.

On July 13, 2011, the Ministry of Science and Technology released the National Program for Scientific and Technological Development during the 12th FYP. Regarding the development of "a new generation of information technology", it declared to implement the industrialization of technologies like new-type display and focus efforts on the R&D of key core technologies such as light-emitting materials of organic light emitting display and TFT arrays.

In November 2011, the Ministry of Industry and Information Technology released the Program for the Innovation of Industrial Technology during the 12th FYP. It classified key devices of new-type panel display as a major development program in electronic information manufacturing industry and proposed to develop key technologies of new-type display such as TFT-LCD and PDP.

In February 2012, the Ministry of Industry and Information Technology released the Program for the Development of the Electronic Information Manufacturing Industry during the 12th FYP, proposing to accelerate the development of panel display and focus efforts on the development of new-type panel display and other strategic emerging industries. As for electronic materials, the priorities of development were glass substrate used in higher-generation of LCD, polarizers, color filters, liquid crystals and other materials, large-size target materials, high-purity electron gas and reagents; while the priorities regarding LCD components are to improve the performance of TFT, raise the carrier mobility and transmission of LCD panel and cut production costs. In particular, in the Program for the Development of Basic Electronic Materials and Key Components during the 12th FYP, a sub-plan, the development of TFT-LCD materials was listed as a major task in the development of electronic materials. It proposed to prioritize the development of materials related to a higher generation of TFT-LCD, which mainly included large-size glass substrate, liquid crystal mixtures, key new-type monomer materials, polarizers, related optical thin films, color filters and related materials, large-size target materials, high-purity electron gas and reagents.

The State Council approved the National Program for the Development of Strategic Emerging Industries during the 12th FYP, proposing to speed up the building of new information technologies. The program emphasized on increasing international competitiveness and new display industry development. It stated that the market should play a fundamental role in resources allocation, and that independent innovation should be strengthened.

In terms of national input, LCD basic research is mainly funded by the National Natural Science Foundation of China and Program 973.[①]

LCD basic research under the Foundation concentrates on the Department of Information Sciences and the Department of Material Sciences (Tables 10-8 and 10-9). The Foundation has attached great importance to the two sectors since 2001. Both general projects and key projects have taken up large proportions in terms of project approval and financial input.

Table 10-8 The number of projects and expenditure of the Department of Information Sciences of the NSFC in the previous years

Year	General Projects		Investment in General Projects		Key Projects		Investment in Key Projects	
	Number	Proportion (%)	Funds (million RMB)	Proportion (%)	Number	Proportion (%)	Funds (million RMB)	Proportion (%)
2001	454	12.88	78	12.21	14	11.29	23	12.78
2002	574	9.89	113	9.78	12	5.77	22	7.15
2003	647	10.17	130	9.85	42	16.67	63	17.47
2004	784	13.41	163	12.64	25	11.16	47	14.14
2005	1,050	15.34	222	12.73	44	14.52	85	17.42
2006	1,102	15.20	263	12.88	24	8.66	46	10.27
2007	792	10.27	208	9.15	42	11.26	81	12.68
2008	957	10.72	295	10.20	45	10.59	90	11.54
2009	1,085	10.78	340	10.28	46	11.76	94	12.98
2010	1,430	10.97	473	10.45	59	13.53	138	14.25
2011	1,611	10.51	955	10.62	66	13.28	189	13.26
2012	1,724	10.21	1,328	10.64	71	13.20	210	13.40
2013	1,646	10.16	128	10.67	82	14.54	237	14.25

Data source: The website of the NSFC.

① There are new display projects in Program 863 too. However, due to insufficient data and the project's focus on application research and commercialization, this project is not analyzed in this section.

Table 10-9 The number of projects and expenditure of the Department of Material Sciences of the NSFC in the previous years

Year	General Projects		Investment in General Projects		Key Projects		Investment in Key Projects	
	Number	Proportion (%)	Funds (million RMB)	Proportion (%)	Number	Proportion (%)	Funds (million RMB)	Proportion (%)
2001	683	19.38	136	21.28	33	26.61	43	23.89
2002	911	15.69	198	17.13	41	19.71	59	18.94
2003	997	15.68	223	16.84	17	6.75	30	18.47
2004	1,170	20.01	267	20.72	33	14.73	51	15.54
2005	1,447	21.14	363	20.83	56	18.48	101	20.89
2006	1,563	21.56	432	21.21	42	15.16	73	16.47
2007	1,118	14.45	356	15.63	61	16.35	109	17.19
2008	1,321	14.81	458	15.88	76	17.89	149	19.14
2009	1,488	14.79	534	16.17	65	16.62	13	17.96
2010	2,078	15.95	779	17.21	63	14.45	148	15.28
2011	2,606	17.00	1,564	17.40	73	14.69	212	14.88
2012	2,729	16.16	2,182	17.49	82	15.24	180	15.28
2013	2,620	16.18	2,096	17.47	82	14.54	248	14.93

Data source: The website of the NSFC.

Between 1999 and 2011, the Foundation subsidized 287 projects on basic research on liquid crystals and their application in display devices (including 6 projects of the National Science Fund for Distinguished Young Scholars and 5 Key Projects), with an accumulated fund of 99.34 million RMB. In recent years, triggered by the development of strategic emerging industries, LCD basic research has become a hot-spot issue, evidenced by the Foundation's subsidies as well as its rapid growth (Figure 10-3).

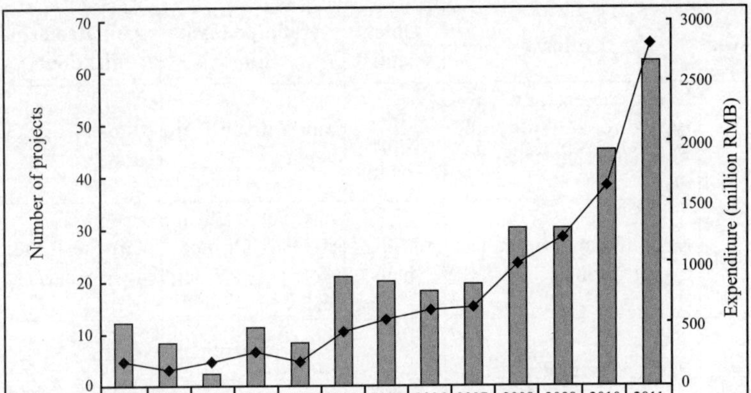

	1999	2000	2001	2002	2003	2004	2005	2006	2007	2008	2009	2010	2011
Number of Projects	12	8	2	11	8	21	20	18	19	30	30	45	62
Expenditure	203	137	200	278	195	435	544	621	639	1,005	1,214	1,649	2,814

Figure 10-3 Projects and expenditure of basic research on LCD industry under the NSFC

Data source: The website of the NSFC.

Of the 10 LCD research projects from 1998 to 2011 under Program 973, chief scientists all came from national key universities and Chinese Academy of Sciences (Table 10-10). Enterprises lack necessary abilities to conduct major fundamental science projects.

Table 10-10 Projects related to the field of LCD under Program 973 (National Key Scientific Research Program included)

Year	Project	Chief scientist	Affiliated institute	Affiliated organization(s)
1998	Research on the structural performance, molecular design, micro-structure design and preparation of photo-electronic functional crystals	Min Naiben	Nanjing University	Ministry of Education, Chinese Academy of Sciences
2002	Research on major basic questions regarding organic/polymeric luminescent materials	Shuai Zhigang	Institute of Chemistry of Chinese Academy of Sciences	Chinese Academy of Sciences

Continued

Year	Project	Chief scientist	Affiliated institute	Affiliated organization(s)
2003	Basic research on new-type micro-display and field emission panel display	Xu Ning-sheng	Sun Yat-sun University	Ministry of Education
2006	Basic research on molecular electronics	Yao Jian-nian	Institute of Chemistry of Chinese Academy of Sciences	Chinese Academy of Sciences
2009	Basic research on materials of organic/polymeric panel display	Wang Lixiang	Changchun Institute of Applied Chemistry of Chinese Academy of Sciences	Chinese Academy of Sciences
2010	Basic research on heterogeneous compatible integration and functional micro-structural system of new-type photo-electronic device	Ren Xiaomin	Beijing University of Posts and Telecommunications	Ministry of Education
	Basic research on the quality optimization and applications of new-type micro-display and field emission panel display	Xu Ning-sheng	Sun Yat-sen University	Ministry of Education
	Structure, performance and preparation of photo-electronic functional crystals	Wang Mu	Nanjing University	Ministry of Education
2011	Applied mathematical research on several major demands in the information fields and related fields	Ma Zhiming	Academy of Mathematics and Systems Science of Chinese Academy of Sciences	Chinese Academy of Sciences
	Research on key basic questions concerning the manufacturing and equipping of high-performance LED	Liu Yan	Research Institute of Tsinghua University in Shenzhen	Economy, Trade and Information Commission of Shenzhen Municipality

In the past 30 years when the LCD industry emerged, Chinese enterprises seldom joined new technology R&D. In the 1980s, guided by the Ministry of Electronics Industry, Chinese electronics industry began to study liquid crystals. But the projects stopped in the 1990s because of the market-oriented reform. The industry relied more on imports and was separated from the international LCD panel industry. Although LCD has already become a mainstream technology for panel display, China switched its support to some so-called "more advanced" projects. LCD, therefore, was not included into the national S&T plan. Things have improved over the years. China pays enough attention to LCD basic research as evidenced by policies and S&T plans in the past decade. However, since LCD industry is a sector with rapidly changing technologies and in need of long-term basic research, research ability does not go hand in hand with industrial technological capability in China.

10.6 Conclusion and discussion

The LCD industry highlights the importance of basic research. Now, the TFT-LCD industry concentrates on Japan, Korea and China in East Asia; Europe and the United States are basically excluded from the industry. The United States was the pioneer of world LCD research and technological innovation, but it failed to succeed later. Japan detected LCD's future, entered the industry and soon became a leader via enterprises-driven industrial basic research and integration among enterprises, universities, and research institutes. Korea picked a similar path: Incumbent firms focused on basic research, and the government planned and arranged innovation-oriented research. Although Chinese Taiwan is a strong force of LCD production, it is weaker than Japan and Korea in basic research in that the role of acquisition in boosting innovation is limited.

The Chinese mainland has entered the industry early

and its scientific community has a strong research capability[①], but it did not make much progress in core technological innovation. LCD panel projects were sought after by many regions, but the industry was driven by investment instead of technological capability, which showed certain negative effects. BOE and other domestic leading enterprises did raise Chinese LCD industry's status and broke foreign capital's monopoly by bringing in and digesting foreign technologies and re-innovating. But China failed to dominate the industry because of insufficient competitiveness and innovation.

In general, China did not invest enough in TFT-LCD basic research. Presently, both Central and local governments have policies and funds in support of the industry, but most of them are for importing high-generation production lines. Next-generation technological R&D does not attract enough attention yet. In the short run, to bring in TFT-LCD technologies and production lines help build domestic capacities and help national and local economies grow. Although the National Program for Scientific and Technological Development during the 12th FYP prioritized new display technology among new-generation information technologies, it did not arrange any TFT-LCD project. The LCD industry was regarded as a mature industry, while new technologies like TFT-LCD had little support (Ouyang, 2011).

Therefore, this industry explicitly shows that basic research is the key to industrial innovation. But the leader might not have the capability to sustain it, so Japan and Korea find their opportunity to catch up and become the new leaders. China failed in the early stage of the industry with less investment in basic research, but now, China is catching up quickly.

① LCD research work of universities, colleges and research institutes in the Chinese mainland is almost at the same level with their international counterparts. The RCA Company in the United States invented LCD in 1968, while Tsinghua University, Shanghai Organic Chemistry Institute and Shanghai Huadong Chemical Engineering College started LCD research in 1969. Later, Changchun Physics Institute, Nanjing No.55 Institute, Huazhong University of Science and Technology and Southeast University studied LCD and PDP (Lu & Cai, 2010).

References

Gao, L., Zhang, D. & Wang, H. What will it bring to the Chinese mainland from the research on TFT-LCD policy in Japan, Korea and Chinese Taiwan?. *Advanced Display*, 2005 (8): 18-23. [高露，张典耀，王红梅. 日、韩、我国台湾地区 TFT-LCD 产业政策研究及对我国大陆的启示. 现代显示，2005（8）: 18-23.]

Liu, X. *Globalization, Pursuing and Innovation*. Beijing: Science Press, 2008. [柳卸林. 全球化、追赶与创新. 北京：科学出版社，2008

Lu, F. & Cai, Y. Strategy and capacity: Seizing the opportunity of China's TFT-LCD industry. Report of the Advisory Committee for State Informatization, 2010. [路风，蔡莹莹. 战略与能力：把握中国液晶面板工业的机会. 国家信息化专家咨询委员会调研报告，2010.]

Ouyang, Z. LCD shouldn't be viewed as a full-fledged industry. *China Electronics News*, 2010-12-07. [欧阳钟灿. 不能把液晶显示当成熟产业. 中国电子报，2010-12-07.]

Ouyang, Z. Intensifying basic research to promote rapid development of new display industry. http://www.cena.com.cn/, 2011-08-30. [欧阳钟灿. 加强基础研究，促进新型显示工业快速发展. 中国电子信息产业网，2011-08-30.]

Tian, M. & Ye, F. *Theory and Technology of TFT-LCD Display*. Beijing: Science Press, 2010. [田民波，叶锋. TFT 液晶显示原理与技术. 北京：科学出版社，2010.]

Hong, J. The Research of development of China's TFT-LCD industry chain. Shanghai: Shanghai Jiao Tong University, 2009. [洪炯星. 中国 TFT 液晶产业链发展研究. 上海：上海交通大学,2009.]

Wang, D. Will China surpass the early starters in developing LCD industry. http://www.sciencenet.cn/, 2008-04-09. [王丹红. 中国液晶面板产业能否后来居上. 科学网，2008-04-09.]

Zhao, Y. Why Chinese LCD leaders invest more despite heavy losses. *People's Daily*, 2011-08-29. [赵永新. 中国液晶面板领先者为何"越亏越投". 人民日报，2011-08-29.]

Chapter 11

Industry-Driven Basic Research: Carbon Fiber Industry

Carbon fibers are a new fiber material with structural feature of carbon materials and morphological characteristics of fibers. Because of its superior mechanical properties, heat resistance, low density, chemical stability, excellent electric and thermal conductivity and abrasion resistance, carbon fibers and carbon fiber-reinforced polymers are widely applied in the areas such as ultracentrifuge drums of oil separation, material of first wall of nuclear reactors, car components and flywheels for energy storage, wind turbine blades, solar panel, parabolic lasers, solar water heaters and electrode material of fuel cells and so on (He, 2004). The preparation process of carbon fibers can be divided into purification and polymerization, spinning, oxidation and carbonization, graphitization and surface treatment. The whole process involves many disciplines including organic chemistry, textile industry, chemical machinery, high polymer chemistry, hydromechanics and so on. The industry of carbon fibers is a high-tech industry requiring complex technologies and multidisciplinary knowledge.

11.1 Basic research and innovation in the carbon fiber industry

The history of carbon fibers can be traced back to 1860, when J. Swan, a British, carbonized slim strands of paper to produce carbon filaments used in lamps. In 1879, Swan immersed cotton yarn into sulfuric acid, then parched and carbonized the product (an alternative method was extruding nitrocellulose through a die orifice and then carbonizing the filaments), and got a patent for the process. Despite the significance of this invention, no practical use was found of it, because at that time, people had not solved the problem of achieving a vacuum environment inside the bulbs yet. Nonetheless, this process laid the foundation for the invention of synthetic fibers afterwards. Another important event that happened in 1879 was that American inventor T. Edison started the research on bulb-used carbon filaments. He made filaments out of the mixture of lampblack and tar, and carbonized them into bulb filaments; afterwards he dissolved cotton fibers in zinc chloride solution, then sprayed the product into a liquid to solidify, and after it dried, carbonized the product in a vacuum to produce bulb filaments. Finally, on October 21, 1879, Edison made a major breakthrough: The bulb with the carbon filament could work for 45 hours straight. However, in 1910, Colidge discovered a way of manufacturing tungsten filaments, which could serve as bulb filaments in place of carbon filaments. This great invention made the carbon bulb filaments obsolete. Soon after WWII, the United States started to carry out research on the invention of rayon-based carbon fibers with the aim of finding suitable thermal insulation materials of strategic weapons and nose cone of missiles. The goals of the United States regarding the R&D of large-scale rockets and satellites as well as overall improvement of the performance of planes gave rise to huge demands for new-type structural materials and ablation-resistant materials. As a result, carbon fibers found a new area of application and witnessed rapid development.

At present, there are three broad raw material systems of carbon fibers, which are rayon-based, PAN-based and pitch-based (He, 2004). Table 11-1 shows the history of innovation and development of the carbon fiber industry.

(1) Rayon-based carbon fibers. In 1891, Cross and Bevan, two Americans, invented rayon fibers. In 1905, Muller used diluted sulfuric acid and sulfate as coagulators, significantly improving the performance of rayon fibers. In 1909, rayon fibers get industrialized, providing materials for later development of rayon-based carbon fibers. In 1950, researchers at Wright-Patterson Air Base in Ohio, the United States started to develop rayon-based carbon fibers as an ablation-resistant material of the rocket nozzle and nose cone of the louder. Abbott, an American textile enterprise, carbonized rayon cloth at the temperature of 1,000 ℃, turning it to carbon cloth used for thermal resistance. This incurred the development of rayon-based carbon fibers by UCC and Hitco (two American enterprises). In 1959, UCC took the lead in launching a kind of general-purpose rayon-based carbon fiber named

"Thornel-25", which was the first commercialized rayon-based carbon fiber in the world. In 1964, UCC again rolled out high-performance rayon-based carbon fibers. This kind of carbon fibers is mainly used as ablation-resistant and thermal-isolation materials. Although the production of rayon-based carbon fibers accounts for less than 1% of the global production of carbon fibers, it is nonetheless an important strategic material by far. The Soviet Union was a major producer of rayon-based carbon fibers, ranking top in the world no matter in terms of basic research, product quality or production scale, while the Belorussian enterprise Svetlogorsk is a typical example in the post-USSR era.

(2) PAN-based carbon fiber. In 1959, Akio Kondo based at Osaka Industrial Laboratory of Japan developed a new way to make carbon fibers from PAN. This significant innovation prompted great development of the carbon fibers industry and has now become the mainstream. However, Akio Kondo did not manage to produce high-performance PAN-based carbon fibers. In 1973, the Japanese enterprise Toho Tenax got licensed to carry out volume production with this method for the first time; but the products were merely low-modulus carbon fibers, while another Japanese enterprise, Toray, started the development of carbon fibers made from civil-use PAN fibers as early as 1962. However, the performance of this kind of carbon fiber was very poor and could not meet the demands of customers. In order to meet market demands, Toray had no choice but to conduct basic research on its own to develop carbon fibers with better performance, and especially to improve the quality of the PAN precursor. The long period from the invention of PAN-based carbon fibers to its raising to the dominant position of this industry is filled with arduous efforts of many researchers from research institutes and enterprise. In 1963, W. Watt and W. Johnson with the Royal Aeronautical Establishment of the United Kingdom applied pressure during the process of pre-oxidation to inhibit contraction of the precursor in the heat treatment process, and by doing so acquired the technology of manufacturing PAN-based

carbon fibers with high performance, which also laid the foundation for modern carbon fiber manufacturing process. Rolls-Royce developed the device capable of continuous production of carbon fibers; the feature of the device was the ability of regulating the drum rotation speed to alter the tensile strength, which was the technological basis of modern carbon fiber production. All such research was generously funded by the British government, and they in turn made the United Kingdom a global center for the R&D of high-performance PAN-based carbon fibers in the 1960s. It is estimated that global demand for PAN-based carbon fibers will exceed 110 thousand tons in 2018. ①

(3) Pitch-based carbon fiber. In 1965, Sukio Ohtani based in Gunma University of Japan invented pitch-based carbon fibers; 5 years later, the Japanese enterprise Kureha launched the general-purpose pitch-based carbon fibers. In 1975, UCC produced the high-modulus pitch-based carbon fiber Thornel-P made from mesophase pitch and quickly commercialized it. Pitch-based carbon fibers can be divided into two categories: One is for general purpose, made from isotropic pitch, while the other is of high performance and made from anisotropic pitch. Currently, the global industrial production of high-performance pitch-based carbon fiber filaments is mainly handed by three enterprises: BP Amoco (the former UCC) in the United States and Mitsubishi Chemical Corporation and Nippon Graphite Fiber in Japan.

Table 11-1 Main development history of carbon fiber industry before the 21st century

Year	Major discoveries and inventions
1860	Swan invented bulbs with carbon filaments.
1878	Swan produced carbon filaments from cotton yarn.
1879	Edison produced carbon filaments from lampblack and tar/cotton yarn and bamboo filaments.
1882	Bulbs with carbon filaments were put into practical use.

① http://www.ocn.com.cn/free/201109/xianwei081429.shtml.

Continued

Year	Major discoveries and inventions
1910	Colidge invented bulbs with tungsten filaments.
1950	Wright-Patterson Air Base of the United States started to develop rayon-based carbon fibers.
1959	UCC of the United States produced "Thornel-25", a low-modulus rayon-based carbon fiber.
	Akio Kondo (Osaka Industrial Laboratory of Japan) developed PAN-based carbon fibers.
1962	Toray of Japan started the production of low-modulus PAN-based carbon fibers (0.5 ton per month).
1963	W. Watt and W. Johnson with the Royal Aeronautical Establishment (RAE) of the United Kingdom invented the manufacturing process of high-performance PAN-based carbon fibers.
1964	British enterprises Courtaulds, Morganite and Rolls-Royce produced PAN-based carbon fibers with the RAE process.
1965	Sukio Ohtani based in Gunma University of Japan invented pitch-based carbon fibers.
	UCC of the United States started to produce high-modulus rayon-based carbon fibers.
1970	Japanese enterprise Kureha produced pitch-based carbon fibers (10 tons per month).
1971	Japanese enterprise Toray mass-produced PAN-based carbon fibers T300 (1 ton per month).
1972	American enterprise Hercules started to produce PAN-based carbon fibers.
	Japan produced carbon fiber fishing poles; the United States produced carbon fiber golf clubs.
1973	Toho Rayon of Japan started to produce PAN-based carbon fibers (0.5 ton per month).
1974	Carbon fiber fishing poles and golf clubs evolved rapidly; Toray of Japan expanded its production to 13 tons per month.
1975	Carbon fiber tennis rackets got commercialized.
	UCC of the United States produced the high-modulus pitch-based carbon fiber Thornel-P made from mesophase pitch and quickly commercialized it.
1976	Japanese enterprise Sumitomo Chemical and American enterprise Hercules jointly set up an enterprise.
1979	Japanese enterprise Toray and Asahi Kasei set up a joint enterprise.
1980	American enterprise Boeing put up demands for high-strength and highly tensile carbon fibers.

Continued

Year	Major discoveries and inventions
1984	Toray of Japan developed high-strength medium-modulus carbon fibers T800.
1986	Toray of Japan developed high-strength medium-modulus carbon fibers T1000.
1989	Toray of Japan developed high-modulus medium-strength carbon fibers M60.
1992	Toray of Japan developed high-modulus medium-strength carbon fibers M70J, with a young modulus as high as 690PGs.

There are four difficulties regarding the basic research and product R&D of carbon fibers. First, it requires the know-how of many industries; second, the manufacturing principle covers knowledge including chemical engineering, polymer synthesis, chemical fiber and materials science; third, the technological integration involves many complex knowledge elements; lastly, there are as many as 19 indexes of qualify control, with a long technological process. Currently, key technologies of carbon fibers are mainly held by a handful of enterprises in the rich world, which enforce block on technologies against China. ① China began to develop carbon fibers and the precursor in the 1960s, but the quality of products was poor and the production was unstable, and most domestic consumption of carbon fibers were imported. In recent years, Chinese carbon fiber enterprises experienced rapid expansion of production driven by global demand, with an overall production capacity of the existing 23 enterprises of 4,000 tons per year. Large carbon fiber projects under construction or to be constructed include Sinosteel, Jinlin Carbon, Jiangcheng Carbon Fiber, Bluestar Carbon Fiber, and CPC Jilin Chemical Carbon Fiber. ② Local governments such as Jilin provincial government also encourage enterprises to accelerate

① http://www.ccin.com.cn/ccin/news/2011/10/13/202999.shtml.

② New materials planning will be released on Sept.7, and market space will be opened, http://www.cnstock.com/, 2011-09-02.

the development of carbon fiber industry.[①] However, China's research efforts in the area of carbon fiber precursor are not enough, which, in addition to the non-disclosure of technology of each enterprise due to market competition, causes the slow development of this area, and the industry can only maintain a small-scale and low-quality precursor production.

To determine the global distribution of basic research and technological innovation capacity in the area of the carbon fiber industry, we carried out an analysis on the SCI and DII of papers and patents related to carbon fiber technology, with the aim of finding out whether basic research has an impact on the capacity of industrial technological innovation.

(1) The top 10 countries that published the most papers in SCI during 1981–2012 are listed in Table 11-2. We can see that China is not lagging far behind the United States, which ranks first (considering the proficiency in English of Chinese scientists), and is ahead of Japan, which shows that China has strong basic research on carbon fibers. The explanations for the good performance in published papers of China are that carbon fiber are an emerging industry that gets much attention from the government and that there is steady development of the material major in Chinese universities and research institutes.

① Jinlin City of Jinlin Province is gradually establishing a complete carbon fiber industrial cluster. It plans to achieve the production capacity of 30 thousand tons of precursors, 12 thousand tons of carbon fibers and 50 thousand tons of carbon fiber products in 2015, with the aim of driving the collective progress of the carbon fiber industry and building a "carbon valley" of China.

Table 11-2 Top 10 countries with the most SCI papers published about key carbon fiber technologies published during 1981–2012

Ranking	Country	Numbers of papers	Global share
1	US	9,507	24.33%
2	China	5,932	15.18%
3	Japan	4,767	12.20%
4	UK	2,920	7.47%
5	Germany	2,361	6.04%
6	France	2,067	5.29%
7	Korea	1,803	4.62%
8	Canada	1,329	3.40%
9	India	1,118	2.86%
10	Spain	1,011	2.59%

(2) In terms of funding for SCI papers, the top 10 funding institutes or enterprises during 1981–2012 are listed in Table 11-3. We can see that the number of papers funded by the National Natural Science Foundation of China is much higher than those funded by the US Natural Science Foundation and the EU. Program 973 also funded much basic research (with 152 papers). The Fundamental Research Funds for the Central Universities, Support Project for New Century Talents in Institutes of Higher Education under the Ministry of Education and Program 863 also funded many papers. This reveals that the Chinese government greatly drives the development of basic research on carbon fibers through various national plans.

Table 11-3 Major funding institutes or enterprises with the most SCI papers published in the area of carbon fiber research during 1981–2012

Funding institute or enterprise	Country	Number of papers	Proportion (%)
National Natural Science Foundation of China (NSFC)	China	927	2.37
US Natural Science Foundation (NSF)	US	325	0.83
Program 973	China	152	0.39

Continued

Funding institute or enterprise	Country	Number of papers	Proportion (%)
EU	—	125	0.32
US National Institutes of Health (NIH)	US	121	0.31
Engineering and Physical Sciences Research Council (EPSRC)	UK	71	0.18
Fundamental Research Funds for the Central Universities of China	China	70	0.18
Chinese Academy of Sciences (CAS)	China	60	0.15
Ministry of Education, Culture, Sports, Science and Technology	Japan	60	0.15
Support Project for New Century Talents in Institutes of Higher Education under the Ministry of Education	China	60	0.15
Program 863	China	50	0.13
Air Force Research Laboratory (AFRL)	US	45	0.12
Natural Sciences and Engineering Research Council (NSERC)	Canada	45	0.12
Office of Naval Research	US	45	0.12
Australian Research Council	Australia	42	0.11
Shanghai Education Committee	China	36	0.09

(3) In terms of industrial patent application, the enterprises and institutes that have the most patent applications in the field of carbon fiber during 1963–2012 are listed in Table 11-4. A large proportion of carbon fiber patents are held by Japanese enterprises, and American and German traditional chemistry enterprises also hold some patents, while the number of patents held by any Chinese enterprise is no more than 100. Surprisingly, two Chinese universities (Donghua University and Harbin Institute of Technology) have more than 200 PCT patents in all.

Table 11-4 Major enterprises with the most carbon fiber patents accumulated during 1963–2012

Enterprise/Institute	Country	Numbers of patents
Toray	Japan	1,307
Mitsubishi Rayon Company	Japan	763
DuPont	US	396
Toho Rayon	Japan	323
Bridgestone	Japan	270
GE	US	252
Teijin	Japan	210
Toyota	Japan	398
Toshiba	Japan	184
Showa Denko	Japan	182
Panasonic	Japan	317
Sumitomo	Japan	177
Kaneka	Japan	150
Asahi Kasei	Japan	144
Osaka	Japan	132
Donghua University	China	126
UBE Industries	Japan	122
Honda	Japan	120
Boeing	US	118
Taiheiyo Cement	Japan	118
Samsung	Korea	117
3M	US	116
Nissan	Japan	111
Honeywell	US	109
Harbin Institute of Technology	China	107
Kuraray	Japan	107

Continued

Enterprise/Institute	Country	Numbers of patents
L'Oreal	France	105
Nippon Denso	Japan	102
Airbus	Germany	101
Voith	US	101
Canon	Japan	99
National Institute of Advanced Industrial Science and Technology (AIST)	Japan	98
Nisshin Kogyo	Japan	94
Yokohama Rubber	Japan	86
Sekisui Chemical Group	Japan	82
BASF SE	Germany	79
Sony	Japan	79

From the above analysis, we can see that Japanese enterprises hold absolute advantages, even monopoly, in this industry, thanks to their perfect combination of basic research and industrial technological innovation. In this regard, American enterprises share the common feature with their Japanese counterparts, and therefore enjoy relative advantages in the carbon fiber industry. By comparison, China shows an imbalanced formula of scientific research capacity and industrial innovation capacity: In the past 20 years, China had paid constant attention to tackling key problems of carbon fibers, but it was without much effect. Most of existing 30-plus Chinese carbon fiber enterprises are in low-level redundant development, and their technologies are not sufficient (Pang, 2012). Currently, production of high-performance carbon fibers is basically unseen in China. Domestic products are equivalent to or inferior to T300 carbon fibers, while T700 carbon fibers are still under engineering research, and T800 and MJ carbon fibers still under development.

11.2 Basic research of the carbon fiber industry in foreign countries

The carbon fiber industry is an industry under the dominance of Japanese enterprises. Since 1959 when Osaka Industrial Laboratory developed the manufacturing process of rayon-based carbon fibers, Japanese enterprises set foot in and gradually built up monopoly in the carbon fiber industry. One of their important lessons is long-term attention paid to basic research by enterprises, which leads to sustained industrial leadership.

Toray is an outstanding representative among Japanese enterprises. It was established in 1926 and started to develop carbon fibers as early as the 1960s. However, because the carbon fibers initially produced were of poor quality, the production was forced to stop for a time. Then Toray enhanced its scientific research and actively participated in international technological cooperation and communication, and soon succeeded in the development of carbon fibers with high strength and high modulus. Toray upholds the philosophy that "R&D is the foundation of the very survival and development of the enterprises" and sets up a basic research institute under the R&D Department to work on exploratory research in the field of carbon fibers. During the past four decades, Toray remained one of the few carbon fiber enterprises around the globe that persisted in conducting basic research on carbon fibers. Apart from the R&D Department (which mainly includes a fiber and textile research laboratory, a basic research laboratory and an exploratory R&D laboratory), it also set up the "Toray Research Center" in 1978 with the main task of conducting basic research on carbon fibers and relevant textiles. This research center is seen by senior management of the enterprise as strategically valuable, so it is equipped with advanced devices and is very powerful (see Figure 11-1). It is exactly because they stick to basic research oriented by industrial demands that discipline Toray became the global leader of the carbon fiber industry after 1980s.

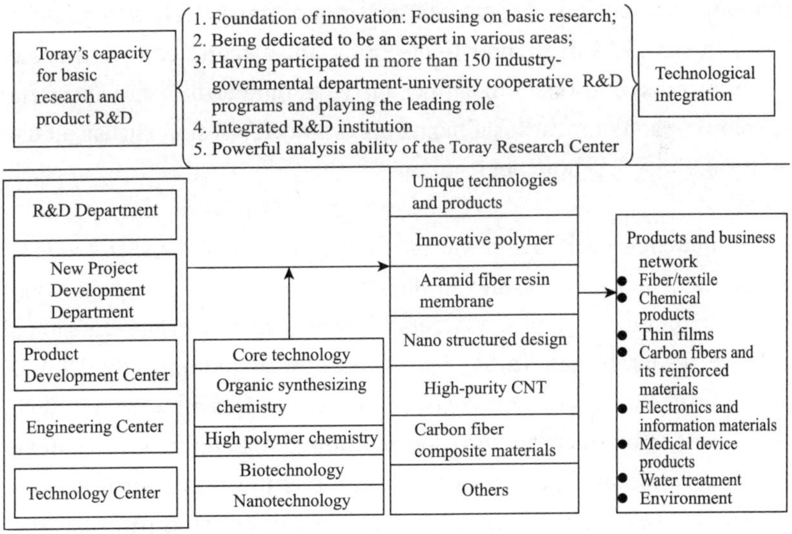

Figure 11-1 Toray's conceptual R&D representation

Data source: The website of Toray.

Toray understands it clearly that the research on carbon fibers is a comprehensive discipline, so the enterprise devotes itself to scientific research in areas of organic synthetic chemistry, high polymer industry, biological technology and nanotechnology and finally turns itself into a globally leading supplier of carbon fiber materials. There are three beliefs that are always being implemented in Toray: First, under no circumstances should the focus of Toray's basic research be driven by the trend of the times; second, the enterprise shall be the expert in a broad array of scientific fields; lastly, the enterprise shall boast of a team made up by a constellation of experts (i.e. the Research Center). Toray attributes its biggest advantage to its superiority in R&D and regards the strengthening of basic research and basic technology as the most important management problem of the enterprise.

Apart from Toray, Mitsubishi Rayon Company, Teijin and several other Japanese big enterprises also attach great importance to basic research on carbon fibers and the composites and have been conducting comprehensive supporting research. These enterprises believe that although the last four decades had seen enormous R&D activities of Japan in the field of carbon fibers, they must keep deepening basic research on carbon fibers to further improve the performance, because the requirements on carbon fibers from the sectors using carbon fibers are becoming ever more demanding.

Founded in 1918, Teijin was the first Japanese enterprise to produce rayon fibers and now remains in the leading position of the production and technologies of synthetic fibers. With no less importance attached to basic research than Toray, Teijin had formulated an overall R&D strategy (including basic research) of the enterprise from early on, with the aim of strengthening cooperation among subsidiaries to achieve practical application of research as early as possible. Teijin values cooperation between enterprises and universities on basic research, and also cooperated from time to time with researchers who are actively involved in the front line of high polymer chemistry, pharmacy, nanotechnology and biotechnology. Besides, Teijin strives to align burgeoning technologies and other advanced technologies with market demands through participation in national projects or joint R&D with universities. Teijin conducts a lot of laborious R&D activities in areas such as the relationship between the micro-structure and performance of carbon fibers, research on rapture failure of carbon fibers and related composite materials. It also attends meetings of national fiber science to exchange on new trends from time to time. Teijin enjoys independence in carrying out basic research, which is featured with a free atmosphere and without the influence of commercialization.

Outside Japan, there are other enterprises that pay long-term attention to basic research, such as American enterprises Zoltek and Hexcel, as well as the German enterprise SGL Group. During the past 20 years, carbon fiber-reinforced composites got the universal attention from all countries in the world. Developed countries including European countries, the United States and Japan all ranked it among the priorities in their national development plans, and the governments encouraged suitable enterprises to strengthen R&D and production capacity of carbon fibers through direct subsidy, tax reduction or exemption and other polices. For example, the US government started a joint research by 3 national laboratories, 15 universities and 50 material suppliers in 2002 and planned to develop a new-generation automobile made of composite materials, which is with less energy consumption and environmentally friendly within 10 years. This move prompted further basic research on carbon fibers.

11.3 Basic research and investment in China's carbon fiber industry

China has always valued the development of carbon fiber technologies. Presently, there are several departments in charge of related activities. However, the decentralized management unavoidably results in low efficiency. To worsen the situation further, domestic enterprises generally do R&D separately and keep their findings secret from each other. As a result, although considerable manpower and money are put into R&D, the allocation of related resources is inefficient, and there lacks a mechanism to encourage cooperation (Pang, 2011). Presently, China is lagging far behind leading countries such as Japan and the United States in terms of the quality of precursors, production technologies and devices, and the actual annual output of carbon fibers of domestic enterprises is only about 40 tons, much less than foreign enterprises, and with much poorer quality (Feng & Du, 2011).

11.3.1 Innovation and development of China's carbon fiber industry

China entered the carbon fiber industry at quite an early stage. Soon after the Japanese developed the process of drafting PAN precursors to make carbon fibers in 1959, China published papers on carbon fiber production in 1962, when the former Jinlin Applied Chemistry Institute had initiated research on PAN-based carbon fibers. Then China started to conduct R&D of PAN-based carbon fiber products, and finished the continuous pilot plant as of the early 1970s. Institutes such as Shanghai Synthetic Fiber Institute and the Institute of Coal Chemistry under the Chinese Academy of Sciences also carried out R&D activities and passed pilot plant test in the middle 1980s.

Chinese enterprises and academic community started to realize the importance of developing PAN-based carbon fiber in the early 1970s. The National Convention of Carbon Fibers and Composite Materials was held during November 13 to 24 of 1975, after which the government organized all parties related to carry out R&D of carbon fibers. In 1975, the State Development Planning Commission allocated 5 million RMB, mapped out a 10-year development plan and appointed 5 teams to carry out research on precursors, carbonization, structural materials, thermal isolation materials and inspection techniques. The plan involved more than 20 institutes and enterprises. For example, the Institute of Jilin Chemical Industry Company, Jilin Liaoyuan Petrochemical Plant, Chemical Fiber Plant of Lanzhou Chemical Industry Company and Shanghai Synthetic Fiber Institute carried out research on PAN precursors using different solvents; Shanghai Synthetic Fiber Institute, four carbon plants in Jilin, Shanghai, Lanzhou and Fushun, Shanxi Combustion Chemistry Institute and the Institute of Chemistry under the Chinese Academy of Sciences were in charge of research on carbonization; applied research of textiles and materials was also arranged. According to statistics, ministries and commissions under the central government invested about 26 million RMB into commercial research on carbon fiber precursors and products from 1975 to the end of 1981 (Li, 2009). China's investment into research on PAN-based carbon fibers started from 1991, while carbon fibers and the composite materials have always been included in China's national scientific research programs, which were successively organized by the Ministry of Metallurgy, the Chinese Academy of Sciences and the Ministry of Chemistry (Chen, 2011).

However, the development of carbon fibers in China was an uphill struggle. Although China got the technology transfer from the British enterprises RK in 1985 and the program "Development and application of carbon fibers and the composite materials" was set up in Beijing by the UN Development Programme and the UN Industrial Development Programme in 1986, it still lacked fundamental breakthroughs in carbon fiber technologies and the industry had not yet been established; therefore, in no way could it meet the demands of national defense. As a result, military industries resorted to imported carbon fibers, which caused a sharp decrease in national investment. From 1996 to 2000, research on carbon fibers in China experienced a sluggish period and almost stagnated (Li, 2009). In 2000, under the vigorous advocacy of Shi Changxu, an academician to both the Chinese Academy of Sciences and the Chinese Academy of Engineering, China once again directed its attention on the research on carbon fibers. During 2001–2005, China concentrated on the research of key technologies of two high-strength PAN precursors and carbon fibers with the aim of providing batches of carbon fibers with good quality to be used in national defense. In January 2003, the National Natural Science Foundation decided to fund two key carbon fiber programs by placing them under the category of "macro-regulation expenses". In 2005, carbon fiber projects were listed in Program 973. Urgent demand from the market and serious attention from the government gave rise to a "carbon fiber craze" in China after 2006: Several carbon fiber projects were listed in Program 863 and Program 973, and copious private capitals were poured into the industry. During 2008–2009, the National Development and Reform

Commission implemented a special project to industrialize technologies related to high-performance fiber composite materials. The priority of the project was the industrialization of production technologies and key devices of carbon fibers, aramid fibers, UHMWPE fibers and related high-performance composite materials, with the aim of breaking the bottleneck restricting the development of China's high-performance fiber and composite industry and improving the design, manufacturing, development and application of high-performance composite materials in China.

Generally speaking, after years of research, the applied technologies of carbon fibers in China are no longer at the mercy of other countries, but the overall level of technologies still needs to be raised and the production capacity is limited. China has successively set up small-scale pilot devices with respective annual output of hundreds of kilograms to several tons and production devices with annual output of dozens of tons. China did not start late in the carbon fiber industry, but the industrial development is very slow, and the production capacity of China can hardly match that of a single enterprise in developed countries or regions. At present, China's high-performance carbon fiber output only accounts for 0.4% of the world's total, with 90% of domestic consumption relying on imported carbon fibers. Besides, because the carbon fibers has important applications in military aircraft, missiles and carrier rockets due to its excellent mechanical properties and light weight, it was regarded as "strategic supplies" by developed countries, which have been enforcing long-term blockade on technologies for China's military products and applications through ban on sales and embargo. [①] This, to some degree, prevents Chinese enterprises from learning advanced technologies from the Western world. Currently, the patents of carbon fiber technologies are predominantly held by Japan and the United States, while Chinese enterprises lack core technologies

① At that time, Western countries signed "COCOM Treaty", which enforced the blockade on China's military technology and equipment, including carbon fiber technology and equipment.

with independent IPR and have not yet fully acquired key carbon fiber technologies. Taking the PAN-based carbon fibers as an example, there is a huge gap between most existing production lines in China and those with international advanced level in terms of production capacity of a single line, production cost, consumption per unit and product performance. Foreign enterprises have developed different models of PAN-based carbon fibers and applied them broadly in many fields, while the industrialization of carbon fibers in China has just started, and the independent production of many high-end products is currently beyond the means of domestic enterprises. The main contributing factor to this situation is the lack of breakthroughs in key technologies (such as PAN precursor technology) and devices, such as saturated steam drawing device and large-caliber high-temperature carbonization device, the absence of which limits the development of China's carbon fiber industry.[①] Carbon fiber products require large investment, and generate high added value afterwards. However, Chinese enterprises are not investing enough in early-stage R&D.

Enhancing basic research oriented by industrial innovation is the ultimate way out for domestic carbon fiber enterprises to achieve international competitiveness. Because the development of PAN-based carbon fibers involves various scientific branches and engineering expertise, Chinese enterprises do not get enough means to take part in the basic research. Regarding the chain of "basic research–technological research–process–application of product and sales", currently the cooperation among participants in each stage (i.e. universities, research institutes and enterprises) is lacking, and an interest-sharing community is absent. National scientific research programs of carbon fibers lack strategic planning, emphasizing one-sidedly on scientific targets but failing to consider the development of the carbon fiber industry; additionally, there are too many state administrations and departments in charge of programs and the funding is dispersed. For

① http://www.miit.gov.cn/n11293472/n11293832/n11294072/n11302450/14470622.html.

example, the National Development and Reform Commission, the Ministry of Science and Technology, the State Administration of Science, Technology and Industry for National Defense, the National Natural Science Foundation, the General Armament Department of PLA and the State Administration of Aerospace Industry for the most part are only concerned with their own scope but do not take efforts to discover principles of key technologies. Besides, there are other reasons including misplaced focus on scientific problems in carbon fiber research programs (for example, focusing solely on carbon filament research but neglecting research on precursors), insufficient participation of chemical fiber enterprises in research programs (which results in disconnection between theoretical research and industrial development) and weak technological capacity of enterprises. Lack of effective basic research and several profound scientific questions seriously limit the development of high-performance PAN-based carbon fibers in China.

11.3.2 Policies and technological plans promoting basic research on carbon fibers

China provides strong support for the development of the carbon fiber industry through national S&T policies and it has been listed into national plans over the years. China entered in and then exited from the carbon fiber industry long ago; in recent years, the Chinese government has sensed the significant development prospect of this industry, so it once again attached great importance to this industry and started to enhance R&D of carbon fibers through the guidance of policies and plans.

On February 9, 2006, the State Council issued the Outline for the National Program for Long- and Medium-Term Scientific and Technological Development (2006–2020). In the section of "Basic materials", it proposed to carry out major R&D activities so as to meet the demands for the development of national basic industries by inventing the manufacturing processes of high-performance composite materials as well as large and super-large composite components.

On October 25, 2006, the Ministry of Science and Technology issued the Outline for the National Key Basic Research Program (Program 973) during the 11th FYP. It pointed out that China's basic research on hi-tech new materials was weak, and it was imperative to strengthen its innovation capacities for key technologies of certain high-end products. The plan set "scientific basis of the structure and shaping control of a new generation of structural materials" as an important research area. In the National Program for the Development of Basic Research during the 11th FYP, which was released in the same year, research on new-type structural materials with high-performance and surface technologies was identified as a major research direction in the field of material science.

In November 2006, the Ministry of Science and Technology issued the National Program for Scientific and Technological Development during the 11th FYP. It proposed to strengthen efforts to tackle key problems in material technologies and the focus was on the research of manufacturing technologies of high-performance composite materials and composite components.

In May 2007, the National Development and Reform Commission (NDRC) released the Program for the Development of High-Technology Industry during the 11th FYP, proposing to accelerate the R&D of aerospace materials and the focus was the R&D of high-performance carbon fibers and relevant composite materials.

On October 10, 2010, the State Council released the Decision to Accelerate the Development of Strategic Emerging Industries, listing the industry of new materials as one of the seven strategic emerging industries and proposing to raise the development of carbon fibers, aramid fibers, polyethylene fibers and other high-performance fibers and composite materials to a higher level.

On April 27, 2011, the National Development and Reform Commission released the Guidance Directory of the Textile Industry (2011). We can see from the directory that the R&D and production of organic and inorganic high-performance fibers and related products including carbon fibers, high-strength and high-modulus polyethylene fibers (UHMWPE), silicon carbide fibers and high-strength glass fibers were all listed as items whose development would be encouraged and highly valued by the nation.

On July 7, 2011, the Ministry of Industry and Information Technology released the Development Guideline of Industrial Key Generic Technologies (2011), which emphasized that carbon fibers, para-aramid fibers and high-strength and high-modulus polyethylene fibers belonged to hi-tech fibers.

On July 13, 2011, the Ministry of Science and Technology released the National Program for Scientific and Technological Development during the 12th FYP. It proposed to vigorously develop advanced structural and composite materials in the field of new material industry, and to implement the industrialization of technologies of high-performance fibers and composite materials. The plan also raised the point of prioritizing the development of key technologies, which can stabilize the production of and lower the cost of mass-production of high-performance fibers, producing a series of high-strength, high-strength and medium-modulus, high-modulus, and high-strength and high-modulus carbon fiber products, and accelerating the development of a new generation of high-performance fibers with independent intellectual property rights.

In November, 2011, the Ministry of Industry and Information Technology released the Program for the Innovation of Industrial Technology during the 12th FYP. Regarding the industry of new materials, it proposed to prioritize the development of manufacturing technologies of high-performance fibers and composite materials such as resin matrix composites, ceramic-based composites, carbon-carbon composites and metal matrix composites.

On January 4, 2012, the Ministry of Industry and Information Technology issued the Program for the Development of New Materials during the 12th FYP, identifying carbon fibers as the key point in the development strategy of reinforced fibers with high performance. It proposed to enhance research on series of carbon fiber products (with different properties such as high-strength, high-strength and medium-modulus, high-modulus, and high-strength and high-modulus) and prioritize the development of technologies key to the industrialization of high-performance PAN-based carbon fibers such as polymerization, spinning, pre-oxidation and carbonization, striving to raise the production capacity of carbon fibers to 12 thousand tons till 2015 to basically meet the demands of aerospace, wind power and transportation equipment. In the section "Key points in the development of high-performance reinforced fibers", carbon fibers were ranked first, before aramid fibers and UHMWPE.

On February 24, 2012, the Ministry of Industry and Information Technology released the Directory of Key Products of the New Material Industry during the 12th FYP, listing general-purpose pitch-based carbon fibers and resin matrix composite materials (including composite blades of wind turbines) as key products for development.

On March 27, 2012, the Ministry of Science and Technology released the Special Program for the Development of Wind Power Technology during the 12th FYP, proposing to carry out research on technologies that apply carbon fibers and other advanced materials to the design of blade structure.

On May 30, 2012, the State Council discussed and passed the

National Program for the Development of Strategic Emerging Industries during the 12th FYP, proposing to focus efforts on the development of new-type functional materials, advanced structural materials and composite materials.

In terms of input from national S&T plans [①],basic research projects of carbon fibers funded by the National Natural Science Foundation (NSFC) are under the Department of Material Sciences. We can see from Table 11-5 that the NSFC started to attach considerable importance to the funding of the Department of Material Sciences since 2001, with numbers and funds of General Projects and Key Projects accounting for a large proportion of all NSFC projects. Ever since January 1988, when the project "Research on the surface texture of low-temperature plasma-modified carbon fibers" headed by Sun Mujin, a researcher with the Chemistry Institute of the Chinese Academy of Sciences (CAS) got the funding from the NSFC, this foundation has continually funded more than 20 projects that are directly involved with basic research of carbon fibers. Figure 11-2 shows that there are in all 91 carbon fiber basic research projects funded by the NSFC during 1999–2011 with aggregated funds of 41.84 million RMB. The amount of funding experienced a huge boost since 2003.

① The "new materials" field also supports carbon fiber projects in Program 863. Taking into account that the program focuses on applied research and industrialization of achievements, coupled with the difficulty in obtaining comprehensive data, we did not analyze this program here.

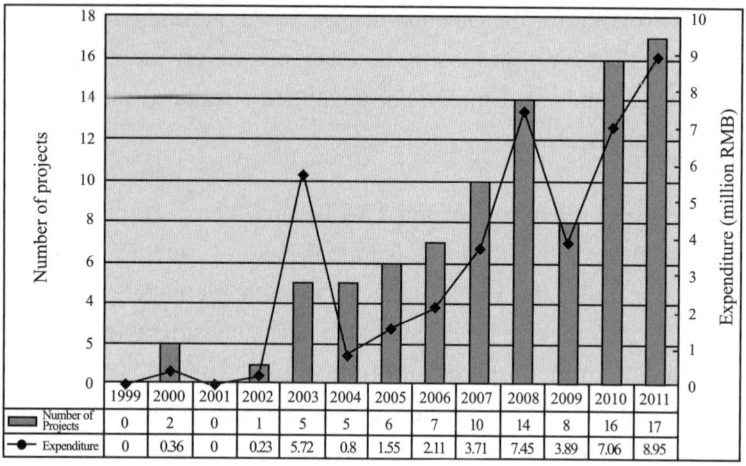

Figure 11-2 Number of projects and expenditure related to basic research on carbon fibers under the NSFC

Data source: The website of the NSFC.

We can also see from the statistical analysis of SCI papers and DII patents in Section 11.1 that there is a huge capacity "gap" between the scientific research and industrial innovation of carbon fibers in China. For the present, most Chinese carbon fiber enterprises tend to value process development while neglecting basic research. However, we can see from Table 11-5 that corporate participation in carbon fiber programs under the NSFC is still scarce, and in the past decade, there was not a single-funded carbon fiber program whose leader was affiliated with an enterprise.

Table 11-5 Carbon fiber technology projects funded by the NSFC during 2000–2011

Year	Project name	Project leader	Supporting organization	Amount of funding (thousand RMB)
2000	Research on the structure and performance of new-type cellulosic fiber precursors of carbon fibers	Shao Huili	Donghua University	170

Continued

Year	Project name	Project leader	Supporting organization	Amount of funding (thousand RMB)
2000	Research on the structural design and absorption mechanism of Xenon-absorbing active carbon fibers	Chen Shuixie	Sun Yat-sen University	190
2002	Research on the electric effect of carbon-fiber-cement matrix composites	Sun Mingqing	Wuhan University of Technology	230
2003	Research on pre-pressed carbon fiber textile-reinforced concrete sheet	Xun Yong	Southeast University	240
2003	Research on the structural regulation and evolution mechanism of PAN-based carbon fiber precursors with excellent mechanical performance	Xu Lianghua	Beijing University of Chemical Technology	1,500
2003	Investigation into the structural formation of high-performance PAN-based carbon fiber precursors and polymers and the controlled pre-oxidation mechanism	Pan Ding	Donghua University	3,500
2003	Surface gas phase oxidation and related catalytic performance of carbon nano-fibers with controlled structure	Dai Yingchun	East China University of Science and Technology	250
2003	Research on the structural design and reduction mechanism of multi-porous carbon fibers used as catalyst-loading material of fuel-cells	Chen Shuixie	Sun Yat-sen University	230
2004	Multi-scale structural regulation in a complex system and the relationship between performance, structure and preparation of related materials	Yuan Weikang	East China University of Science and Technology	0

Continued

Year	Project name	Project leader	Supporting organization	Amount of funding (thousand RMB)
2004	Controllable preparation and dual-pore fractal structure of new-type active carbon fibers with dual-pore size distribution	Wu Qilin	Donghua University	230
2004	Research on the seismic performance of smart high-strength concrete columns with carbon fiber casing	Wang Zhenyu	Harbin Institute of Technology	240
2004	Mechanical performance of super-high-performance concrete beams with CFRP and pre-pressed tendon	Fang Zhi	Hunan University	250
2004	Research on the synthesis of carbon fiber sizing material and its synergistic effect on composite interface	Bai Yongping	Harbin Institute of Technology	80
2005	Research on the Pd-carbon nano-fiber catalyst with controllable crystal orientation and direct synthesis of hydrogen peroxide	Zhao Tiejun	East China University of Science and Technology	230
2005	The law of degradation of mechanical performance of carbon fiber composite-reinforced corroded RC components	Zhang Weiping	Tongji University	240
2005	Nanodisk electrode ultrahigh temporal-spatial resolution research of membrane fusion dynamics	Zhang Rongying	Huazhong University of Science and Technology	280
2005	Research on the application of durable composite materials in long-span cable-stayed bridges and the design methods	Xie Xu	Zhejiang University	260

Continued

Year	Project name	Project leader	Supporting organization	Amount of funding (thousand RMB)
2005	Basic research on the solid-liquid high-pressure filtrating-squeezing near-net shape technology of light high-performance Csf/Mg composite material	Qi Lehua	Northwestern Polytechnical University	280
2005	Research on the structural properties and design methods of new-type composite timber structure	Liu Weiqing	Nanjing University of Technology	260
2006	Fire-resistant performance and design method of carbon fiber cloth-reinforced concrete flexural members	Zheng Wenzhong	Harbin Institute of Technology	300
2006	Micro-structural evolution of PAN precursors in the process of pre-oxidation and carbonization	Wang Chengguo	Shandong University	280
2006	Research on the mechanism of bonding between CFRP and steel plate	Lu Yiyan	Wuhan University	300
2006	Association study on the preparation of new-type carbon fibers by PAN electro-spinning	Liu Jie	Beijing University of Chemical Technology	310
2006	Multi-field coupling mechanism of carbon fiber smart layer and its field domain diagnosis	Li Zhuoqiu	Wuhan University of Technology	280
2006	Mechanical performance of carbon fiber-reinforced nano-clay-modified epoxy composite material under low temperature	Fu Shaoyun	Technological Institute of Physics and Chemistry of CAS	350
2006	Basic research on preparation of CNT-reinforced PAN carbon fibers through in situ polymerization	Dong Yanming	Xiamen University	290

Continued

Year	Project name	Project leader	Supporting organization	Amount of funding (thousand RMB)
2007	Preparation of ethylene glycol through hydrogenolysis of biomass by orderly structured carbon nano-fiber-borne Ru-based catalyst	Zhou Jinghong	East China University of Science and Technology	270
2007	Shear property and design method of non-anchored carbon fiber cloth-reinforced continuous beam	Zhou Chaoyang	Central South University	320
2007	Research on the mechanical performance of concrete beams with CFRP tendon	Zhang Peng	Guangxi University of Science and Technology	220
2007	Impact of the arrangement of carbon fibers on the cathode spot movements of electrode material	Zhang Chengyu	Northwestern Polytechnical University	190
2007	The induction mechanism of pyrolytic carbon deposit by carbon fiber's surface status	Yu Shu	Central South University	200
2007	Research on carbon fiber-reinforced foamed aluminium	Yao Guangchun	Northeastern University	340
2007	Research on the seismic performance of carbon-PAN fiber concrete bridge piers	Wang Kehai	Research Institute of Highway of the Ministry of Transport	300
2007	Basic research on the highly efficient energy transformation process based on reaction of catalytic pyrolysis of methane	Li Yongdan	Tianjin University	1,500
2007	Research on the enhancement of Mg-Li-based composite material interface by radiation-modified carbon fibers	Li Junqing	Harbin Engineering University	170
2007	Research on the structural form and structural behavior of pre-tensioned tethered rail bridges	Li Fangyuan	Tongji University	200

Continued

Year	Project name	Project leader	Supporting organization	Amount of funding (thousand RMB)
2008	Research on the quantitative image analysis of irregular porous structure in active carbon fiber material	Zhu Yu	East China University of Science and Technology	200
2008	Configuration and application of new-type highly sensitive electrochemical sensors based on biological catalysis	Zheng Jianbin	Northwest University	320
2008	Carbon fiber composite materials cutting technology and the testing & rating theory and technology of surface quality	Zhao Fuling	Dalian University of Technology	290
2008	Research on the synergic physical chemistry - electrochemistry controllable purification mechanism of underground water with a high concentration of arsenic	Zhang Xu	Tsinghua University	350
2008	Basic research on the preparation of inert anode made of nickel-ferric spinel particle/metal/fiber multiphase composite material	Yao Guangchun	Northeast University	1,800
2008	Research on the realization mechanism of synchronizing toughened and strengthened CFRP between layers of polysulfide/CNT nano-fiber membrane	Yang Xiaoping	Beijing University of Chemical Technology	300
2008	Research on the advanced technology of synchrotron radiation small angle X-ray scattering and several applications	Wu Zhonghua	Institute of High Energy Physics of CAS	2,200
2008	Structural control of V2O5/CNFs/cordierite catalyst and its low-temperature de-NOx behavior	Wang Yanli	East China University of Science and Technology	180

Continued

Year	Project name	Project leader	Supporting organization	Amount of funding (thousand RMB)
2008	Research on the rational utilization of new-type tensile spatial structure of high-performance carbon fiber cables and the construction technology	Tu Yongming	Southeast University	200
2008	Research on basic questions in the preparation process of CNT-reinforced new-type cellulose-based carbon fibers	Shao Huili	Donghua University	320
2008	Preparation of the carbon fiber and control of its morphology and structure as an absorbing material	Li Guang	Donghua University	350
2008	Drilling defects in the integrated assembly of carbon fiber composite material/titanium alloy and its suppression strategy	Gao Hang	Dalian University of Technology	380
2008	Research on the growth pattern of the spiral carbon nano-fiber which is catalytic synthesized by nano-copper	Du Fanglin	Qingdao University of Science and Technology	360
2008	Basic research on the improvement of fatigue performance of welded joint by carbon fiber composites	Chen Tao	Tongji University	210
2009	Atomic radial distribution function and structural evolution of short range order of PAN-based carbon fibers	Yu Meijie	Shandong University	200
2009	Research on the protection mechanism of concrete cathode by functional carbon fiber cement-based material covering	Yao Wu	Tongji University	350
2009	Research on the SR-SAXS of the internal structure of high-performance carbon fibers	Xu Jian	Institute of Chemistry of CAS	2,200

Continued

Year	Project name	Project leader	Supporting organization	Amount of funding (thousand RMB)
2009	Research on the mechanical behavior of C/SiC composites under extreme conditions	Suo Tao	Northwestern Polytechnical University	260
2009	Research on the analysis of mechanical performance and optimization design of cable-stayed bridges made of carbon fiber composite materials	Mei Kuihua	Chang'an University	200
2009	Electro-spinning rheological feature and the mechanism of high-temperature carbonization of precursors of liquefied wood carbon fiber	Ma Xiaojun	Tianjin University of Science and Technology	180
2009	Research based on high current electron beams of metal-carbon fiber composite cathodes	Liu Lie	National University of Defense Technology	300
2009	Basic research on the preparation of foamed aluminium with high-energy absorption capacity	Cao Zhuokun	Northeastern University	200
2010	Assembly of polymer brush in the carbon fiber/epoxy resin surface and construction of polymer brush in reinforced and toughened surface structure	Zhou Xiaodong	East China University of Science and Technology	380
2010	Research on membrane electrode components of fuel cells based on carbon nano-fibers with in situ growth diffusion layer	Zheng Junsheng	Tongji University	190
2010	Research on the inherent laws of the micro-structure, surface property and electro-catalytic activity of carbon nano-fibers	Zhang Xinsheng	East China University of Science and Technology	360

Continued

Year	Project name	Project leader	Supporting organization	Amount of funding (thousand RMB)
2010	Research on moisture absorption and aging mechanism of carbon fiber composite surface in the seawater environment	Zhang Fuhua	Shanghai Maritime University	190
2010	Aspherical shaping of thin lens made by carbon fiber-reinforced polymeric composite materials	Zeng Chunmei	Suzhou University	360
2010	Catalytic oxidation of HCHO under room temperature by active carbon fiber-reinforced metal phthalo-cyanine-activated molecules	Yao Yuyuan	Zhejiang Sci-Tech University	200
2010	The diffusion mechanism of carbon fiber composite grout in rock fractures under high dynamic water pressure	Yang Ping	Tongji University	200
2010	Evolution of multi-scale carbon layers and amorphous structure in PAN-based carbon fibers	Wu Gangping	Institute of Coal Chemistry of CAS	200
2010	Seismic strengthening and design method of carbon fibers in the non-ductile-reinforced concrete frames	Wang Zhenyu	Harbin Institute of Technology	380
2010	Preparation of ball-rod-shaped short carbon fiber-reinforced composites and materials	Liu Lei	Shanghai Jiao Tong University	380
2010	Research on the micro-structural evolution mechanism during the formation of unidirectional electro-spinning PAN-based carbon fibers	Liu Jie	Beijing University of Chemical Technology	380
2010	In situ study of PAN precursor's oxidation process by method of synchrotron radiation small angle scattering	Li Zhihong	Institute of High Energy Physics of CAS	420

Continued

Year	Project name	Project leader	Supporting organization	Amount of funding (thousand RMB)
2010	Surface reinforcing mechanism of CNT-grafted carbon fiber-reinforced resin composite materials	Li Yibin	Harbin Institute of Technology	230
2010	Research on structural soundness monitoring based on the sensing ability of CNT and guided waves	Li Fucai	Shanghai Jiao Tong University	460
2010	Research on in situ electrochemical preparation of nano nickel by putting chelate polymer-grafted carbon fibers in spent chemical nickel plating liquid	Jin Guanping	Hefei University of Technology	360
2010	Influencing mechanism of the interface with nano-scale component on the performance of carbon fiber/epoxy composite materials	Huang Yudong	Harbin Institute of Technology	380
2010	Research on nitrite coupling agent-modified, carbon fiber–reinforced resin-based composite materials	Hu Yanhong	East China University of Science and Technology	360
2010	Research on the influence of carbon fiber's surface self-assembly on the atomic oxygen resistance of the composite interface	He Jinmei	Harbin Institute of Technology	200
2010	Research on the carrier emitting mechanism of carbon nano-fiber/polyethylene composite semi-conducting materials	Hao Chuncheng	Qingdao University of Science and Technology	350
2010	Structural performance of extra long span concrete cable-stayed bridges based on high-performance materials	Fang Zhi	Hunan University	380
2010	Research on new-type carbon fiber velvet cathodes	Fan Yuwei	National University of Defense Technology	350

Continued

Year	Project name	Project leader	Supporting organization	Amount of funding (thousand RMB)
2010	Relevance between micro-structure and macro-mechanic properties of fibers	Bian Wenfeng	Harbin Institute of Technology	350
2011	Research on the ultrasonic phased array inspection technique of aircraft structural parts made of carbon fiber composites	Zhan Xianglin	Civil Aviation University of China	280
2011	Research on the theory and method of sparse pulsed eddy current array imaging of carbon fiber composite members	Ye Bo	Kunming University of Science and Technology	260
2011	Research on the design and microwave absorbing property of SiC nano-wire arrays growing in situ on the surface of carbon fibers	Wu Renbing	The Second Artillery Engineering University of PLA	250
2011	Regulation and construction of the catalyst used for cleaning soot particles emitted by diesel engine with the method of carbon nano-fiber template	Wu Qiang	Shanghai University of Electric Power	250
2011	Research on mechanical performance and absorption mechanism of carbon fiber-reinforced composite sandwich structure under intense dynamic loading	Shu Xuefeng	Taiyuan University of Technology	650
2011	Research on construction and mechanism of CNT/carbon fiber synergic resilient bismaleimide resin composites	Qiu Jun	Tongji University	600
2011	Research on carbon fiber surface modification based on super-critical fluid technology and its functional mechanism	Meng Linghui	Harbin Institute of Technology	600
2011	Research on key technologies of imaging detection by infrared thermal wave radar facing the side of CFRP laminate defect and the layer	Liu Junyan	Harbin Institute of Technology	550

Continued

Year	Project name	Project leader	Supporting organization	Amount of funding (thousand RMB)
2011	Research on the mechanism of accelerated decomposition of carbon fiber-reinforced amine cured epoxy resin by the combination of hydrogen-donor solvent and catalyst	Liu Jie	Changchun Applied Chemistry Institute of CAS	250
2011	Surface modification of carbon black and the preparation of silicon rubber with high conductivity by mixed filling of carbon black and carbon fibers	Li Qiuying	East China University of Science and Technology	250
2011	Development of new-type high-performance nano-catalyst and its application in the reaction of synthesizing ethylene glycol by CPPO	Li Chunping	Inner Mongolia University of Technology	560
2011	Key technologies and mechanism of efficient recycling of used carbon fibers/epoxy resin composites by using super-critical normal propyl alcohol	Jing Deqi	Institute of Coal Chemistry of CAS	250
2011	Research on the preparation of the graphene-modified carbon fiber by electro-spinning technology and its application as the negative electrode material of lithium batteries	Guan Hongyu	Northeast Normal University	250
2011	Dynamic evolution model and cooperative control in the forming process of carbon fibers	Ding Yongsheng	Donghua University	290
2011	Research on the structural performance and monitoring technology of pre-pressed smart carbon fiber plate-reinforced steel	Deng Langni	Guangxi University of Science and Technology	250

Continued

Year	Project name	Project leader	Supporting organization	Amount of funding (thousand RMB)
2011	Research on the preparation of the CNT-reinforced carbon fiber based on electrostatic spraying technology and its reinforcing mechanism	Chen Bowen	Tianjin Polytech University	550
2011	Influence mechanism of force-thermal coupling on the drilling of carbon fiber composites and its countermeasures	Bao Yongjie	Dalian University of Technology	250

Data source: The website of the NSFC.

As Table 11-6 shows, there are 9 carbon fiber research projects under Program 973 during 1998–2011, among which 3 are undertaken by two enterprises as the supporting organization of the chief scientists. However, these two enterprises are both large state-owned enterprises (Sinopec and Aviation Industry Corporation of China) rather than specialized carbon fiber enterprises. Currently, the scale of Chinese carbon fiber enterprises is not large enough to manage basic scientific projects with such significance.

Table 11-6 Projects related to the field of carbon fibers under Program 973 (National Key Scientific Research Program included)

Year	Project name	Chief scientist	Supporting university or institute	Supporting organization(s)
1999	Basic research on performance optimization of commodity polymer materials	Yang Yuliang, Qiao Jinliang	Fudan University, Beijing Research Institute of Chemistry Industry	Sinopec, Science and Technology Commission of Shanghai Municipality

Continued

Year	Project name	Chief scientist	Supporting university or institute	Supporting organization(s)
2003	Research on the multilayered structure and high performance of advanced polymer-based composites	Han Zhichao	Institute of Chemistry of CAS	Chinese Academy of Sciences
2005	Basic research on the multilayered structure and high performance of polyolefin	Yang Yuliang	Fudan University	Sinopec, Ministry of Education
2006	Basic scientific problems of rayon-based and PAN-based carbon fibers with high performance	Xu Jian	Institute of Chemistry of CAS	Chinese Academy of Sciences
2010	Research on basic scientific problems related with aerospace application of advanced composite materials	Yi Xiaosu	Beijing Institute of Aeronautical Materials of AVIC	Aviation Industry Corporation of China
2011	Research on the scientific foundation and common problems of PAN-based carbon fibers with super high performance and low cost	Xu Jian	Institute of Chemistry of CAS	Chinese Academy of Sciences
2011	Basic research on major questions related to high-performance carbon fibers	Yang Yuliang	Fudan University	Ministry of Education
2011	Basic research on the effective preparation and service of high-performance carbon-carbon composite materials	Xiong Xiang	Central South University	Ministry of Education
2011	Key scientific problems in the preparation process of high-performance aramid fibers	Yu Muhuo	Donghua University	Ministry of Education, Science and Technology Commission of Shanghai Municipality

Data source: The websites of the Ministry of Science and Technology of China.

11.4 Conclusion and discussion

The carbon fiber is a very promising strategic new material, and its industrial innovation and development are under great influence of basic research. China began to develop high-performance carbon fiber in the mid-1970s and has made great progress. However, China lacks strength in this industry. Presently, 95% of the global markets are taken up by foreign giants. Domestic industry is quite immature. There are at best six small precursor producers in China. China imported 90% of precursors. Its products are mainly for general, basic and low-end use. Many enterprises invest a lot in expansion but little in basic research. For example, no domestic enterprise develops oiling agent products to increase fiber strength.[1] Due to insufficient basic research and deep-level issues yet to be addressed, Chinese high-performance polyacrylonitrile-

⑩ Difficulties for Chinese carbon fiber industry, http://www.dragon-raja.com.cn/201203/tanqianweichanye211424.htm.

based carbon fiber's development is constrained. Therefore, Chinese leading enterprises should strengthen research and experiments. The government should grant easier approval and special funds to support carbon fiber precursor and carbon fiber preparation, structural form and performance research and to break through technological obstacles that prevent rapid industrial development and core competitiveness. ①

Innovation and development of biomedicine, LCD panel and carbon fiber industries show that scientific research and industrial competitiveness will mismatch if basic research is not equipped with innovation. Enterprises should be a major innovative force; the government should better input allocation by enhancing innovation-orientated input of both self-initiated and planned basic research and engaging enterprises in research through tax and financial policies. Corporate participation proved highly efficient in promoting technological innovation. The government may also advance commercialization by integrating enterprises, universities and research institutes.

Case analysis shows that leaders of the three industries—he United States (biomedicine and carbon fibers), Japan (LCD panels and carbon fibers) and Korea (LCD panels)—share a common characteristic and that is industry-driven basic research. China's subsidies to basic research have long been constrained to self-initiated purely theoretical

① For example, on October 31, 2010, China Carbon Fiber and Its Composite Material Technology Innovation Strategic Alliance was founded in Beijing, with members including National Carbon Fiber Engineering Technology Research Center, National Engineering Laboratory for Carbon Fiber Preparation and Engineering, National Engineering Laboratory for Carbon Fiber Preparation Technology, Beijing University of Chemical Technology, Jiangsu Hengshen Fiber Materials Co., Ltd, National New Material Industry Development Strategy Advisory Committee, Weihai Tuozhan Fiber Co., Ltd, Zhongfushenying Carbon Fiber Co., Ltd, Institute of Coal Chemistry of the CAS and Jilin Petrochemistry of Petrochina. In June 2012, Dalian Carbon Fiber and Its Composite Material Industry Alliance were founded. Seven carbon fiber enterprises and research institutes signed a strategic cooperation agreement. The alliance was devoted to technological innovation, resource pooling via market, breaking through technological bottlenecks through cooperation and industrial development.

research or national planned research. Although some enterprises have invested into application or even basic research, few Chinese enterprises support basic research because it is public and takes long to pay off. Large SOEs like Petrochina, Sinopec, China Mobile and State Grid have enough fund and research institutes and labs to conduct medium- and long-term research, but the majority of Chinese enterprises lack the ability to do so.

Basic research in China is normally carried out by state-sanctioned universities and research institutes driven by scientist curiosity and national needs (national defense for example), and its commercialization is too difficult. The mismatch between industrial and national needs should be addressed through specific and industry-oriented research. Universities and research institutes conduct research according to national needs. Research has to go through project approval, planning, application and assessment and normally takes 3–5 years or even longer to generate results. The industry, on the contrary, is in urgent need of basic research. Scientists may not have interest in industrial basic research. Presently, scientists work for research institutes and are devoted to vertical subjects backed by the government. But they hesitated to work on horizontal subjects backed by enterprises because they may be despised by peers due to the for-profit nature of the research. Such mindset is a constraint to the development of talent strategy of enterprises. Under new circumstances, scientists abandon the old mindset and work on all types of scientific research to maximize their values (Liu, 2012). An entrepreneur of a domestic enterprise of large-scale wind power generation facilities said, “To improve generator’s performance is the top issue in our industry. The generator blade contains resin, whose upgrading depends on carbon fibers. Japanese carbon fiber technologies are already mature, and they have obtained almost all relevant patents. We have to pay them for using the technologies. If China wants to have a share of the world manufacturing and scientific research, its input into basic research has to be diversified." [①] This quote shows that Chinese enterprises have realized basic research's importance to their development.

The last three chapters' analysis shows that Chinese current basic research system fails to support key technological breakthroughs. The National Natural Science Foundation of China, Program 973 and so on are mainly targeted at academic exploration and national needs but fail to meet industrial and market needs. An input system to tackle key technological bottlenecks is yet to be put in place. Basic research is supported by the United States, Japan and Korea because it has a direct bearing on industrial key technology capability and is the key to economic growth. But it is the weakest part for China. In recent years, China has made progress in basic research of many fields, but the results hardly bring about remarkable progress in technological competitiveness. Apart from the three industries mentioned, nanometer is another typical example. In the past decade, with the support of Program 973, basic research in the nanometer industry has developed fast, but the Chinese nanometer industry does not have a clear development path. Up till now Chinese enterprises are only capable of making low-end nanometer products (Cheng et al., 2011).

① Speech by Wu Gang, President of Jinfeng Science and Technology Incorporated Company, at the Business Starting and Innovation Forum of an alumni meeting of Management School, CAS on November 27, 2011.

References

Chen, Y. Chapter 15: Carbon fiber composite materials//*Report on the Development of China's New Material Industry (2010)*. Beijing: Chemical Industry Press, 2011. [陈亚莉．第 15 章 碳纤维复合材料 // 中国新材料产业发展报告（2010）. 北京：化学工业出版社，2011.]

Cheng, P., et al. Basic research and China's industrial technology development. *Management Review*, 2011, 23 (12): 46-55.[程鹏，等．基础研究与中国产业技术创新．管理评论，2011，23（12）：46-55.]

Feng, J. & Du, X. Review on China's carbon fiber technology patents. *China Intellectual Property News*, 2011-12-07. [冯洁，杜岩希．我国碳纤维技术专利状况浅析．中国知识产权报，2011-12-07．]

He, F. *Carbon Fiber and Its Application Technology*. Beijing: Chemical Industry Press, 2004. [贺福．碳纤维及其应用技术．北京：化学工业出版社，2004．]

Li, K. The past and present of China's carbon fiber research. *Advanced Materials Industry*, 2009 (10): 101-105. [李克健．中国碳纤维研究的过去与现在．新材料产业，2009（10）：101-105．]

Li, Y. New materials planning will be released on Sept. 7 and market space will be opened. http://www.cnstock.com/, 2011-09-02. [李雁争．新材料规划 7 日发布　市场空间被打开．中国证券网，2011-09-02．]

Liu, X. Policies should support enterprises to participate in basic research. *Chinese Science News*, 2012-01-09. [柳卸林．企业参与基础研究亟待政策支持．中国科学报，2012-01-09．]

Pang, L. How to keep to the direction to make China's carbon fiber technology breakthroughs. *China Chemical Industry News*, 2011-08-11. [庞利萍．中国碳纤维技术攻关如何把握方向．中国化工报，2011-08-11．]

Pang, L. Carbon fiber: A promising industry starts successfully. *China Chemical Industry News*, 2012-04-24. [庞利萍．碳纤维：希冀产业成功起步．中国化工报，2012-04-24．]

Chapter 12

Conclusion and Suggestions

12.1 Conclusion of the research and discussion: Inspirations for China

Since the 20th century, it has always been a problem for countries around the world regarding how to effectively support the technological innovation of domestic industries through scientific research. As to technologic policy; how to encourage more breakthroughs of core industrial technology to be based on basic research is a topic of much attention. Both in developed countries that took the lead in accomplishing industrialization, such as America, Britain and Germany, and latecomers succeeding in catching-up innovations, such as Japan and South Korea. Most Chinese enterprises choose an innovation path different from those of Japanese and South Korean counterparts. After the introduction, they tend to focus on rapid imitation, cost reduction and process improvement. With the increasingly fierce competition in the market, enterprises lack the resources and energy to conduct core technology innovations through basic research.

Although during the recent decade, China has become a major R&D country through quickly mobilizing tremendous resources for science and technology development. The rapid increase in R&D investment didn't lead to the corresponding rise in innovation capacity of core industrial technology, which is still relatively weak. Realizing this problem, the Chinese government has been advocating the new strategy of enterprises' independent innovation, and enterprises have become the dominant player of R&D investment and implementation. However, most resources are used to reduce production costs, improve technological process and the development of application and research centered on accelerating manufacturing speed. Except some highly-targeted fields (such as nanotechnology), there is still a huge gap between the small share of basic research , large share of technology development and production improvement (OECD, 2011).

There is another important background for this book: China now faces the challenge that its economy only has a large size instead of high quality. There are also other problems such as the lack of homegrown brands, and the energy-intensive and heavily-polluting manufacturing mode. We think that the core of the problem lies in the weakness of enterprises' independent innovation capacity. Although the country has strengthened the investment, in and support for, independent innovation, the technological capacity of enterprises still did not receive due attention.

One of the basic starting points of this book is that, only when Chinese enterprises focus on basic research, applied research and original innovation, can their innovation capacity be really improved. Relying on imitation enables rapid manufacturing of products, but can cause the lack of product design capacity, which reflects the R&D ability of enterprises.

But the basic situation of China's investments is that the economic growth didn't drive more investments from the government and enterprises in basic research or the large-scale construction of enterprise technology centers. Thus, here is another starting point of this book: stopping imitation, mastering fundamentals, building technologic centers and enhancing the innovation capacity through research ability.

China is now in a catching-up situation almost in all industries. which brings two associated questions in terms of basic research: first, is the insufficiency in basic research an inevitable phenomenon in the catching-up process of China's industrial technology? Second, which kind of basic research should China's independent industrial innovation stick to? How to balance the curiosity-driven, industry-driven and government-driven basic researches? How to balance the role of governmental planning and market mechanism in resource allocation of basic research?

As to the first question, we think the answer is no. In fact, Japan and Korea attached great importance to basic research in their technological catching-up process with the basic research intensity never less than 10%. In the mid-1950s, when Japan's R&D intensity was 1%, its basic research intensity reached 20%, and even 30.3% in 1965, down to 12.9% in 1965 and picked up afterwards, peaking at 35%. In 1983 when Korea's R&D intensity was 1%, its basic research intensity reached 18%. Although then the number declined, the lowest level reached was 12.5% in 1995. Also in the stage of transformation, China's basic research intensity has always been under 6%. That's mainly because in the technological catching-up process, China overemphasized market orientation, which caused both the government and enterprises to tend to focus on experimental development, and even think it's appropriate to put experimental development before basic research. However, this conception has been proven to be seriously wrong and very dangerous. The low basic research intensity is not an inevitable phenomenon of high economic growth period. Basic research has a significant position no matter in technological catching-up or technological leapfrogging.

As to the second question, we think that China's independent innovation should pay more attention to strengthening industry-driven basic research, which is the basic technological R&D based on the specific needs of industrial innovation and market judgment. It serves to solve the key bottleneck of industrial development and transform basic scientific theory of innovation to the core technological capacity of enterprises. The market nature of enterprises determines that only the basic research driven by industrial needs is the direction of enterprises (Pasteur's Quadrant instead of Bohr' Quadrant). The government-driven and curiosity-driven basic researches are not in line with the industrial needs and can't provide enterprises with much benefit. In history, developed countries sustained their industrial leadership through industry-driven basic research. Although now American industry has changed the innovation model which relied on major companies to build large-scale R&D labs, its basic research intensity still remains in an ideal range. Japanese enterprises which had completed the catching-up process have also increased their investments in industry-driven basic research during the recent decades, which helped Japan to succeed in scientific and industrial competitiveness (e.g. Japanese scientists have won Nobel Prize for several times in recent years). In an environment of globalized technology and industrial competition, the traditional model of Chinese industries, that is wining market by advantages of cheap land and low cost, has become unviable. That's not only attributed to the rapid rising costs of China's raw materials, land and labor, but also because countries with even lower costs, such as Cambodia and Vietnam, have become more attractive for the manufacturing industry. Without the support of strong industry-driven basic research, it's difficult for China's industrial independent innovations, especially breakthrough innovations, to succeed.

At present, the attention paid to China's resource investment in basic research for industrial technology innovation is still not enough. Influenced by the planning model of the Soviet Union for a long time, China has put a large amount of basic research resources in universities and public research institutes, allocating resources through science and technology programs. Other reasons include multinational corporations' fight for Chinese market shares after reforms and opening-up. The choice of Chinese industries' developmental paths, is commonly lacking in foresight and impetus for basic research among Chinese enterprises and no effective incentive mechanisms for enterprises to engage in basic research in national science and technological programs. Most Chinese enterprises mainly focus on innovations aimed at reducing costs. The technology reserve can only cover at most 2–3 years, while innovative companies including DuPont, IBM, GE all take the lead in the basic research of their own industry. As the representatives of enterprises succeeding in basic scientific research, DuPont even has extended its basic research plans to the year of 2090. According to the findings of a specific survey on corporate innovations by China's National Bureau of Statistics on October 22nd, 2006, Chinese enterprises have tended to invest most of their R&D funding in the improvement of existing products and technology, and very few in the development of new products and basic research. 33% of R&D investment was spent on "improving process to raise efficiency and reduce costs"; 31% on "development and research aimed at perfecting existing technology and expanding the applications of products"; 24% on "completing new products/developing new process (expected to be commercialized within 3 years)"and less than 9% for "basic research (not to be commercialized within at least 3 years)". [①] This pattern of resources allocation apparently can't support enterprises in mastering core technology. Foreign enterprises which entered China through FDI have been conducting basic or breakthrough research. They take full advantage of China's human resources of science and technology and preferential R&D policies, constantly strengthening their industrial technology advantages, in some degrees further suppressing the impetus and possibility of Chinese enterprises engaging in basic research.

---

① Only one-third of R&D funding of Chinese enterprises was spent on innovation. Xinhuanet, October 22, 2006.

For latecomers it is not easy to realize the transformation from imitation to innovation, and not all of them have the opportunity and ability to succeed in the technology catching-up. Although China has achieved rapid economic growth by taking the advantage of emerging globalization, its scientific and innovative capacity is still relatively weak compared to its strong manufacturing capacity. Simply building factories with foreign investments can't help to improve the domestic technology capacity, and domestic enterprises still haven't completely gotten rid of the vicious cycle of "introducing—falling behind—reintroducing—falling behind again". Therefore, for Chinese enterprises, how to break up the shackles of core technology is still an important problem concerning China's leap from catching up to taking the lead in technology. Cheng et al. (2011) noted that breaking the restraints of industrial core technology through industrial basic research is one of the sources for knowledge accumulation in the catching-up process. Against the background of globalization, in recent years many multinational enterprises have moved their R&D departments to China, which however doesn't mean the high values added were also transferred to China. It is always quite difficult for China to benefit from the value-adding process of the innovation value chain. Insufficient invention patents and their limited values directly lead to the relatively low status of our country in the global innovation field. Most enterprises conduct innovations only to maintain their existing market shares instead of pursuing technological leadership. Large amounts of R&D investments were spent on low-cost and large scale technical transformations for maintaining profitability. [①]

The starting point and conclusion of this article is: one of the major roots causing the serious insufficiency of China's industrial core technological innovations is that enterprises lack the powerful scientific capability. Basic research targeted at industrial technology innovation is the core of China's independent innovation strategy. The development

① What is Missing for "Innovative China"? *Science Times*, December 2, 2011.

of China's science and technology policies should take into account how to provide resource support for industry-driven basic research. In many Chinese industries, the core technical problems haven't been well solved. Chinese enterprises lack the basic research orientated at technology innovations, and it's also hard for public basic scientific achievements to be applied in innovation activities of enterprises. Before the 1980s, pure basic research was valued in American science and technology policies. However, with the increasing pressure from competitors such as Japanese electronics and automobiles, American's policies funding scientific research witnessed obvious changes, which required basic research to be connected with national objectives and closely integrated with the development of production and technology, and tend to prefer research programs with economic application value. Scientists and policy researchers had to endow basic research with new understanding to include application (Liu, 2007).

This book proposed that China increase the investment of resources in corporate basic research, which bears no intention to deny the leading role of the government in basic research. Actually, in most countries, the dominant investors of basic research are still governments, and the focus of industrial R&D is technology development (experimental development). The comprehensive reform of China's science and technology system has introduced market factors to science and technology factors and promoted the connection between R&D of public R&D departments, universities and industrial and economic development. The establishment of science parks, transformation of R&D institutes and projects integrating enterprises, universities and research institutes has also facilitated industrial technology innovations. However, due to the drop (relatively speaking) in support for public R&D, and the brain drain to developed countries and multinational enterprises, China's basic research suffered from some adverse impacts (OECD, 2011). More importantly, a large amount of scientific research doesn't care about the practical needs of industrial development, leading to more papers and yet less applicable achievements. The basic research oriented at industrial technology innovations is particularly hurt. China's investment in basic research is relatively low as a whole, and the industry lack the basic research capacity. As an instrument guiding public resources to first achieve national science and technology objectives, China's present basic research plan has its advantages, but the shortcomings are also evident. For example, the plan is often devised in a top-down way through "selecting winners". That is the government determines the program plan and sets privileged rights, while enterprises and other stakeholders find it difficult to participate in the process. In the past, China's funding for basic research was mainly driven by the national plan. As to research driven by industrial needs, influenced by the concepts such as "the achievements of basic research is a kind of information of public nature", "enterprises can't acquire all benefits from their own basic research no matter through technical secrets or intellectual properties", there was never a sustained and stable supporting mechanism. In fact, because the industrial basic research is targeted at market needs and especially some are closely connected with industrial technology development, it is not enough to simply rely on governmental plans in allocating basic research resources (Cheng et al., 2011).

At present, Chinese universities and research institutes haven't formulated the "basic research community" with enterprises. Many enterprises are unable to engage in major basic research due to its own insufficient capacity. According to statistics from Ministry of Science and Technology, the proportion of science and technology plan projects conducted by enterprises has been increasing year by year; over 95% of the projects in the country's science and technology support plan are participated by enterprises. Enterprises account for 30% of supporting institutions of 863 plan subjects; more than half of national major special subjects were led by enterprises, but the initiative of enterprises and the support of projects for the enterprises' innovation development still need to be improved. Although many projects are now participated in by enterprises, the "academic mindset" of research projects hasn't been completely changed. ① China has begun to carry out an independent innovation policy since 2006, but now our economic growth still relies too much on foreign technology and enterprises with a low key technological self-sufficiency rate. Because of unsatisfactory innovation performance, limited profits for domestic enterprises and the expensive intellectual property fees paid to multinational enterprises, there has been a paradox of rapid economic growth with low innovation capacity, and economic opening up with increasing dependence on foreign technology. Besides, governments and enterprises are short-sighted and still prefer experimental development and applied research.

Therefore, we think that China should ensure basic research can get sufficient resource investment and maintain balance among industry-driven basic research, free-exploration basic research and government-driven basic research. First, Chinese enterprises should strengthen the investment in basic research, which is the most direct measure. Second, if Chinese enterprises lack the investment ability the country should offer guidance and coordination; such as further strengthening the basic research of universities, the integration of enterprises, universities and research institutes, and the rapid transformation of basic research achievements to productivity. Third, look for balance between competitive subsidy and stable subsidy to public research institute. The government should encourage basic research targeted at industrial technology innovations to narrow the gap between basic research and industrial needs. The theoretical questions of industrial basic research actually should be mainly proposed by enterprises for they have a direct understanding of the industrial development trend. Of course, it is unrealistic to let all Chinese enterprises become IBM, Microsoft, Si, and build labs with superior research capacity. In the transformational stage and facing fierce international competition, Chinese enterprises can hardly pay attention to redevelopment of technology; instead they focus on engineering existing technology and realizing commercial

① http://finance.ifeng.com/roll/20110119/3234442.shtml.

values. Generally Chinese enterprises lack the accumulation of technology and knowledge and it's not easy for them to find basic scientific questions connected with industrial core technology. So at this time universities and research institutes are needed to take the initiative and conduct the search in the market. And actually, government is not able to see the development trend of basic research in all fields. With the lack of participation by enterprises, the direction of industrial basic research cannot be well devised. A balance must be reached between the government and market, and it will be unrealistic to emphasize either one.

12.2 Policy suggestions

As technology rapidly develops and industry profoundly changes, basic scientific research faces many problems that need to be solved. It is an important challenge for China's basic research to adapt itself to global trend and meanwhile provide long-term support for the country's economic and social development. ①

Basic research holds the key to the breakthroughs of core technology. Although China has caught up on a number of industries through incremental innovation, yet second-fast advantage tends to be replaced by first-mover advantage as patent and standard become main indicators of industrial development. It, therefore, becomes far more

① National Center for Science & Technology Evaluation (NCSTE), Report on Comprehensive Evidence for Contributions from China's National Natural Science Foundation and for International Assessment on Management Performance, 2010.

difficult for companies to catch up on technology since breakthroughs based on scientific innovation become the new rule of market competition. The solution is the attention paid to basic research. To this end, China should substantially adjust its basic research methods, increase investment in industry-driven research and optimize resources allocation so as to improve the input mechanism for basic research.

1. Government planning and market mechanism should complement each other to better allocate R&D resources. Chinese enterprises should be encouraged to increase investment in basic research and efficiency of resources allocation.

China's industrial innovation model based on scientific projects is facing mounting challenges. Due to planned economy and the influence of the Soviet Union, attention paid to science far outweighs that to the market, thus innovation is led by research institutions and universities in China. Science and technology is regarded as a scarce resource and secured through specialized systems. This causes technology resources to be channeled to research institutions and universities, hindering companies' capacity for innovation.

Over the past 30 years, China's innovation is advanced by scientific and technological progress, including application of research findings and cutting-edge technology, and technology transfer completed through scientific projects. The launch of the atomic bombs, hydrogen bombs and artificial satellites has awakened people to the important role of technology in national security (highlighting the need of national security over that of industries and market). The impact of this concept is still far-reaching even today.

Depending on the government for technological research was the mindset of the era of planned economy. Despite China's achievements in technological innovation in recent years, problems of the innovation model remain, including low efficiency of technology transfer and innovation, slow growth of innovation capacity, lack of impetus for application of research findings, conflict between various interests, etc. These problems can largely be solved by China's National Plan for Medium and Long-term Technology Development, which focuses on enhancing companies' innovation capacity and advancing major technology projects represented by large aircraft.

Traditionally, China's basic research is made possible by resources provided by governmental and scientific projects. This fails to serve the actual requirements of innovation. Current evaluation system of scientific projects also fails to help companies participate in basic research. That is why policies of fiscal subsidies, taxation incentives and base construction should be adopted to make market system and government plans complement each other and to channel resources to more areas so as to diversify the sources of research resources.

Through guidance and coordination, an inclusive mechanism of research input including enterprises, universities, and non-profit organizations should be established (Zhou , 2008). Governments should guide, not dominate or replace the basic research of industries in terms of technology innovation and development policies. To improve the efficiency of using governments' R&D resources, the Ministry of Science and Technology, the Ministry of Education, Chinese Academy of Sciences, and National Natural Science Foundation of China should cooperate with each other (OECD, 2011).

2. China's universities and research institutes should strengthen industry-driven basic research while enhancing academic research capacity.

China's universities and research institutes face two disadvantages in basic research: first, some professors and researchers over commercialize basic research. They prefer R&D and experiment development that can quickly be turned into products to the boring and unpredictable basic research on scientific theories. These people fail to fulfil their due responsibilities. Second, despite the ever-increasing input by the government, the innovation and application systems remain unchanged over the past decade. Because research results are evaluated based on the number of patents and articles published, professors and researchers prefer profitable pure research to applied research. This leads to a new round of "back to the ivory tower" phenomenon. It also makes research on applied science and experimental development paper-targeted, thus scientific research fails to serve actual requirements of companies.

China ranks second in terms of the number of scientific papers included in the Science Citation Index (papers of certain disciplines such as nanotechnology already rank the top), but few of those papers have been applied. If China's universities and research institutes focus solely on the back end of R&D (research and experiments), or decouple basic research from actual requirements, China's core technology development will be inhibited by foreign competitors, and independent innovation is hard to be realized. For instance, China has leapfrogged in high-speed railway depending on the importation of foreign cutting-edge technology, but if the latest foreign technology comes from nowhere, China will be overwhelmed in technology development, because it lacks the basic scientific knowledge of the industry.

How to balance basic research and the application of research findings is a global problem. Chinese enterprises, different from foreign ones, are still weak in basic research. Universities and research institutes remain the trendsetters and main suppliers of findings. That is why fast application is a main target for policymakers. Because industrial innovation is not a priority of the evaluation system, China's government-subsidized research projects fail to increase international competitiveness of industries despite an increasing number of papers.

Pure research which is not in line with the requirements and trends of industrial development is unlikely to enhance the innovation capacity for core technology (although it is the cornerstone of national scientific development). Public research institutes should play their role in basic research, applied research and strategic technology development, and bear the risks of research in order to distinguish their role in innovation of market entities. They should reduce overlap of research fields among universities, especially in basic research. They can draw upon the experience from Germany's Fraunhofer Society. This Society has conducted basic research and applied research that serves innovation requirements, thus narrowing the gap between universities and Max-Planck Society on basic research and industrial research. Its main function is to lead the most promising innovation projects and to pave ways for applied industrial innovation (OECD, 2011).

3. Chinese enterprises should be encouraged to dominate industry-driven basic research through policy guidance and optimizing external environment.

Generally speaking, the advantages of Chinese enterprises are low cost and market segmentation. Most enterprises are at the low end of global value chain and do not have the need for basic research. With China's accession of WTO, companies face more fierce external competition. Even if they have the awareness of the importance of basic research, many of them continue to lower their cost and tap the market, driven by the mindset of rapid commercialization of technology and increase of profit. They lack the impetus and energy of R&D.

One of the key targets of state policy of independent innovation is to lend such impetus and cultivate companies' capacity for applied research. This requires a new system of the government. Although government-backed research findings have a spillover effect on R&D, these findings cannot be expected to automatically solve problems of core technology. This is true both at home and abroad. In terms of technological innovation and industrial development policies, the government should guide companies to conduct basic research that serves industrial need through fiscal policy, economic leverage, and management innovation, because only companies themselves are aware of the trend of technology development.

In the past, most R&D projects on applied technology were submitted by universities and research institutes, and approved by relevant government agencies. This has been changed. After companies invite the bids from universities and research institutes, the government provides necessary conditions and policy support for those projects based on their importance. In terms of industrial application, an innovation system dominated by companies should be set up and resources should be channeled to industrial upgrading and market-oriented innovation.

Enterprises should say farewell to imitation. Imitation can help enterprises survive, but at the expense of low profit, no independent brands, IPR, or core technology, making them lag behind. Chinese enterprises have often been sidetracked in getting rid of imitation. Because they are producing a large number of goods, but are not aware of the scientific knowledge, they can only imitate others' products. Once the technology is upgraded, they have to continue to import and imitate. To solve the problem, they must build their own R&D system, and improve capacity for product design and development, and for application of research findings.

Chinese enterprises should strike a balance between basic research, development research and experimental development. Progressive innovation and technology pursuits are still essential for Chinese enterprises, but in the landscape of global innovation, second-fast companies face more difficulties in catching up as patent and standard lead industrial development. Currently, most Chinese enterprises fail to prioritize innovation in their strategies. Their needs for scientific resources are quite limited (OECD, 2011).

Companies' interference in basic research has two advantages: first, gaining core technology and IPR to improve competitiveness; second, better absorbing foreign technology, promoting technological transfer, and providing necessary support for economic development. That is why the government should increase their policy support for investment in basic research and give preferential policy for venture capital investment and taxation (Ma & Zeng, 2007).

4. Chinese enterprises can adopt an open innovation model led by independent R&D, in order to improve research capacity by integrating global know-how.

For Chinese enterprises, globalization opens the door of learning online from world leading enterprises, but open innovation should be based on independent R&D. Open innovation without domestic basic research will not help absorb the latest technology or know-how; instead, it will help MNEs with global innovation and geographical distribution of industries. Chinese enterprises should avoid such model, since knowledge is gained through knowledge bases within enterprises.

If China's enterprises focus solely on technological transfers and the back end of R&D (research and experiments), China's core technological development will be inhibited by foreign competitors and independent innovation will be hard to be realized. While building internal R&D capacity and balancing basic research, development research and experiments, they should enhance cooperation with universities and research institutes and fully utilize global innovation resources to expedite R&D and product innovation.

References

Liu, L. *Theory and Practice of Basic Research Policy*. Beijing: Tsinghua University Press, 2007. [刘立．基础研究政策的理论与实践．北京：清华大学出版社，2007．]

Ma, B. & Zeng, G. "Kealey's problem" and the reason of government-funded basic research. *Science & Culture Review*, 2007, 4 (5): 90-101. [马佰莲，曾国屏．"基莱问题"与政府资助基础研究的理性．科学文化评论，2007，4（5）：90-101．]

OECD. Reviews of innovation policy: China, 2011.

Zhou, J. *Foundation and Source of Innovation: Investment, Evaluation and Coordination of Basic Research*. Beijing: Science Press, 2008. [周寄中．创新的基础和源泉：基础研究的投入、评估和协调．北京：科学出版社，2008．]